T0094395

Leadership and Women in Statistics

Leadership and Women in Statistics

Edited by

Amanda L. Golbeck
University of Montana
Missoula, Montana, USA

Ingram Olkin
Stanford University
Stanford, California, USA

Yulia R. Gel
University of Texas at Dallas
Dallas, Texas, USA

CRC Press
Taylor & Francis Group
Boca Raton   London   New York

CRC Press is an imprint of the
Taylor & Francis Group, an **informa** business

A CHAPMAN & HALL BOOK

Cover painting: Jungle Gym (Tuner)© by Molly Barker

CRC Press
Taylor & Francis Group
6000 Broken Sound Parkway NW, Suite 300
Boca Raton, FL 33487-2742

© 2016 by Taylor & Francis Group, LLC
CRC Press is an imprint of Taylor & Francis Group, an Informa business

No claim to original U.S. Government works

Printed on acid-free paper
Version Date: 20150623

International Standard Book Number-13: 978-1-4822-3644-6 (Hardback)

Visit the Taylor & Francis Web site at
http://www.taylorandfrancis.com

and the CRC Press Web site at
http://www.crcpress.com

To Craig and Dan

To Anita, Vivian, Rhoda and Julia,
the women in my life

To my parents and to Alex and Lena

To all the past generations of women who have opened doors for
future generations

Contents

Preface   **xi**

About the Editors   **xv**

I   Fundamentals of Leadership   **1**

1   Outlook on Statistics Leadership   **3**
Ronald L. Wasserstein

2   Persuasion, Presence, and Patience: Important
Characteristics of Leadership for Women in Statistics   **9**
Laura J. Meyerson

3   Domains of Leadership Behavior in Organizations   **19**
Sim B. Sitkin

4   Four Leadership Principles for Statisticians: A Note on
Elizabeth L. Scott   **31**
Amanda L. Golbeck

II   Fresh Opportunities and New Challenges for the Statistics Discipline   **53**

5   Robust Leadership in Statistics   **55**
Jon R. Kettenring

6   Leading Significant Change in Official Statistics: A Woman's
Place Is in the Counting House!   **69**
Cynthia Z.F. Clark

7   What Makes a Leader?   **85**
Lynne Billard

8   To Be or Not to Be Bold: Effective Leadership in Various
Individual, Cultural, and Organizational Contexts   **103**
Kelly H. Zou

III Project Leadership   129

9 Statistical Challenges in Leading Large-Scale Collaborations:
   Does Gender Play a Role?   **131**
   Bhramar Mukherjee and Yun Li

10 Leadership in Statistical Consulting   **145**
    Duane L. Steffey

11 Women Leaders in Federal Statistics   **161**
    Marilyn M. Seastrom

IV Leadership Competencies   191

12 Competencies Needed for Statistics Leadership from an
    International Perspective   **193**
    Motomi (Tomi) Mori and Rongwei Fu

13 Organizational and Business Acumen: Observed
    and Latent Attributes   **205**
    Sally C. Morton

14 Leadership: An Untold Story   **217**
    Sallie Keller and Jude Heimel

15 Leadership and the Legal System   **229**
    Mary W. Gray

V Leadership Development Platforms   245

16 Professional Organization Membership   **247**
    Lee-Ann Collins Hayek

17 Leadership Development in the Workplace   **259**
    Gary R. Sullivan

18 Research Team Experience as a Platform for Leadership
    Development   **273**
    William A. Sollecito and Lori A. Evarts

19 How Statisticians Can Develop Leadership by Contributing Their
    Statistical Skills in the Service of Others   **307**
    Sowmya Rao

VI Individual Strategies 321

20 Lessons from an Accidental Leader 323
   Arlene S. Ash

21 Practical Suggestions for Developing as an Academic Leader 339
   Charmaine B. Dean, Nancy Heckman, and Nancy Reid

22 The Many Facets of Leadership 353
   Jacqueline M. Hughes-Oliver and Marcia L. Gumpertz

23 Leadership and Scholarship: Conflict or Synergy? 365
   Roy E. Welsch

VII Institutional and Network Strategies 371

24 The Value of Professional Champions and Mentors in an
   Academic Environment 373
   Katherine Bennett Ensor

25 Mentoring: It Takes a Village. Personal Story 383
   Sastry G. Pantula

26 "If You Would Consider a Woman ..." 395
   Daniel L. Solomon

27 Fostering the Advancement of Women in Academic
   Statistics 413
   Judith D. Singer

Contributors 429

Index 443

Preface

Leadership and Women in Statistics explores the role of statisticians as leaders, with particular attention to women statisticians as leaders. Borrowing from the Harvard Business School's communal definition of leadership, we define a statistics leader as a statistician who advances the quality of research methods or outcomes, influences others to go down a better path, and makes sure her positive impact on both people and science lasts in her absence.

The time has come to explore the role of statisticians as leaders. The field of statistics, which has tended to be a crucial yet inconspicuous science, is changing. It is expected in the future, with the overwhelming increase in data collection in all fields, that there will be an increasing number of proactive roles for statisticians. With this growing expectation, we are noting growing support within our professional associations for the development of statisticians as leaders, manifesting in initiatives such as the American Statistical Association (ASA) president's workgroup, under president Nat Schenker and chaired by Janet Buckingham, on developing training in statistical leadership.

The time has also come to openly explore the role of women statisticians as leaders. Women in statistics have tended to be in the background in professional statistical associations and settings, even though they make major contributions to the statistics profession in their workplaces. Statisticians have been talking about women's issues, but mainly in the back room. At the same time, women in other fields have been visibly leading change all around the world. There is every reason that statistics women should be among these visible inspirational leaders.

For these reasons, we have come together to develop this book in the areas of "statistics leadership" and "women in statistics." This book, which is a first, offers advice for emerging and existing leaders of both genders, and with diverse backgrounds, on converting their passion for statistical science into visionary, ethical, and transformational leadership. It illuminates fresh and engaging approaches and identifies opportunities to foster creative outputs and develop strong leadership voices. This book aims to break the silence about women's issues in statistics and advance the leadership of women in statistics.

Leadership and Women in Statistics includes contributions by experts in leadership, together with contributions by exemplar statistician leaders at various career stages and in various sectors. The book starts out by exploring "Fundamentals of Leadership." Ron Wasserstein, executive director of the ASA, presents an outlook on statistics leadership from the perspective of the largest professional organization of statisticians and urges all statisticians to step up to leadership in order to maximize the difference they can make for themselves, their organizations, their profession, and science in general. Laura Meyerson points out that great leaders, including great statistician leaders, masterfully exercise persuasion, executive presence, and patience/humility, and that the exercise of these three fundamental Ps can present challenges for women. Sim Sitkin delineates six fundamental behavioral elements of effective leadership that provide a basis for individuals to improve their leadership; these are thought of as a pyramid that includes foundational domains, action-oriented domains, and an integrative domain. Finally, Amanda Golbeck recommends that statisticians should regularly practice natural leadership and should aspire to organizational leadership; they should reflect on their leadership and seek out leadership development opportunities; they should practice evidence-based storytelling; and they should exercise influence by sharing their intellectual wealth.

Part II provides "Fresh Opportunities and New Challenges for the Statistics Discipline." Jon Kettenring argues that the statistics discipline needs robust leadership, that is, the profession needs to attract diverse talent and maximize the potential of all statisticians, including women, if it is to embrace its potential in the 21st century information age. Cynthia Clark illustrates the essential role of the statistics leader in formulating, communicating, amassing support for, and implementing vision, as well as other factors that are significant in catalyzing and enabling program and structural change processes in organizations. Indeed, an effective leader is one who is able to successfully lead change, and Lynne Billard discusses various principles that make leadership effective, while the statistician is urged to embrace the art and science of inclusive transformational leadership. Kelly Zou discusses the complexity of leadership in its many contexts — individual, cultural, and organizational — and illuminates existing leadership challenges that persist under the glass and bamboo ceiling metaphors as well as some emerging leadership challenges we are facing in the 21st century globalization era.

Part III, "Project Leadership," focuses on leadership challenges within the unique role that statisticians play in interdisciplinary research and policy environments. Bhramar Mukherjee and Yun Li examine the underrepresentation of and challenges for women in leadership of large-scale collaborations, including how to get started and stay involved, how to advance to formal leadership roles, and how to provide the best leadership. Duane Steffey outlines the role of leadership in the different kinds of statistical consulting practices, including how to provide principle-centered leadership such that exceptional value for clients is created, and development of a strong new generation of statistical consultants is assured. Marilyn Seastrom addresses the question of how women can best position themselves to help solve some of our nation's greatest challenges, with examples of how women in statistics have impacted government by creating new cooperative connections among professionals, raising important new questions, promoting new technologies, applying new project management concepts, and addressing new policy and legislation challenges.

Leadership competencies are skills and behaviors that lead to excellent leadership performance. There are many of these. Part IV, "Leadership Competencies," zeroes in on four such competencies in relation to statistics leadership, namely, cultural intelligence, acumen, preparedness, and law. Motomi Mori and Rongwei Fu point out that the field of statistics is becoming increasingly globalized and diverse, and statistics leaders need to have high cultural intelligence — they need to be culturally aware and culturally sensitive — in order to be successful. Sally Morton argues that statistics leaders need to have strong organizational and business acumen — they need to have fluent understandings of the observed and latent aspects of the big pictures of their organizations — in order to succeed. Sallie Keller and Jude Heimel discuss that leaders need to be prepared for unforeseen catastrophes and life-changing events; they need to expect the unexpected. Mary Gray reasons that statistics leaders need to know what help is available to them through the legal system when all other avenues of corrective assistance have failed; they are advised to pursue this help with caution.

All people can develop their leadership, regardless of their leadership skills level at baseline. Part V, "Leadership Development Platforms," discusses some ways to develop leadership. Lee-Ann Collins Hayek describes how statisticians at any career level can benefit from developing and practicing leadership as members of professional associations. Gary Sullivan describes how individuals can develop leadership as employees within workplaces, and suggests that individuals choose the leadership competencies they most want to strengthen and work on these through observation, discussion, reading, and, most important, coached practice. William Sollecito and Lori Evarts take a close look at how statisticians can learn, develop, and master leadership skills by participating as members of work teams, with special attention given to experiential learning as members of clinical research teams. Sowmya Rao presents a case study of learning and exercising servant leadership while doing pro bono statistical work, in this case in a developing country.

There are many barriers to leadership, particularly for women, and Part VI of this book, "Individual Strategies," illuminates some of the strategies that individuals can exercise in order to maximize the likelihood that they will become leaders. Here there are four contributions that span the United States and Canada, and that include practical experience in statistics and other departments in comprehensive universities as well as biostatistics departments in medical schools. Arlene Ash presents advice from an "accidental leader"; Charmaine Dean, Nancy Heckman, and Nancy Reid present advice for developing as an academic leader; Jacqueline Hughes-Oliver and Marcia Gumpertz present advice on facets of leadership; and Roy Welsch presents advice related to leadership and scholarship.

The final part of the book, "Institutional and Network Strategies," illuminates some networking strategies, as well as some strategies that organizational leaders can exercise, to overcome the many barriers to leadership. Katherine Ensor discusses how women and others in statistics should seek out both mentors and champions. Advice given by mentors can be highly instrumental in overcoming barriers to leadership, and professional champions are even better because they proactively connect individuals to career advancement and leadership opportunities. Sastry Pantula presents a personal story describing the influence of the many effective mentors (a village of mentors!) who contributed to his highly successful career. Dan Solomon delves into the important issue of diversity in the workplace, including why it is an economic, academic, and business imperative, and presents many concrete examples of how to achieve successes in diversity and inclusion. Judith Singer looks at six specific lessons that can be used to support women in academe and other knowledge organizations; these lessons involve relationship building, policy and practice improvement, evidence sharing, work-life program support, institutional accountability enforcement, and career initiative backing.

By paying special attention to women's issues, this book aims to provide a clear vision for the future for women as leaders in statistics. It aims to provide a vision of how women can incorporate leadership into the multifaceted paths of their diverse careers while simultaneously enjoying satisfying lives outside of work, and of how they can optimize their lives within constraints.

The new metaphor for a career is the jungle gym. This metaphor fits women's careers better than the traditional metaphor of a ladder. The jungle gym allows multiple trajectories, and multiple entry and exit points, and thus fits the challenges faced by women in real life. By showcasing perspectives of women statisticians in various leadership roles, this book provides examples that will help to inspire young women statisticians to "lean in" to their careers.

We hope this book will be beneficial to all statisticians and especially to those who are ready to consider leadership as an important element of their careers, and also those who are already leaders but want to deepen their own perspectives on leadership. We expect this book will also be beneficial to individuals in other science and technology disciplines and beyond. We anticipate this book will be useful for leadership training and development, and for personal professional development.

The editors wish to thank the outstanding contributors for their strong and thoughtful chapters and for their enthusiastic support of the book throughout the development process. We also wish to thank John Kimmel of Chapman and Hall/CRC Press for his expert guidance and patience, and the University of Montana School of Public and Community Health Sciences for its support of Professor Amanda Golbeck in this project. Finally, we wish to thank Molly Barker for permission to use her brilliant painting on the book cover; Ron Wasserstein and the American Statistical Association, and Elyse Gustfasson and the Institute of Mathematical Statistics, for permission to use their galleries of photographs; and Nancy Geller and Brian Phillips for permissions to use individual photographs.

About the Editors

Amanda L. Golbeck is professor of biostatistics at the University of Montana and former vice president for Academic Affairs at the Kansas Board of Regents. She served in a variety of leadership roles from department chair to vice president and academic dean at four academic institutions. She currently teaches courses in public health administration and management and public health leadership, and she is a member of the American Statistical Association President's workgroup on developing training in statistical leadership. Dr. Golbeck is an elected fellow of the American Statistical Association, an elected member of the International Statistical Institute, a past president of the Caucus for Women in Statistics, and a country representative to the International Statistical Institute Committee on Women. She earned a BA degree from Grinnell College, followed by MA degrees in anthropology and statistics and a PhD in biostatistics from the University of California at Berkeley. She also earned certificates in educational management at Harvard University's Graduate School of Education, leadership at the American Association of State Colleges and Universities' Millennium Leadership Initiative, negotiation at the John F. Kennedy School of Government, and women in leadership at the American Academy of Neurology. Grinnell College, the Kansas Board of Regents, and the American Statistical Association have presented awards to her in recognition of her leadership.

Ingram Olkin is professor emeritus in statistics and education at Stanford University. He earned a doctorate from the University of North Carolina, and before moving to Stanford was on the faculties of Michigan State University and the University of Minnesota. He has written a number of books, and has served on editorial boards for statistical, educational, and mathematical journals. He has served as chair of the Committee of Applied and Theoretical Statistics, chair of the Committee of Presidents of Statistical Sciences (COPSS), and president of the Institute of Mathematical Statistics. He has been a Guggenheim Fellow, an Alexander von Humboldt Fellow at the University of Augsburg, a Lady Davis Fellow at Hebrew University, an Overseas Fellow at Churchill College, and received an honorary DSci from DeMontfort University. He received a Lifetime Contribution Award from the American Psychological Association, and a Wilks medal and a Founders Award from the American Statistical Association. For many years, he has been involved in the furtherance of women in statistics, for which he received the COPSS Elizabeth L. Scott award.

Yulia R. Gel is professor in the Department of Mathematical Science at the University of Texas at Dallas. In 2004–2013 she was assistant/associate professor at the Department of Statistics and Actuarial Science at the University of Waterloo, Canada, where she started upon completion of a postdoctoral fellowship at the University of Washington. She earned her PhD degree in mathematics from Saint-Petersburg State University, Russia. She held visiting positions at Johns Hopkins University, George Washington University, and University of California, Berkeley, USA. She was vice president for membership and outreach of the International Society on Business and Industrial Statistics (ISBIS) and is currently the treasurer of the International Environmetrics Society (TIES). She is also a member of the Committee on Women in Statistics of the International Statistical Institute. Yulia is a fellow of the American Statistical Association and received the Abdel El-Shaarawi Young Researcher's Award in environmental statistics.

Part I

Fundamentals of Leadership

1

Outlook on Statistics Leadership

Ronald L. Wasserstein

American Statistical Association

Too few statisticians think of themselves as leaders. Too few PEOPLE think of themselves as leaders. Perhaps this is because their notion of leadership does not reflect the full breadth of the term. Maybe they think only "bosses" are leaders. Or perhaps they simply do not give themselves credit for what they do. Whatever the reason, the result is lost opportunity. The opportunity loss is not just personal. Organizations and even whole professions may languish for lack of leadership.

Statisticians can be leaders. Statisticians should be leaders. Statisticians must be leaders. By the time you complete this book, you will be convinced this is true, and you will be inspired by the examples of statistical leaders, enlightened by their experiences, and empowered by their advice.

Bob Rodriguez, 2012 president of the American Statistical Association (ASA) and an extraordinary leader, stated succinctly the reason why statisticians should be leaders:

> Leadership ability is a prerequisite for the growth of our field because statistics is an interdisciplinary endeavor and our success ultimately depends on getting others to understand and act on our work.

Rodriguez' statement is spot on. A good deal of the work statisticians do is work that serves the needs and interests of others. Statistics makes science better. What would crop science be, for example, without the contribution of statistics to agricultural experimental design and data analysis? And how much poorer statistics would be without the methods developed because of the needs of agricultural experimenters! Examples abound, as renowned statistician and scientist John Tukey memorialized with his famous statement about statisticians getting to play in everyone's sandbox (*The New York Times*, 2002; Brillinger, 2012). In industry, in government, in medicine, and in manufacturing, in the courtroom and in the classroom, statisticians fundamentally are collaborators, partners in the proper use of statistics to gain new insight and make better decisions.

Leadership skills — especially the ability to listen well, to communicate well, and to effectively influence decisions — are necessary for statisticians to be successful collaborators. First of all, they are needed in many instances just to get a seat at the table, a chance to weigh in on the statistical and scientific work being done. (Virtually all of us have witnessed the results of statistical work being done without any involvement of professional statisticians. Not pretty.) However, once we get to the table, leadership is required to maximize the contribution that statistics makes to the endeavor.

These relationships are the foundation of future leadership opportunities. My first job in academic administration came in this way. After an administrative shakeup at Washburn University, a faculty member in psychology was appointed as interim dean. She said she would only take the job if she could hire me as assistant dean! The reason for this: we had served together on a committee to review faculty salaries, and she was impressed with the power of statistics and my ability to communicate it. She said she had spent a career working with deans who did not seem to care about data, and she wanted to be different.

That job, and the mentoring she provided me, launched my career in leadership. Rodriguez noted that mentoring is another form of leadership, a thought that might not have occurred to many mentors. "Along with influencing others, leadership is about helping others succeed," Rodriguez says. "A university professor coaching a student on how to give an effective presentation is a statistical leader, as is an experienced industrial statistician guiding a younger colleague through multiple revisions of a paper." I have had the good fortune of having multiple wonderful mentors in my career. As executive director of the American Statistical Association, if I have been a good leader, then others will have been mentored well by me.

Rodriguez sees statistical leadership opportunities everywhere. "We need many kinds of leaders, not only in positions of prominence, but in any area of our profession in which there is opportunity to influence the acceptance of statistical contributions," he says. "A statistician writing an editorial letter advocating for data-based policy decisions is a statistical leader, as is a manager working with human resources staff to define the job responsibilities of statisticians in a company or government agency."

Leadership opportunities, then, manifest themselves to statisticians in their roles as collaborators and mentors. They arise in situations where statisticians can advocate or educate. In the past year, however, the leadership of the ASA has learned from professionals on the front lines that the secret to "making it to the top" is to "make it to the middle."

ASA leadership has been spending time with statistical leaders in "big data" industries. It has become clear from these conversations that, broadly speaking, in many organizations, there are two largely distinct groups: those people who are highly technically skilled, and those deft at business or policy. The people who make the most important contributions to the organization are those who are eventually able to operate in the middle, serving as a bridge between the two groups, having the ability to effectively communicate and work with both. Those people are enormously valuable and, by making it to the middle, they are often the ones who make it to the top, becoming increasingly valuable to their organizations and gaining in leadership responsibility and influence as a result.

It is not hard to visualize this concept in many other types of organizations as well. In government, for example, there will be technical geeks and policy wonks, and these people may be very good at what they do, but if they cannot effectively communicate with and influence their opposite numbers, their effect on the success of the agency may be marginal. Katherine Wallman, Chief Statistician of the United States, is a good example of the kind of statistical leadership in government that results when a person has mastered the technical side and the ability to communicate with the policy people. As chief statistician, Kathy works in the White House Office of Management and Budget, providing strategic guidance to the vast federal statistical system. She provides oversight and facilitates coordination of statistical policy, standards, and programs for the system, and represents the US government in major international statistical entities, such as those in the United Nations and the Organization for Economic Cooperation and Development. Lisa LaVange, director of the Center for Drug Evaluation and Research (CDER) Office of Biostatistics at the Food and Drug Administration, is another excellent example. CDER statisticians and other scientists make certain that safe and effective drugs are available to the American people. LaVange leads a large team of statisticians and is responsible for developing policies and procedures for statistical review of regulatory submissions and coordinating research to evaluate drugs and biologics.

In academe, senior leaders like Dan Solomon, dean of the College of Sciences at North Carolina State University; David Madigan, executive vice president and dean of the faculty for the Arts and Sciences at Columbia University; Judith Singer, senior vice provost for faculty development and diversity at Harvard University; Sallie Keller, formerly dean of the College of Engineering at Rice University; Xiao-Li Meng, dean of the Harvard University Graduate School of Arts and Sciences, and Sastry Pantula, dean of the College of Science at Oregon State University, have risen to major university leadership roles, in no small part because of their abilities to speak the language of a broad range of disciplines.

In industry, leaders are making a difference in their companies by successfully demonstrating and communicating the importance of statistics to the bottom line of their companies. Three examples:

- Christy Chuang-Stein, vice president and head of the Statistical Research and Consulting Center at Pfizer, provides scientific leadership for statistical advice to Pfizer projects, collaborating with researchers to develop innovative methodology that can help increase the efficiency of pharmaceutical product development.

- Martha Gardner, chief scientist and global quality leader at General Electric (GE) Global Research, helps GE teams integrate the use of quality improvement tools and methods into their research and product development programs. In addition, she leads GE-wide initiatives, such as the GE Analytics Engineer program and the GE Reliability Council.

- Theresa Utlaut, principal engineer at Intel Corporation in the Logic Technology Development group, is responsible for the research, development, and deployment of statistical methods that are used throughout Intel's manufacturing environment. She supports the research and development groups through consulting, developing, and deploying statistical training, and statistical computing.

In education, statistical leaders have had a significant impact on the quality of teaching of statistics at the K–12 level by effectively connecting the statistical community with the educational community.

- Richard Scheaffer, professor emeritus of statistics at the University of Florida and former ASA president, launched a nationwide quantitative literacy program and is considered by many to be the "father of Advanced Placement (AP) statistics." He has been the "statistical education guru" for several decades.

- Christine Franklin, senior lecturer Lothar Tresp Honoratus Honors Professor at University of Georgia, is a leader in US efforts to improve the teaching of statistics and the preparation of teachers at the K–12 level.

- Roxy Peck, associate dean emerita and professor emerita of statistics at California Polytechnic State University in San Luis Obispo, is among the most respected leaders in improving undergraduate instruction of statistics and has also been instrumental in advancing AP statistics.

Not surprisingly, all three are recipients of the ASA Founders Award for their outstanding service to the association and the profession.

All of these people are statistical leaders, and, thankfully, there are many more like them (with apologies for not listing them). Unfortunately, far too few statisticians are making it to the middle and fully realizing the leadership opportunities that await them, indeed, that NEED them.

Over the next few years, the ASA hopes to change that. To begin with, every statistician who wants to develop leadership skills should view her or his professional society as a place to do so. All of the people listed have had significant leadership roles within the ASA. Four of them (Wallman, Keller, Pantula, Scheaffer) have been ASA presidents. The ASA provides several hundred leadership opportunities each year through its committees, councils, chapters, sections, and other entities. Opportunities to mentor people new to our field are on the increase within the ASA as well.

However, the ASA has been providing these kinds of leadership opportunities for decades, and has realized in the last few years that it needs to do more. To that end, the association has introduced a personal skills development program to complement the professional skills development that takes place through traditional continuing education courses. The personal skills development program is or soon will be providing opportunities across four domains:

- Communication: speaking, presentation, consulting, listening, writing, and other communication skills

- Collaboration: also including team building, teamwork, understanding personality types

- Career planning: finding a challenging and rewarding position, goal setting, career advancement, negotiation, strategic planning

- Leadership: influence, conflict resolution, creative problem solving, and other leadership skills

Providing the chance for statisticians to sharpen these "get to the middle" skills is every bit as important as providing opportunities to learn new technical skills.

I am grateful for the opportunity to write the first chapter in this book on leadership for women in statistics. In my seven years at the ASA, I have been mentored by and marveled at the leadership and integrity of five outstanding women who were elected by their peers to be president of the largest statistical community in the world.

- Sallie Keller, who now leads the Social and Decision Analytics Laboratory at the Virginia Bioinformatics Institute, hired me, and provided both tremendous advice and a great start, leading the ASA toward a new era of involvement in science policy.

- Mary Ellen Bock, professor and former chair of the Purdue University Statistics Department (one of the largest statistics departments in the world), coached me through my first board meeting as executive director. Her calm, thoughtful professionalism helped me immensely, and her personal commitment to the success of the ASA was inspiring. Indeed, Kellers and Bock's commitment and dedication led to their receiving the ASA Founders Award, the ASA's highest recognition for service.

- Sally Morton, chair of the Biostatistics Department at Pittsburgh University, raised the professionalism of the association even higher, launched our strategic plan, and brought the perspective of a senior leader in industry to the board in ways that made all of us better.

- Nancy Geller, director of the Office of Biostatistics Research at the National Heart, Lung, and Blood Institute, was fearless and relentless, particularly about making sure everyone was heard and no one marginalized. She cared deeply about preparing the next generation of statisticians.

- So did Marie Davidian, William Neal Reynolds Professor of Statistics at North Carolina State University, who led with endless energy, with a vision for increasing the visibility of our profession, and a remarkable ability to connect with scholars in other disciplines.

What an extraordinary set of women role models I have been blessed with!

Of course, I have just listed a few current women leaders in the statistics profession. There are many others, and there have been many remarkable women leaders through the years. Fourteen women have served as the president of the ASA. The first was Helen M. Walker of Teachers College, Columbia University, a renowned statistical educator; she was ASA president in 1944 and later served as president of the American Educational Research Association. Dr. Walker helped increase professional visibility for women in statistics at a time when most prominent academic positions were held by men. The 1980 ASA president Margaret Martin was a leading government economist and the first executive director of the National Research Council Committee on National Statistics. She was influential in developing the Current Population Survey. And then there is the incomparable Gertrude Cox, a leading researcher in experimental design, president of the ASA in 1956, an eminent statistician, founder of the Statistics Department at North Carolina State University and director of statistics at what is now RTI International. Her contributions to statistics and to professional leadership continue to inspire young statisticians.

Every day, the executive director of the ASA sees how much difference statistical leaders make. We hope this book will help you take your place as a leader, in whatever roles that may entail. If you have never thought of yourself as a leader, we hope you will have a new mind-set, starting now!

Bibliography

Brillinger, D.R., "John W. Tukey: His Life and Professional Contributions," *The Annals of Statistics* 2002, Vol. 30, No. 6, 1535–1575.

Rodriguez, R. (2012). Amstat News President's Corner, February, 2012, http://magazine.amstat.org/blog/2012/02/01/statisticalleadership/.

2

Persuasion, Presence, and Patience: Important Characteristics of Leadership for Women in Statistics

Laura J. Meyerson

Biogen Idec

When I was asked to write this chapter, I had some hesitation. I am a woman and currently lead a staff of approximately 200 that includes statisticians, data managers, medical writing scientists, and data analysts at a Fortune 500 pharmaceutical/biotechnology company. I have been with the same company in this role for more than 10 years, and have quadrupled the size of my department with the growth of the enterprise. Yet, I had some trepidation. What could I impart on this topic of leadership? Did I have a unique point of view? Or, would I just reiterate what others have said about women and leadership? Is there something unique about the field of statistics that makes leadership especially challenging? As I started to ask myself these questions, I started to dig deep and ponder what it was that I had learned on my journey that was important to me with respect to leadership. I thought about what are the specific challenges related to being a woman leader. I also thought what was challenging about the field of statistics and leadership. It occurred to me that I might be able to impart a unique perspective on women, leadership, and specifically leadership challenges in the area of statistics.

My role in industry started as an individual contributor working on a team as a lead statistician with other members from various disciplines. These teams typically manage a clinical development program encompassing several studies or experiments designed to gather evidence for or against the marketing of a medicine. Products do not move forward either because they have a safety problem or they do not work as well as planned. Team members are highly trained in their fields and almost all have advanced degrees. Because I was typically the only member trained in statistics, I needed influencing skills in order to persuade others to consider statistical principles and concepts. As I took on more responsibility, moving into roles responsible for multiple products where other statisticians reporting to me were individual contributors, I eventually managed people not only from statistics, but also from other disciplines, such as statistical programming, data management, and medical writing. I sit at the decision table now with the leaders of those other disciplines that sit on the aforementioned teams (i.e., regulatory, medical, clinical operations, etc.). In this role, I am called to be a good people manager and also need to influence my peers who represent other-than-statistics disciplines. I not only need to persuade others using statistics as it applies to drug development, but also how to persuade others to allocate positions and money to certain areas. In all of my roles, from individual contributor to the leader of a large, diverse organization, I have found persuasion to be a critical element of effectiveness.

The points I make in this chapter are my own opinions that I have gathered from my own experience and are not those of my company or any other company. I hope I can impart some insight on this topic for many reasons. One is that there is a great need for statistical leadership, given the magnitude of data that is now and continues to be available via the internet. Data analysts with very little statistical training will be able to analyze the data using software tools. The 2011 McKinsey report (Manyika, 2011) predicts by 2018 the United States alone will need 150K data mining skilled professionals. Currently, the American Statistical Association has about 18,000 members. A quick calculation shows there will be a need for 10 times the number of people with data mining skills as there are statisticians. Each statistician will have to have incredible leadership skills to influence those mining the data to minimize reporting of false signals. Secondly, I am a woman, and there are certain advantages and disadvantages in being a woman leader that I would like to impart. And, thirdly, statistics by its very nature of being detail oriented and assumption based with a lot of unfamiliar terms makes it especially difficult to influence others. Business executives want to know your position, not always the details of how you got there. Too many assumptions and caveats become confusing and are not persuasive to those untrained in statistics.

In this chapter, I cover three characteristics of leadership. I believe these three characteristics have been the most important for me as a statistician and in particular a woman statistician in industry. Most great leaders possess these three characteristics: 1) excellent *persuasion* skills, 2) executive *presence* and 3) *patience* and humility. I call them the 3 Ps: persuasion, presence and patience.

As a leader of a large organization and people manager, I should mention the fourth P of leadership, *people*. You must like people to interact with them, to help them grow, and reach their potential. Tom Peters, an American writer on business management practices best known for *In Search of Excellence* (coauthored with Robert H. Waterman, Jr.), in an interview with McKinsey (Peters, 2014), says:

> If you are a leader, your whole reason for living is to help human beings develop — to really develop people and make work a place that's energetic and exciting and a growth opportunity. If you are in leadership, you do people. Period. It's what you are

paid to do. People, period. Should you have a great strategy? Yes, you should. How do you get a great strategy? Hire a great strategist, not by being the world's greatest strategist. You do people. Not my fault. If you don't get off on it, do the world a favor and get the hell out before dawn.. . .

Persuasion

Statisticians and all great leaders must possess excellent persuasion skills. Whether you are trying to impact a team as the sole statistician or convince those working for you to change, you must be able to utilize techniques of persuasion. For example, a good lead statistician must convince the team to consider good statistical principles such as bias minimization when interpreting data summaries and analyses. Additionally, executives must influence employees to follow them in times of uncertainty and change. Also, you must be able to influence others in a committee discussion regarding new investments in resources. All of these activities require excellent persuasion skills.

Techniques of persuasion or influence are not easy to develop, even though it is a relatively old concept. The first recorded reference to persuasion techniques was in BC Aristotle's rhetoric on ethos, logos and pathos (Rhys Roberts, 2004). Ethos, or ethical appeal, is creating a convincing argument based on the characteristic of the messenger. Logos, or the appeal to logic, is to create a convincing argument based on reason. Finally, pathos, or emotional appeal, is to convince your audience based on its motives, feelings, and attitudes. One of the most useful courses covering persuasion is that given by Gary Orren at Harvard University JFK School of Government. Below are some examples from his Orren's LEAP framework (Orren, 2005).

Examples of Ethos, Logos, and Pathos:

ETHOS

I will end this war in Iraq responsibly, and finish the fight against al Qaeda and the Taliban in Afghanistan. I will rebuild our military to meet future conflicts. But I will also renew the tough, direct diplomacy that can prevent Iran from obtaining nuclear weapons and curb Russian aggression. I will build new partnerships to defeat the threats of the 21st century: terrorism and nuclear proliferation; poverty and genocide; climate change and disease. And I will restore our moral standing, so that America is once again that last, best hope for all who are called to the cause of freedom, who long for lives of peace, and who yearn for a better future.

This example is a quotation from the Democratic presidential candidate acceptance speech by Barack Obama on August 28, 2008. He is appealing to the ethics of his audience in this quote. He speaks about partnerships, moral standing, hope, and a better future. All these attributes have ethical appeal.

LOGOS

> However, although private final demand, output, and employment have indeed been growing for more than a year, the pace of that growth recently appears somewhat less vigorous than we expected. Notably, since stabilizing in mid-2009, real household spending in the United States has grown in the range of 1 to 2 percent at annual rates, a relatively modest pace. Households' caution is understandable. Importantly, the painfully slow recovery in the labor market has restrained growth in labor income, raised uncertainty about job security and prospects, and damped confidence. Also, although consumer credit shows some signs of thawing, responses to our Senior Loan Officer Opinion Survey on Bank Lending Practices suggest that lending standards to households generally remain tight.

This quotation is from The Economic Outlook and Monetary Policy by Ben Bernanke on August 27, 2010. It has logical appeal by using numbers, trends, and reasoning. This can be very convincing to a statistician, but often fails to persuade a general audience.

PATHOS

> I am not unmindful that some of you have come here out of great trials and tribulations. Some of you have come fresh from narrow jail cells. And some of you have come from areas where your quest — quest for freedom left you battered by the storms of persecution and staggered by the winds of police brutality. You have been the veterans of creative suffering. Continue to work with the faith that unearned suffering is redemptive. Go back to Mississippi, go back to Alabama, go back to South Carolina, go back to Georgia, go back to Louisiana, go back to the slums and ghettos of our northern cities, knowing that somehow this situation can and will be changed.

This example is from the "I Have a Dream" speech by Martin Luther King Jr. August 28, 1963. It has emotional appeal by speaking to one's faith and redemptive suffering. Emotional appeal was very important in the civil rights movement. It would have been very difficult to move this initiative forward with only a logical argument.

Statisticians often fail in leadership because the logical, coherent argument frequently fails to persuade (for e.g., Clinton health care plan, bush v. Gore). Statisticians may tend toward the logical appeal, since we are comfortable with numbers and trends. Have you ever felt like it is so logical, why do they not get it? Logic is definitely a necessary component of what we do, but not sufficient. If you want to have impact with statistics, you must not only appeal to the logical argument, but also think about the motives and attitudes of your audience.

Aristotle also considered Agora, a Greek word for a gathering place, as an important element of persuasion. In other words, context is important. Where are you when you are trying to persuade someone? Do you catch them in the elevator? Or, do you have a meeting? Or, do you invite them for coffee? It depends on the relationship, but it matters. Also, the timing of when you try to persuade someone is important. It must be done in a manner that is not interrupting of other agenda items.

The last element that Aristotle considered important for persuasion is called Szygy, which in Greek means the rare alignment of the celestial bodies, such as the sun, moon, and earth during an eclipse. He believed that in order to persuade effectively, you must have a perfect balance between ethos, logos, pathos, and agora. I think it is too often that statisticians rely on logic and neglect the other pieces that are important, thereby not achieving Szygy in leading others. Table 2.1 lists the elements of the LEAP framework as noted by Orren.

Is persuasion really that important — we are statisticians, we think logically, and impart our logic. Dr. Orren speaks to the three factors of social influence. They are power, negotiation (payment), and persuasion. Dwight D. Eisenhower said that leadership is about getting others to do something **you want** done because **they want** to do it. This takes persuasion. Sometimes statistical arguments are very complex. Statisticians must think of simple, yet credible ways to make their statistical points.

Agora can follow simple formulas. This often depends on the rules of an organization. Who has the decision-making power? What are the forums where decisions are made? Whereas, ethical and emotional appeal may be more challenging. Women and minority leaders may have a special challenge in regard to ethical appeal. Because this is about the character of the messenger, it can be difficult when others have not experienced that type of leader before. "This is called implicit bias" in Olkin (2014). He illustrates "implicit bias" when faculty are considering the admission of students, hiring of faculty, or giving an award. The committee wants to choose the "best" candidate. At face value, this is a most reasonable requirement. He states that in the shadow the

Table 2.1
Five Elements of Persuasion (Orren, 2005).

LOGOS	Message	Content of the argument, reasons, data
ETHOS	Messenger	Character, credibility, plus other characteristics
AGORA	Context	Where, when (setting, rules, timing)
PATHOS	Audience	Emotions, plus other predispositions
SZYGY	Alignment	Reasonable balance among all other elements

definition of "best" often means "just like us," and, of course, the "us" are men. What Olkin is referring to in the business world is called "executive presence," and although it is a requirement to get to the corner office, it is more challenging for those that are different, such as minorities or women.

Presence

Presence is that elusive quality that is perhaps why you are not given a seat in the corner office. It is part of ethical appeal or ethos as described in Aristotle's rhetoric and seems to be related to three things: gravitas (most important), how you communicate, and how you look. Presence is easier for men than for women. As Olkin says in his article in *Amstat News* on tenure and women is true in industry and government. When it is the men who are deciding on who sits in the corner office, and to date all corner offices are inhabited by men, then the best person for the corner office must look like a man. The bar for executive presence is higher for women than for men.

Sylvia Hewlett (Hewlett, 2014) wrote a wonderful book with a lot of examples and detailed descriptions and examples of executive presence (EP) and how to attain it. In her book, she describes gravitas as the core characteristic. Some 67 percent of the 268 senior executives surveyed nationally said that gravitas is what really matters. Signaling that "you know your stuff cold," that you can go "six questions deep" in your domains of knowledge, is more salient than either communication (which got 28 percent of the senior executive vote) or appearance (which got a mere 5 percent).

Gravitas was one of the four Roman virtues. It indicates seriousness, dignity, a certain depth of personality, and authority. This comes easily to some but is more difficult for others. However, I believe if one wants to learn this, they can. Image is one of the elements of gravitas. Image is about what you are telegraphing to others. The wrong message and the wrong messenger can destroy careers no matter the substantive reality. Do you signal to others that you are a star and have what it takes? Do you act like you know your material six questions deep? You want to create an indelible impression.

A recent study underscores the importance of image in the world of music. In a piece published in the *Proceedings of the National Academy of Sciences*, the University of London researcher Chia-Jung Tsay, working with a sample audience of one thousand, reports that people shown silent videos of pianists performing in international competitions picked out the winners more often than those who could also hear the sound track (Chia-Jung Tsay, 2013). The study concludes that the best predictor of success on the competition circuit was whether a pianist could communicate passion through body language and facial expression. This evidence from the world of music underscores the tremendous power of image: How musicians present themselves creates an indelible impression. We might like to think that we're evaluating a performance of Bach or Shostakovich based solely on what we

Table 2.2

Gravitas Elements (Hewlett, 2014)

Element	Women	Men
Grace under fire	79%	76%
Decisiveness/showing teeth	70%	70%
Speaking truth to power	64%	63%
Demonstrating emotional intelligence	61%	58%
Right-sizing your reputation	57%	56%
Vision and charisma	50%	54%

hear, but in reality we're profoundly conditioned by the visuals. Judgments are made before the first note sounds in the concert hall. It is no different in the workplace.

This is most complicated for women and minorities. Questions around conformity versus authenticity arise for these historically underrepresented groups. How much should you fit in? How much should you stand out? In Hewlett, all of those interviewed said they struggled with this, as they all need to fit into a specific organizational culture, but for the historically underrepresented groups, it was especially painful, since they are required to "pass as straight white men." This is because this continues to be the dominant leadership model. The majority of those who sit in corner offices on Wall Street and Main Street are looking this way. It should be that over time, this authenticity struggle will get easier because more women are getting into the corner offices, and with age and experience they will be able to bring more of themselves to work.

Hewlett describes the top aspects of gravitas as being 1) grace under fire, 2) decisiveness and showing teeth, 3) speaking truth to power, 4) demonstrating emotional intelligence, 5) right-sizing your reputation, and 6) vision and charisma. Hewlett and her research team conducted a national survey that involved nearly 4,000 college-educated professionals — including 268 senior executives (both male and female) — to find out what coworkers and bosses look for when they evaluate EP. They also conducted forty focus groups and interviewed a large number of leaders. Table 2.2 shows the survey results on the importance of each of these elements. The percentages, listed by gender, refer to the number of senior executives that found the element important. Although the differences appear small from a sample of only 268, the weight of evidence from all the interviews and focus groups confirm that on most elements, women are held to a higher standard.

Many believe that gravitas cannot be learned. In Hewlett's book on EP, she breaks it down into its elements and gives examples in a way that it can be learned. It is a process, and it takes experience and patience to learn these elements and gain gravitas.

Patience

I have included patience as an important element of leadership, because it is the thread that binds us in both persuasion and presence. To be persuasive, you must be patient to have the right time and place for your argument. You must patiently wait for your audience to be ready to accept the concepts you are presenting. And to have EP, you must have experience. You must experience a mistake or accident to have grace under fire. You must have opportunities to show that you have teeth and can make those tough decisions. It takes time to demonstrate gravitas, but you can get an edge in if you have good communication skills and a polished appearance.

Concluding Remarks

I would like to impart some of my personal challenges and solutions to persuasion, presence, and patience so that you can understand my path and why I have chosen these specific characteristics. Persuasion has been something of a challenge my entire career. Early in my career, I worked as an assistant professor. Of course, teaching is a form of persuasion that I found challenging, but not impossible. There is a certain amount of leverage a professor has over his or her students that makes persuasion easier. I believe the most difficult part of being a woman professor had to do with my presence, which illicited a bias in finding more senior professors that would work with me and help me with my research. Almost all the senior professors were male, and I felt there was an unspoken discomfort for them to mentor females. I really needed this in order to gain tenure, and felt it was too difficult so moved to the pharmaceutical industry.

I started as an individual contributor, where my biggest challenge was to persuade other team members as to the importance of statistical principles. They were not trained in statistics, and they were not my students, so I had to find a way for them to see why the statistical points were important. One of the most effective ways to do this was to better understand their fields (typically medical) and then explain the statistical point in their terms, so that they could see how it would affect their experiment. I often felt my job was to ensure there was no bias in the experiment or interpretation of the results. Was there sufficient randomization, what decisions were made as the trial was ongoing and did they affect the study conduct, what was pre-specified, how was missing data handled, was any modeling biasing the results, etc.? It was sometimes difficult to convince the nonstatistician why these were important questions.

As my career progressed I excelled in project management, and being a statistician on a clinical program in industry takes a lot of project management. I enjoyed working with multiple statisticians and statistical programmers creating the most efficient and high-quality statistical presentations for regulators to examine. In one submission, there are thousands of presentations of data, programs, and datasets that need to be delivered. I enjoyed leading and organizing this and was recognized for being good at it. I did find that once I was recognized for being good at this, I was asked to do it over and over again and not given new opportunities, such as leading a department. I wanted to be a line manager and saw that my male colleagues were asked to do this even though they may not have been as good at project management as I was. I found a company that was willing to give me a line management position and moved. I have made three major company moves to get to the position I am in today. Each move was with increasing responsibility, and I have had success in each of the roles I have taken.

It is not clear whether the opportunity at my current company existed at the time that I was ready for a new opportunity, or if it was that I was not viewed as having the "presence" necessary for that new opportunity. If I may, I will speculate that once a woman proves that she is good at something, she is more often asked to keep doing that rather than looked at for a higher position or another opportunity. This might be another expression of "implicit bias" towards woman. It never took me long to get a new position at a new company. So, that was what I did. I rarely was not offered a position after an interview. My main experience on EP has been from feedback from certain male bosses. The feedback feels very much to me like they want me to act more masculine in the boardroom or at executive meetings. Sometimes when I show passion, I am told to not show too much passion (females look like they are going to "lose it"). I have even been told not to be "too authentic." My take on all this is that femininity is still not comfortable in the business arena.

Being a woman presents many challenges with respect to leadership positions in statistics and other disciplines. In this chapter, I have not discussed work–life balance and how it can be more difficult for women in the workplace. As male roles in the family change, work–life balance will become less gender specific. I have spoken about the tools I learned during my journey of increasing roles in leadership. I hope that these words leave you with something to think about as you follow your path.

Bibliography

Hewlett, S.A. (2014). *Executive Presence: The Missing Link Between Merit and Success*. HarperCollins.

Manyika, J., Chui, M., Brown, B., Bughin, J., Dobbs, R., Roxburgh, C., and Hung Byers, A. (2011). Big data: The next frontier for innovation, competition, and productivity In *McKinsey Report*, May.

Olkin, I. (2014). Where have all the tenured women gone? *Amstat News*, January.

Orren, G. (2005). Persuasion: The Science and Art of Effective Influence. In *A Presentation from the JFK School of Government*, Harvard University.

Peters, T. (2014). Interview with Tom Peters on Leading the 21st Century Organization. In *McKinsey Quarterly*, September.

Roberts, W.R. (2004). *Aristotle's Rhetoric* (translated). Dover Publications.

Tsay, C.-J. (2013). Sight over Sound in the Judgment of Music Performance.

3

Domains of Leadership Behavior in Organizations

Sim B. Sitkin

Duke University

In this chapter, I will discuss a relatively new model of leadership that draws broadly on the social science research literature. It conceptualizes leadership in terms of specific clusters of behaviors that have distinguishable effects on those a person is trying to lead. While there are many theories of leadership and voluminous writing on the subject, this research-based framework is intended to synthesize that work into a systematic, intuitive, and grounded set of behavioral guidelines by which anyone can consciously work to improve their leadership effectively. Because of the systematic and scientific attributes of this model, we have found in our work with scientists, executives, and others that this approach makes leadership to be less of a fuzzy-but-nice topic based on weak research, and more of a set of specific, challenging targets to pursue. To our early surprise, these attributes led to the model being particularly valuable to those who came from technical fields (medicine, financial analysis, etc.), and thus it may be useful for statisticians.

With this volume's focus on women in statistics, this approach to leadership may provide some additional benefits. Sometimes, the route to effective leadership is perceived to be paved with technical excellence, gaining power positions, and increasingly fancy titles. All of those are, of course, valuable and should be equally accessible to women in any field. But they are not a sure route to effective leadership, much as they can complement it. Instead, I would suggest those desirable outcomes are made more likely by the ability to execute effective leadership behaviors — and my goal in this chapter is to very briefly provide an overview of what those behaviors are and what their effects are.

Before I describe our model and the behaviors it implies, let me first clarify what I mean when I refer to leadership.

A Different Way of Thinking About Leadership —
Putting Leadership in Context

The term "leadership" is used in many ways. As an academic, I want my students to play a "leadership role" in their field. I also want them to be viewed as a "thought leader" in their chosen area of specialty. Finally, I want them to aspire to and achieve administrative "leadership positions" in their organizations. But if leadership is to be a substantively meaningful term that provides guidance for effective behavior, we need to be more precise in how we deploy it.

I will distinguish the term leadership from other similar and related terms, acknowledging that it is fine (and inevitable) that everyday usage of the term will be far less precise. But in the absence of a clear sense of the distinction, and especially in technical fields like statistics, my experience is that when leadership is used in a way that is not distinguished from other constructs, the effect is that a focus on actual leadership gets lost. So I hope my nit-picking will be tolerated.

What do we really mean when we refer to a "leadership role," "thought leader," or "leadership position?" The term "leadership roles" is typically used to mean accepting managerial responsibility for getting something done through a group of other people. "Thought leadership" is a lay term that scientists typically do not use, but is widely used in business and the popular press and coincides with the notion that a person is recognized as a highly regarded technical expert who is called upon to give advice, make presentations, lead panels, and edit books. Administrative "leadership positions" involve being given formal power and authority to direct the activities of others in our field or related activities, and involves (sometimes grand) titles, including becoming a department head, and perhaps even a university dean or provost or a corporate or government senior executive. Although these roles are themselves quite distinct — manager, technical expert, and administrative authority — none necessarily imply the presence of outstanding leadership. They are often simply referred to as "leadership" even though we can all think of people with titles or managerial responsibility or great expertise who are terrible leaders. Similarly, we can think of people we have worked with or encountered in everyday life who are wonderful leaders even though they may not be good managers, be the best technical expert, or have a fancy title. So if leadership is clearly distinct from these other constructs, what is leadership?

Based on the work I have done with my colleague, Allan Lind, I define leadership as "behaviors that influence others in the pursuit of goals." So leadership is about engaging in specific behaviors that exercise influence. It is not about power or titles or even expertise, though each of those factors can certainly make it easier to effectively exercise influence. But why "the pursuit of goals"? Because leadership is exercising influence with a purpose, rather than merely to control others or be viewed as a big shot. The goal can be an organization's goal or your goal in leading those others, or it can be helping others achieve their own personal goal. It can involve a technical goal, such as solving a challenging problem. It can involve an administrative goal, such as improving a department's productivity. Or it can involve a person goal, such as helping someone achieve the next steps in their career ladder or learn a new skill. But leadership is engaging in behaviors that foster the pursuit of a goal.

From this way of thinking about leadership, we all have the opportunity to exercise leadership regardless of our formal position or our personal style. We can not only lead subordinates (if we supervise others) but can also "lead up" or lead laterally toward peers or "out" toward those who use our professional services (e.g., clients).

Creating a New Framework for Understanding and Testing Leadership Effects

In creating a new model of leadership, we did not want to ignore what had been done before, so we carefully drew not only on the leadership literature, but also from work on leadership in organizational behavior, social psychology, sociology, and political science. What emerged from this review was six distinct areas of leadership behavior, which we refer to as "domains of leadership," each with its own distinctive behaviors and effect on followers.

We have examined these domains in our research and in our teaching and consulting in many countries around the world, in different industries and occupations, and a wide range of levels of the organization hierarchy. While the way one implements the behaviors can vary (for example, letting others know what level of expertise you bring to a task needs to be very indirect under some circumstances but can be quite direct under others), the sets of behaviors and effects have thus far received consistent support.

The Leadership Pyramid

The six domains provide a framework to address the full range of skills that are needed to exercise leadership. The three foundation domains include letting others know who the leader is and what competencies, values, and style she offers (personal leadership), conveying to others that you understand and care about them and will treat with respect and fairness (relational leadership), and helping others to make sense of the situation and their role in it (contextual leadership). These foundation elements are intuitive — one is about the leader, one about the follower, and the third about the situation. Building upon this foundation are the action-oriented parts of leadership — communicating that others should strive for achievement and be optimistic and enthusiastic about the opportunity (inspirational leadership) and offering a clear sense of the scope for effective action through honest feedback, encouragement of initiative, and protection when a reasonable action does not succeed (supportive leadership). The final integrative domain of leadership involves fostering a sense of balance, ethical principles, and sense of personal ownership of the mission (responsible leadership). As described below, each domain represents a cluster of specific behaviors that result in distinct effects on those being led.

Personal Leadership

The effect of personal leadership is credibility. What builds a leader's credibility? Of course, it involves having the capability to lead. But having such a capability, we have found, is not enough. The key is to let others know that you have that capability — otherwise, you may merit credibility but not be perceived as credible. So one set of personal leadership behaviors involves communicating to others what your skills, knowledge, and other relevant capabilities are. I am from New York but live in North Carolina. In New York, I can be quite direct in letting people know what I am and am not good at. In North Carolina, I need to be much more indirect in my communication. But in order for your personal leadership behavior to be effective in creating credibility, it is not enough that you yourself know, you must ensure that they know too. Put differently, you may not need to blow your own horn, but the horn must be blown for others to hear the music.

Similarly, for the other aspects of personal leadership behavior, being (e.g., being competent) is not enough; one must take action to convey in a personally and situationally appropriate manner (e.g., acting to let people know about a competence that you have that they may not be aware of). These other aspects include your beliefs and values, and your level of dedication to the organization and mission (versus self interest). And finally, personal leadership involves that familiar buzzword — authenticity. While our technical and managerial roles do not require that people have a sense of who we are as people, as professionals, including what we believe and value, to lead others they need to feel like they have a sense of who it is who is leading them.

This is not about bragging or making it all about you. But people follow people and not abstractions, so to lead (as opposed to manage or analyze), you need the courage to reveal just a bit of yourself. They do not need to know everything about you (leadership is not reality TV), but they do need to have just enough of a sense if you want the credibility for your advice and your analyses to be given the credence they deserve.

Why does this matter? Allan Lind and I have suggested that an effective personal leader must be seen as authentic, with knowledge, insights, and values that are worthy of respect, and whose dedication to the team is clear. And credibility is important because it makes others more willing to accept the leader's direction.

Relational Leadership

Relational leadership emphasizes interpersonal ties with others. You do not have to like the other person or have a personal relationship with him or her to exercise effective relational leadership. It is not about becoming friends, or even about being friendly — but about establishing a very specific type of connection through your actions that convey, that you understand and care about the other person, and that you respect him or her and will treat him or her fairly. Think about those you might be trying to influence. Do they believe that you understand them as people and as professionals, what they value, and what they are trying to achieve? Do they believe you are just using them and their talent, or that you really are concerned with their welfare independent of its benefit to you or the organization (for example, do you prepare them to be well-positioned for a promotion or to leave for another job because it is better for them, even though not necessarily so great for you). If you have conveyed that you understand them, including their weaknesses, do they know you respect them — if so, how do they know that?

Relational leadership complements personal leadership by making the leader–follower relationship become two–way; it is about the leader (personal) but also about the leader's connection to the follower (relational). There are four clusters of behavior that comprise relational leadership — each implied by the foregoing questions — understanding, respect, concern, and fairness.

The first involves demonstrating that you really understand the person you are trying to lead. This may mean that you let them know you see potential in them or recognize their strengths or weaknesses that they were unaware others could see. It also involves conveying that you know them as a person. You do not have to be close friends with someone to know what their hobbies are, know about their family, and if there are current life circumstances that might be of concern or joy. It may seem silly or tangential to your responsibilities as a leader, but they are fundamental to leading effectively, because much of your other leadership efforts hinge on whether I believe you have a clue as to who I am and what I am capable and not capable of doing.

The second involves conveying that you respect the person you wish to lead. If you understand me, you know my weaknesses — which leads me to worry that you will no longer respect me as a professional or as a person. So I need to know that you do — and feeling respect for me is not enough, you need to find a way to let me know. Note that having established a sense of understanding is critical for respect to be meaningful. It is meaningful that your family respects you as a person, but less so that they respect you as a professional, unless they have expertise in statistics so that their judgment is based on relevant understanding.

The third type of actions involve conveying that you are concerned about me and are not just using me to achieve your or the organization's ends. I take personal pride in having "lost" many assistants over the years as I encouraged them to go back to school, advocated for them for promotions, or supported them in moving to other organizations to advance their careers or address personal concerns. I make assignments that give them the chance to develop and demonstrate a readiness for those next moves — sometimes in cases where it slows down some of our work but allows them to develop. And I let them know why I am doing it. I ask the team to pitch in and cover for someone who has a sick relative. I lead up by protecting those above me from distractions and by giving them a chance to safely vent about issues or express personal angst about a failed initiative or personal setback when I can see they need an outlet. Relational leadership is not just about leading those below you but also leading up and leading peers. When leading up, does that individual know you care about their success and feelings, or only that you come to them for information, resources, or approval?

I was recently at a well-known high-tech company that studied what differentiated their highly successful technical managers from those that were less successful. They expected to find that the successful managers were technical wizards, problem solvers, or gave clear direction. Those things mattered, but what emerged as the most important factor was that the manager let the employees know that they knew and cared about them through little things, such as remembering the names of their children or asking about an ill relative, as well as work-related issues like project interests or special individual abilities or aspirations. Successful managers got to know those they wished to lead, whereas unsuccessful ones did not. In short, successful managers were better relational leaders.

The effect of strong relational leadership is that followers trust you. Trust is defined as a willingness to be vulnerable to another under conditions of uncertainty. If I am to trust you in leading (i.e., influencing) me, it makes sense that I would want to know that you respect me and care about my welfare (after all, I am making myself vulnerable to you) and that you know me well enough to not inadvertently hurt me out of ignorance (e.g., assure me that I am capable of doing this new, highly visible assignment, but I very publicly fail because you misjudged my relevant knowledge). Not only is it intuitive, but our research supports that effect.

Contextual Leadership

Contextual leadership is based on a very simple idea: People look to those who lead them to help them make sense of the world. This sensemaking function of leadership seems so obvious and basic that ironically it is easily overlooked. In fact, in our research and in working with individuals and organizations from around the world, we found that this dimension of leadership is almost universally the weakest aspect of leadership. It is viewed as very high in importance, yet is done especially poorly. Why? I suspect it is because it seems too obvious and not because it is difficult to do once recognized. One striking finding is that this weakness is even more true of technical organizations and fields. In technical organizations, it is as though the leadership is thinking, "if I have to explain this to you, maybe you are not as good as I thought you were." Which also means that when those you lead do not understand, they would not want to ask or let you know, as it could damage your respect for them.

This sensemaking function has several specific behavioral components. The first is simplifying and prioritizing. We live in a complex world and deal with complex tasks that are only getting more complicated over time. And change is so constant that even if I knew our priorities quite clearly yesterday, I may need assurance that they remain intact today. Clearly, rookies need such guidance. But counter-intuitively, sometimes very knowledgeable and experienced pros need it, too, since their in-depth knowledge can sometimes make it hard to see the forest for the trees. That is where contextual leadership comes in.

The second aspect of contextual leadership is providing a sense of coherence in how things get done. This involves helping others understand how our organization works, how an analysis was done or what its implications are, and seeing where my tasks fits into the bigger picture of a project or a broader strategy. The low scores result from the assumption that people know and understand — but they look to their leaders to help, and it appears that clarification is far from sufficiently given.

Contextual leadership leads to a sense of shared community, where people understand what is important and how the pieces fit together.

Inspirational Leadership

Inspirational leadership may seem like the most familiar aspect of leadership, as it is written and talked about often. But it is misunderstood in that it is often associated with the trait "charisma," often drawing on nonscientific notions or weak research. Charisma may be a trait that leads people to think a person is cool or appealing, but that is not leadership, which aims to have influential effects on others rather than merely encouraging them to admire you from afar.

So what is inspirational leadership behavior, and what effect does it have? The effect is to get those you lead to raise their aspirations and aim for more than they (or you) thought was possible previously. This is accomplished when leaders convey three things: an expectation of excellence and innovation, optimism, and enthusiasm. Even some of these behaviors (specifically optimism and enthusiasm) are often misconstrued, thus requiring some explanation.

Research suggests that when we convey our expectations of others, they tend to meet those expectations. Inspiring others thus involves, in part, having high expectations both for the excellence of the outcome and the ability to be creative when confronted with obstacles in pursuing that outcome. Whereas the managerial function involves enforcing standards by setting goals, monitoring goal achievement and then rewarding or punishing effort and outcomes, the leadership function here is not to require but to encourage and expect. The purpose here is to elicit a sense of ownership and desire to achieve (leadership) rather than to demand minimum standards are met (management).

Optimism is the second component of inspirational leadership, and we can almost hear leaders try to inspire us with statements like, "I know you can do that" or "We're the best, and I am confident our great team can accomplish this." Such assurances are certainly a part of optimism, but they overlook a critical component. To inspire others, we need to be more than just cheerleaders, we need to convey to them why we believe it is possible, even likely, that we can achieve an especially tough objective. When you wish to inspire (and this could be inspiring your boss to let you or the team take on a challenging task that is high risk, or inspiring your boss to personally take on an aspiration set of tasks for the group), do not just assert that you are confident. It is much more effective to also explain the basis for your confidence and paint a picture of a pathway by which success can be achieved.

The third component of inspiration is enthusiasm. This is not about general positive affect. Mere cheerfulness is not inspirational. Instead, enthusiasm needs to be very targeted to the mission or goal at hand. If optimism is about conveying why the person or group is the right one to be pursuing a challenging goal, enthusiasm is about conveying why that goal is worth pursuing to begin with. Inspirational leaders like Martin Luther King or John F. Kennedy conveyed why we should be excited about pursuing civil rights or a moon landing. Solving an especially tough problem or getting two noncooperative groups to coordinate can be daunting, and part of inspiring others involves helping them to feel like those daunting tasks are worth taking on to begin with.

Effective inspirational leadership gets those you lead to raise their expectations for themselves and for the team. Our research suggests that even for individuals who start out with very high aspirations, leaders who expect excellence and convey optimism and enthusiasm are able to influence those individuals to raise their aspirations even higher.

Supportive Leadership

Supportive leadership is also often misconstrued as merely providing positive and warm encouragement that leads others to feel appreciated, safe, and perhaps even complacent. Making someone feel valued and appreciated is certainly one aspect of this dimension of leadership. But it is more. Supportive leadership is about giving those you lead the resources they need to be able to take effective initiative. If you want others to learn to fish, you certainly need to teach them how to fish, but you may also need to give them a fishing pole, transport them to a lake, and make sure they know that if they fail to catch a fish, other food will be available to ensure their survival. The components of supportive leadership are: efficacy, security, and blame control.

Efficacy involves giving honest feedback about where someone has performed well and where they have performed poorly, where they have strong capabilities and where they are weak and need to either develop more or get help. This may mean letting someone who lacks confidence know that you believe that they are underestimating their knowledge or special abilities. But we also work with individuals who have oversized egos or are impatient for promotions and may overestimate their readiness. It may mean letting some air out of their balloons.

Security implies that if someone deploys their ability appropriately and puts in their best effort, that they will not be punished. In the absence of this your best people would be foolish to take on critical but risky tasks or try unproven but high potential solutions. Accompanying this is blame control, which can poison any organization. As a leader you want to promote a sense of accountability, but eliminate the common tendency for finger pointing when things go badly.

The effect of good supportive leadership is that people will be more willing to take appropriate initiative, rather than just waiting for direction. If people have a realistic sense of their capabilities, and know if they do their best they will be supported and not blamed for failure they could not control, then they will be willing to take action.

Responsible Leadership

Effective responsible leadership involves being a steward of the core values and missions of one's institution. This entails three kinds behaviors. First, responsible leadership includes personally exhibiting and advocating for a sense of balance. Second, effective responsible leadership need to personally embody and articulate a sense of responsibility and ethicality for the whole. Finally, responsible leadership implies a public role that involves others looking to the leader to personify the institution.

We all face competing demands and interests. Are our employees, those who use our services, or our top brass given priority? Are long-term goals or short-term issues given priority? When raises, promotions, or the most attractive project assignments are being doled out to a staff group, who gets what? When we face a need to get work done but also want to develop people, how do we strike the right emphasis on each element? When we need to make tradeoff choices involving the precision, comprehensiveness, and timeliness of our work projects, who and how do we settle on the right mix? The question underlying each of these decisions is, "What is the right balance?." One key aspect of responsible leadership is to properly balance among competing demands, but also to be sure to make clear how those balancing choices are being made so that the leader's approach can help guide others. This same logic applies to ethical role modeling. When it comes to ethical tradeoffs, leaders also serve as the template that others will use to judge what behavior is appropriate and what is not. So leadership involves both enacting the organization's and the leader's ethical principles, but also articulating them so that others both know what they are and also to reinforce them as operating principles as we do our work.

The final aspect of responsible leadership is *the public role of leadership*. Leaders represent their team and organization. Whether by choice or not, if one is exercising influence and looked to by others as a leader, then one has stepped into the spotlight. It simply comes with the territory. When acting as a leader, it is not effective to merely hide behind the technical and hope to avoid scrutiny. If others are allowing you to influence them, you should expect that they will wish to scrutinize your behavior and alignment with more general organizational and societal principles.

The effect of strong responsible leadership is not gaining the admiration of others (nice as that would be to have). Instead, it is about your leadership's effect in influencing them to feel a sense of stewardship. If you succeed, they should feel they own the project, the team, the organization — and thus will act in the broader best interests. That is the ultimate effect of strong responsible leadership.

Conclusion

What I have tried to do in this chapter is to provide a brief overview of the behavioral elements of effective leadership. As I noted at the outset, leadership is not about position or title, and being effective can be aided by personal traits (e.g. intelligence, oratory ability, social skills), but we can overcome our limitations and leverage our strengths by knowing what behaviors are required to obtain the effects of good leadership.

Whether you are trying to advise others to accept your analysis, you are serving as a team leader, or trying to be a good team member, these leadership behaviors apply. They apply at all levels of the hierarchy, in all types of organizations, and across national cultures. You can make them work for you as an introvert or an extrovert, but you must execute them in a way that fits your own preferences and style. And, finally, they apply whether you are leading up, down, or laterally in a hierarchy. In fact, they apply not only at work, but in family life and in our community work.

A last word about how this model relates to gender. There are stereotyped images of male leadership and female leadership, in which men are stronger at personal, inspirational, and responsible leadership whereas women are stronger at relational, contextual, and supportive leadership. We have not found that to be supported by the data. Individuals differ, but gender groups do not. But even if it were true, our world is moving in a direction where relational, contextual and supportive leadership are increasingly critical aspects of leadership for success. So the key is to use an understanding of the range of leadership behaviors required and then develop and deploy those behaviors to aid your leadership and have positive effects on those you wish to lead.

Some Suggested Readings

Much of the literature on the topic of leadership is a dangerous combination of being voluminous and of questionable value due to poor theory and methodology. I suggest a few readings that relate to the model I discuss in this chapter (Sitkin and Lind, 2007; Hernandez et al., 2014; and Janson et al., 2008), a few that provide overviews of areas of the field I think are more valid (DeRue et al., 2001; Eagly and Karau, 2002; and Van Knippenberg et al., 2004), and one that represents a cautionary critique of the field (Van Knippenberg and Sitkin, 2013). The goal is to reflect both this model and some diversity of approaches.

Bibliography

DeRue, D.S., Nahrgang, J.D., Wellman, N., and Humphrey, S.E. (2011). Trait and behavioral theories of leadership: An integration and meta-analytic test of their relative validity, *Personnel Psychology*, 64, 7–52.

Eagly, A.H., and Karau, S.J. (2002). Role congruity theory of prejudice toward female leaders, *Psychological Review*, 109, 573–598.

Hernandez, M., Sitkin, S., and Long, C. (2014). Cultivating follower trust: Are all leader behaviors equally influential? *Organization Studies*, in press.

Janson, A., Levy, L., Sitkin, S., and Lind, A. (2008). Fairness and other leadership heuristics: A four-nation study, *European Journal of Work and Organizational Psychology*, 17, 251–272.

Sitkin, S.B. and Lind, E.A. (2007). *The Six Domains of Leadership: A New Model of Developing and Assessing Leadership Qualities*. Chapel Hill, NC: Delta Leadership, Inc.

Van Knippenberg, D. and Sitkin, S. (2013). A critical assessment of charismatic-transformational leadership research, *Academy of Management Annals*, 7, 1–60.

Van Knippenberg, D., Van Knippenberg, B., De Cremer, D., and Hogg, M.A. (2004). Leadership, self, and identity: A review and research agenda, *Leadership Quarterly*, 15, 825–856.

4

Four Leadership Principles for Statisticians: A Note on Elizabeth L. Scott

Amanda L. Golbeck
University of Montana

Introduction

Statistics leaders advance the quality of research methods or outcomes in a broad spectrum of disciplines from medicine to astronomy to economics, influence others to go down a better path, and make sure their positive impact on both people and science lasts in their absence. One of the greatest examples of a statistics leader by this definition was University of California — Berkeley ("Berkeley") statistics professor Elizabeth L. Scott (1917–1988). Her legacy continues in the form of the Elizabeth L. Scott Award, administered by the Committee of Presidents of Statistical Societies (COPSS). While Scott at times held titled administrative positions in the university or professional associations, she was at her best when she exercised natural leadership based on pure influence toward solving societal problems, especially problems surrounding the status of women in academe. In this chapter, I discuss four leadership principles based on the qualities and behaviors of Scott that made her an effective natural leader and prominent role model for future generations, and I note the potential for all statisticians to develop their leadership and increase their positive impact in their workplaces and communities.

Background

I met Scott in 1977 at Berkeley when my research interests in applied stochastic processes and a market study catalyzed me to move to the doctoral program in biostatistics from the doctoral program in anthropology. Biostatistician Chin Long Chiang (1914–2014), whose specialty was applied stochastic processes, immediately advised me to meet with Scott to develop a plan to earn a master's degree in statistics along the way to the PhD in biostatistics. I found my way to Scott's office in Evans Hall on the Berkeley campus and looked around. Her desk, tables, and file cabinets were piled high with papers and files. Impression: She was serious and busy. Her office walls were covered with beautiful images of nebulae and other astronomical objects. Impression: She was into astronomy in addition to statistics. She sat behind her desk and asked me about my background and interests. Impression: She was a no-nonsense, even formidable, figure. I left her office feeling self-conscious about my red nail polish, not from anything she said, but from her demeanor. It was the last time I wore red nail polish. Bottom line: This was a good meeting. Scott was clear about what I needed to do, and she was encouraging. She held the door wide open for my entry into statistics, and I walked away from her office knowing that I was making a large leap into the right future.

Scott supervised my master's thesis, which I chose to do on a birth distribution problem. I presented the final product to her with typical young female timidity, to which she countered: "I think you have undervalued what you have done!" And she was right, because I was readily able to publish the thesis in a highly respected journal (Golbeck, 1981). Right after I completed the thesis, I decided to get married and took a leave of absence from my doctoral program. This must have made Scott cringe, because she had over the prior decade been conducting research on women's issues in academe, and one of the recurring charges from male faculty members was that women were not good investments in doctoral training, precisely because of actions like mine! But I knew two weeks into my employment that I would return to my doctoral program, and so my spouse and I made a two-year plan to return to Berkeley. In the middle of my leave of absence, I made a trip back to the campus to confer with Chiang and Scott about returning to my graduate program. Even though she was on sabbatical, Scott said it would be nice to see me in Berkeley and took me to lunch at the Faculty Club with some other women graduate students in statistics, and we discussed their activities and my plans. Scott was again holding the door wide open, this time for my return to biostatistics. I always felt she was there for me. As an example, she came to my dissertation defense even though she was not on my doctoral committee and her purse had been stolen from her office only minutes before. Her presence was a support, and it was a gift.

And so, even though I was not at the time aware of all of the details about Scott's contributions to and impact on the profession, it was no surprise to me that many others had the same high regard for Scott, and COPSS decided to establish an award in her name. The Scott Award is one of only five awards given by COPSS, which is comprised of the presidents of a number of eminent statistical professional organizations: American Statistical Association (ASA), Eastern and Western North American Regions of the International Biometrics Society (ENAR/WNAR), Institute of Mathematical Statistics (IMS), and Statistical Society of Canada (SSC). The operating principles for the Elizabeth L. Scott Award Committee state the following about the purpose and history of the award (COPSS, 2014):

> This award shall recognize an individual who exemplifies the contributions of Elizabeth L. Scott's lifelong efforts to further

the careers of women in academia. An astronomer by training, she began to work with Jerzy Neyman in the Statistical Laboratory at Berkeley during World War II and had a long, distinguished, career as a professor at Berkeley. She worked in a variety of areas besides astronomy including experimental design, distribution theory, and medical statistics. Later in her career, Dr. Scott became involved with salary inequities between men and women in academia and published several papers on this topic. In addition to her numerous honors and awards, she was president of the Institute of Mathematical Statistics (IMS) and the Bernoulli Society, vice-president of the American Statistical Association (ASA) and International Statistical Institute (ISI), and elected an honorary fellow of the Royal Statistical Society (RSS) and fellow of American Association for the Advancement of Science (AAAS). In recognition of her lifelong efforts in the furtherance of the careers of women, this award is granted to an individual who has helped foster opportunities in statistics for women.

Scott became involved in women's issues as a senior professor. She and others struggled for women to achieve equal academic rights. Scott became a leader within the academic women's movement on the Berkeley campus, for the Carnegie Commission on Higher Education, on national science committees, for the American Association of University Professors (AAUP), and within the national higher education community. She was an advocate.

In her time and within the academy, Scott argued for the replacement of antinepotism rules with conflict of interest rules. She argued for the constitution of a faculty that was representative of those who were trained. She pushed for promotion of women faculty members at the same rate as men. She argued for paid maternity leave. She campaigned for the appointment of women faculty members to major policy-making committees in representative numbers. Discovering in the late 1960s that at Berkeley only 2 percent of full professors were women, and there were only 5 women professors in the sciences, Scott was the first at Berkeley to articulate research questions on the status of academic women, especially,

"Why are there so few women on the faculty?"

Also, Scott asked that departments admit women to graduate school and recommend them for fellowship awards and teaching assistantships on an equal basis with men. She argued for the improvement of the status of women in research units, especially that women should be allowed to apply for grants as principal investigators in their own right. She pushed for equity in insurance and other benefits. She recommended that there be one faculty club that would serve the needs of faculty of both genders. Having been regularly denied access to The Faculty Club at Berkeley because she was a woman, and even though she was a full professor and department chair, Scott planted the seed that led to the opening of the club's membership to women.

But perhaps Scott's largest social contribution was in the area of pay equity. Many women faculty members over the years at many different higher education institutions benefitted from her nationally disseminated salary evaluation kit. Scott was not the first to use multiple regression to show that salaries were different for women and men, but she and her team were the first to use linear regression with interaction effects and, when possible, with productivity controls, to estimate the actual magnitude of these salary differences. This estimation approach allowed many women to successfully argue to their institutions and the courts for adjustment of their salaries.

Methods

I have been conducting historical biographical research about Scott over the past six years with funding from the Phoebe W. Haas Charitable Trust and the Philadelphia Foundation. My work draws heavily from information in archival records, most located at The Berkeley Bancroft Library, where the university archivist made available to me the very large, uncatalogued Elizabeth Scott Collection that contains all of Scott's professional files. I also accessed other collections at Bancroft, the Berkeley Academic Senate Office, the University of California at Santa Cruz Special Collections and Archives, and the Oakland Public Library Local Area History Section. My research is also using original, face-to-face, telephone, or email interviews, visits to the observatories where Scott did astronomical measuring, existing oral histories, and articles published in professional journals or on the Internet.

Four Leadership Principles and Recommendations

There are many self-evident truths about leadership that apply in any setting. Here I present and discuss four principles that I believe will particularly benefit statisticians, based on the professional example of Scott and some 30 years of my own experience as an academic leader.

(1) **Leadership is based on influence, not position. Statisticians should practice natural leadership each and every day and not be afraid to aspire to organizational leadership.**

Leadership is influence. If you are an influencer, you are a leader, whether or not you have an organizational title that presupposes leadership (Maxwell, 1998). Thus, managers may provide little or much leadership, and leaders may or may not be managers. While both management and leadership are enormously complex, difficult, and essential to organizations, they are also radically different. The work of managers is to keep their organizations running smoothly, and to ensure the consistent and ongoing quality of their organization's products or services. Management includes operations like budgeting, planning, structuring and staffing jobs, measuring performance, and solving problems. The work of leadership, on the other hand, is to set a vision for useful change and then find and exploit strategic opportunities to move the organization forward toward that vision. Leadership involves getting people to leap into the right future, no matter how large the leap or how quickly they need to make it (Kotter, 2013). Scott is an interesting case study for leadership, because at some times she behaved primarily like a manager, and at other times she acted like a natural leader.

Scott in early to mid-career, as a faculty member and before she held titled administrative positions, developed a strong advocacy background. She argued for academic and political freedom, civil rights, community assets, and students. Specifically, as a new PhD, Scott spoke out to the California governor in the anticommunist McCarthy era against the action of the Board of Regents that all employees should sign a loyalty oath saying they were not communists. A few years later, she argued to the IMS leadership that the association's bylaws should be amended to prohibit racial segregation. In the early 1960s, she helped raise money to support Dr. Martin Luther King and the civil rights movement, and she advocated that a city aquatic park be saved. In the mid-1960s, she fought for students who were arrested during the Berkeley Free Speech Movement. In these activities, Scott worked with others on these issues and could best be described as an advocacy collaborator. She had yet to emerge as an advocacy leader.

Scott, as a senior faculty member, held two administrative titles within the university. One was assistant dean of the Berkeley College of Letters and Science from 1965 to 1967. It can be presumed that her duties were typical for this kind of position and involved handling and resolving individual student issues that arose in the college. The other was chair of the Berkeley Statistics Department from 1968 to 1973. Again, typical duties can be presumed, and they would have involved orchestrating meaningful collective activities among the faculty, building consensus and cooperation on unit decisions, and balancing and communicating the needs of the department with those of the college and university. All evidence points to the fact that Scott saw these two titled administrative positions as primarily managerial rather than leadership positions.

From one standpoint, Berkeley Statistics Professor Jerzy Neyman, who was the founding chair of the department, was always in the background of whoever was serving as the department chair and led the department from the background as a *pater familias* (Yang, 1999). From another standpoint, Scott had an extremely strong family military background, which clearly influenced many of her behaviors and decisions (see principle 2). It is very likely that Scott viewed these two positions in analogy with the role of a noncommissioned officer in the US Army whose duty it was to primarily manage, and viewed the higher-level positions of chancellor, provost, and probably also dean as analogous with the role of a commissioned officer, where there was an expectation to exercise strong leadership. In any case, there is no clear archival evidence of whether Scott practiced affirmative action advocacy leadership within these titled administrative roles. We do know she acted as a unit manager, and she always worked within channels and the chain of command at this time (Susan Ervin-Tripp, personal communication, 2010).

It was not from within these titled administrative roles that Scott emerged clearly as a leader: Instead, it was when she became an advocate for affirmative action for women in the academy that she emerged as a natural leader based purely on her influence. Advocacy leadership involves identifying and seizing strategic opportunities to improve our social landscape and inspiring others to leap forward. Indeed, Scott's most effective leadership was advocacy leadership. Her mission was to improve the workplace for women in the academy. Her work toward this mission began in 1968 when the University of California decided it must be engaged in finding solutions to the so-called urban crisis, characterized by race riots and violence. Scott was selected to a women's group that was formed at Berkeley to help offer solutions and ideas to the crisis, a group that decided to become an advocacy group for academic women. Her work toward the mission, which she found to be exhausting, continued for twenty years until her death. Her influence continues to this day in the form of affirmative action and salary equity standing committees and initiatives at Berkeley and other universities across the nation that strive to use solid evidence and statistical methods in their work. The principles that Scott used to make her advocacy effective are described in a separate section below (see advocacy leadership). Scott was a good role model for how to function in the academic environment. I saw from her example that I could use my natural tenacity to good advantage and "lean in" to effect positive change in the academy.

I was recruited to titled organizational leadership early on in my academic career, in 1987. I had been a faculty member for only four years at San Diego State University, and had just been awarded tenure, when I was asked by the chair of my very large mathematical sciences department to succeed Charles "Chuck" Bell, Jr. (1928–2010), who specialized in nonparametrics and stochastic processes, as the coordinator for the statistics division. I leaned in and immediately said "yes." The statistics division had eight tenure ladder positions, and I was the first woman among them and therefore the first woman coordinator, and the statistics program consisted only of a small master's degree. Women statisticians at that time were in short supply. I recognized the potential to expand the division and wrote a five-year plan for the division that won two new tenure ladder positions, to which I recruited biostatistician Kung-Jong Lui and applied statistician and consultant Duane Steffey, both of whom are now fellows of the American Statistical Association. I led the enlarged division to add a bachelor's degree in statistics, a master's degree concentration in biostatistics, and a collection of new courses (biostatistical methods, survival analysis, statistical computing, categorical data analysis, and Bayesian statistics were all added under my leadership) to support these two new degrees. All of this took vision, energy, influence, tenacity, and an understanding of how the university and the university system worked (new degrees required system approval). I did not see my role as managerial, although I did carry out the managerial responsibilities. I saw my role as a leader. I believed in the law of victory — finding a way for the division to win — and would not accept defeat for the advancement of statistics (Maxwell, 1998). I had to bring both the faculty and the administration to my vision for the division, which sometimes was not easy, given that the statisticians were already carrying a heavy teaching load with assignments in both the department of mathematical sciences and the business school, and given that the pure mathematicians ruled the department of 75 or so tenure ladder positions, of whom only four were women.

Leaving behind a strong foundation in the statistics division, I went on to the position of associate dean for undergraduate studies at San Diego State University, where there were some 15,000 undergraduate students across 7 colleges. This was a university wide position. Again, I saw myself as more of a leader than a manager and formed visions for change. I provided leadership to all of the university wide diversity initiatives, including a state-funded Faculty–Student Faculty Mentoring Program for students who were underrepresented in higher education. This was a good fit, given my graduate training in anthropology and my orientation toward social activism that I developed as a student at Grinnell College and then Berkeley. Being a statistician, I immediately realized that I needed to form a relationship with the institutional research office on the campus so that I could access new breakdowns of data to use in my diversity efforts. I learned from the data that the group of first-year students who had not declared majors was like the group of underrepresented students in that their retention was relatively low. I acted on this information by adding a new section of the mentoring program for these undeclared students. Also, in the position of associate dean, I provided leadership to university wide assessment activities, including chairing the university assessment committee. This was an excellent fit, given my master's level training in statistics and doctoral training in biostatistics. Being a statistician, I saw the need for administrators and others to have better access to institutional research data, something that Scott pioneered in her career. And so I developed plans for an executive information system that would make it easier for administrators and others to access and use institutional research data, and that would in turn make it easier to evaluate inequities between groups, for example, males and females.

While still the statistics coordinator, my spouse and I decided to have a child and became adoptive parents of a newborn. A good friend reassured me that this would work out. She had her children while in graduate school and told me they fit right in: She had her book and yellow highlighter (this was before tablet and notebook computers), and her children had their books and yellow highlighters. Her advice turned out to be good, and even though I was a minority as a woman in my department, I found my colleagues to be generally tolerant about having my son sit quietly in the back of a room doing his quiet activities while we were having a professional meeting, and I found them to be generally tolerant about me scheduling my classes during the day instead of the evening. To this day, I am grateful for having had an environment that permitted me to advance in both my professional and personal life, especially as I understand that my experience was above average. To this day, I am also grateful to have had a spouse who runs half of our home, another experience that I understand to be above average. What I did not have was a good school situation for our son. The schools in the city we lived in had a negative reputation, and so we placed our son in an expensive preschool that required a thirty minute commute each way in stop- and- go traffic. This precipitated a move to the Midwest, to be closer to our families, and to a place where our son could attend a good public school under a reasonable commute.

(2) **Leadership competencies determine leadership effectiveness. Statisticians should understand their leadership influences, and they should seek out opportunities to develop their leadership.**

Leadership competencies are behaviors, skills, and abilities that are critical to successful performance as a leader. These include both innate and learned characteristics. One of Scott's sets of characteristics derived from her strong family military background. Scott came from a family of military leaders. Scott's maternal grandfather, John Charles Waterman, graduated from the United States Military Academy at West Point in 1881; served on frontier, border, international, university, and other duty; and then retired as a colonel in 1919 after some 40 years of military service. It can be noted that his first service was as a cavalry officer on frontier duty in the Dakota Territory, where he participated in the Wounded Knee Massacre and the engagement at White Clay Creek: There he saw both how good senior leadership could build relationships to attain positive outcomes, and how poor senior leadership could act on grudges and inexperience to achieve disastrous outcomes. There was ample opportunity for Grandfather Waterman to pass military leadership lessons down to Scott, especially as he lived with college-age Scott for the last few years of his life in Berkeley (West Point, 1939).

Scott had at least five other relatives who were also military. Two also graduated from West Point, including maternal uncle John Julius Waterman, who in 1910 retired as a colonel after a distinguished career as a field artillery officer in the American Expeditionary Force in World War I (*Military Times*, 2014); and Scott's youngest sibling (among three males), Loxley Radford Scott, retired as a captain in the United States Army in 1961 after serving in World War II and the Korean War (Scott, R.W., 2014). Scott's father, Richard Christian Scott graduated from the United States Naval Academy in Annapolis in 1911 and retired as a major in the United States Army after serving as a field artillery officer in World War I (Scott, R.W., 2014). Her oldest sibling, Richard C. Scott Jr. also served in the military in World War

II, although little is known other than that he sailed on the USS Wasp, which was lost in 1942 (United States Navy, 1942). And her middle brother, John W. Scott, entered the army as a private, served in World War II and the Korean War, and eventually became an officer, serving as a lieutenant colonel in the United States Army Air Force (United States Department of Veterans Affairs, 2014).

With all of these military officers and influences in Scott's family, it is no surprise that the leadership principles of the military rubbed off on her. Traditional United States Armed Forces leadership principles have been codified into the following eleven statements (Army JROTC, 2014):

1. Know yourself and seek self-improvement.

2. Be technically and tactically proficient.

3. Develop a sense of responsibility among your subordinates.

4. Make sound and timely decisions.

5. Set an example.

6. Know your people and look out for their welfare.

7. Keep your people informed.

8. Seek responsibility and take responsibility for your actions.

9. Ensure assigned tasks are understood, supervised, and accomplished.

10. Train your people as a team.

11. Employ your team in accordance with its capabilities.

Evidence exists that Scott embodied these leadership principles. She continuously worked to: mitigate her weaknesses, expand her methodological toolkit, communicate standards, reason clearly under pressure, be a role model, build trust among those around her, generate information and explain reasons, find and accept new challenges, employ sound judgment toward mission accomplishment, benefit the team, and increase the capabilities of those around her.

Scott had a commanding presence ... a fearlessness ... and considerable organizational acumen. She understood the concept of chain of command and operated within channels. She exhibited both the character and the behavior of a leader in the military sense. Her strong military-influenced leadership competencies contributed significantly to her effectiveness (Golbeck, forthcoming).

I would love to have had such a family military background to inform my leadership. My father served as a technical sergeant in the United States Army in World War II, serving in North Africa, Italy (Anzio), and the liberation of

Rome. There were no West Point graduates in my family, nor any officers of field artillery. Like many of my father's contemporaries, my father preferred not to talk about what had happened to or around him during the war. So I did not have a military influence on my leadership development.

But what I did have was an intentional decision to practice leadership in junior high, and then high school and beyond. When I was in grade school, one of my teachers took my class over to the junior high school to watch the awards ceremony for the graduating 9th graders. I watched a girl win the award for outstanding extracurricular activity, and she became a five-minute role model: I decided that when I graduated from junior high school I was going to be the winner of that award. When I got to junior high school, I became editor in chief of the school newspaper, president of the art club, an actor, a backstage manager, etc. I did win that outstanding extracurricular activity award, but more importantly, I developed leadership competencies along the way.

I did not know about formal leadership development programs, and so I did not have any formal leadership training as a statistics coordinator, just leadership experience that I had gained along the way. Although I had considerable success by any measure in my position as statistics coordinator, I could still have benefitted from programs that are targeted at department chairs in general, such as the American Council on Education's Leadership Academy for Department Chairs. This workshop "focuses on the chair not only as a unit leader but also as an academic leader in service to the institution and its mission" (ACE, 2014). Or I could have benefitted from a program targeted at statistics department chairs in particular, such as the American Statistical Association's annual Workshops for Chairs of Programs in Statistics and Biostatistics. This program "is designed to provide information and discussion for leaders of statistics and biostatistics groups and departments. In particular, it will try to stimulate discussion between new and experienced chairs on a range of topics from the scientific diversification of our field and its implication for hiring, education, and departmental organization to better attracting talented students to our discipline" (ASA, 2014). As associate dean for undergraduate studies at San Diego State University, I felt very prepared as a statistician to lead the assessment responsibilities in the position, but I thought I could benefit from learning more about mentoring to better lead the diversity responsibilities. And so I sought out and found a national conference on the subject of mentoring leadership, which influenced and improved my leadership.

Then I discovered executive education and made a lifelong commitment to intentionally develop my leadership with the help of organized programs. There are so many good formal leadership development institutes and workshops now to choose from. I have had the privilege of being selected to attend five. Each of these formal offerings was unique — one was on management, one on fund-raising, another on executive leadership, yet another on negotiation, and, most recently, one on women and leadership.

I was recruited from my position as associate dean for undergraduate studies at San Diego State University to a position as vice president and academic dean at Bethany College, a small, liberal arts college in Lindsborg, Kansas. I was the first woman to hold this position. This was a full-time administrative position, with personnel and budget authority, and where I was responsible for overseeing the academic program, the library, the career services office, the academic support center, the enrollment services office, and the computer center. I had provost responsibilities, and was in charge of all college operations while the president was away. Yet I made sure I continued my research program by working long hours in the evenings and weekends, even meeting with collaborators on the weekends. While I had many broad management responsibilities in the vice president and academic dean position, I saw myself primarily as a leader. My vision was for a college that was more efficient, effective and updated, had more of an emphasis on scholarship, and had improved retention. I led a number of initiatives to these ends. In this position, I benefitted from attending Harvard University's Institute for Educational Management. For senior leaders in higher education, this is "an intensive, total immersion experience that provides a rare opportunity to assess your leadership skills, renew your commitment to higher education and develop tangible strategies for long-term institutional success" (Harvard Graduate School of Education, 2014). I also benefitted from attending the Council for Advancement and Support of Education (CASE) session on Development for Deans and Academic Leaders. This is billed as "the required curriculum for deans, presidents, provosts and other academic leaders with fundraising responsibility. You will strengthen your partnership with your advancement officer and learn to engage with potential donors in productive, meaningful and authentic ways — ultimately leading to a new level of comfort and greater success with fundraising" (CASE, 2014).

My next position was as chair of a public health sciences department at Wichita State University. This move brought my son into urban schools, where the children were raised to be less discriminatory (my son is biracial). Then I was recruited to a very challenging and rewarding position as vice president for academic affairs at a state board of regents (a position I discuss in principle 3). It was a position I loved and only left because I was unable to be at home during the week, and my son needed me more at home, as he was entering high school. Next came a position as associate dean for research at the University of Kansas School of Medicine — Wichita. After that, I was offered a dean position at a university and a director position at a higher education coordinating board, but I turned them both down because our son was still at home and there was not a position for my spouse. Now I am a full professor at the University of Montana, where my spouse also has a position. My spouse and I have always tried to optimize both of our careers simultaneously. This has not always been easy, but we have somehow managed. In the end, my career has looked less like a ladder and more like a jungle gym.

I continued to seek out executive education over these years. As vice president for academic affairs at the board of regents, I was selected to participate in the Millennium Leadership Institute offered by the American Association of State Colleges and Universities. This is "a premier leadership development program that provides individuals traditionally underrepresented in the highest ranks of higher education the opportunity to develop skills, gain a philosophical overview and build the network needed to advance to the presidency" (AASCU, 2014). As associate dean for research, I decided to further develop my leadership in a program on Mastering Negotiation: Building Sustainable Agreements offered by the Harvard University John F. Kennedy School of Government. This "addresses the challenges of negotiating across cultures, organizations, and sectors in a world of various economic, political, and social problems, where sustainable solutions require consensus among multiple stakeholders" (Harvard, 2014). Now as a full professor focusing on natural leadership, I just participated in a course on Women in Leadership offered by the American Academy of Neurology: "Current research indicates that women's strengths and styles are more necessary than ever for institutions to be successful, and balancing women's and men's voices at the table often creates stronger results" (AAN, 2014).

Each of these opportunities was illuminating and highly valuable in its own way toward my development as a leader. Each was important toward developing my critical thinking skills and building functional networks for my leadership. Statisticians should consider the value of formal professional development of their leadership and include it in their plans for lifelong learning.

(3) **Leaders inspire action through compelling stories. Statisticians should add evidence-based storytelling to their professional toolkits.**

Storytelling is a useful tool that leaders can use to communicate challenges and move people toward a new vision. Stories can touch hearts and minds in ways that reason with facts and figures cannot. Scott as a statistician lived in a world of evidence-based reason, but she nonetheless understood the efficiency and effectiveness of a good story. Scott had a number of compelling stories that she used to inspire action. One that stands out is the Big Telescopes Story.

The background to the Big Telescopes Story is that there had been a long-standing differentiation in roles between men and women in astronomy. Historically, at the major observatories in this country, it was thought the men should use the telescopes, and women should do the long, detailed computations on the data produced using the telescopes, computations that would then be published by the men. The supposition was that men were more capable of doing hardy and creative work, and women were more capable of doing tedious and meticulous work analogous to stitching and embroidery. As time went on, more and more women became unhappy with the traditional role differentiation and began striving themselves to become fully engaged in the science of astronomy. More and more women tired of watching the men and wanted to do their own observing using the telescopes, including the big telescopes (Lankford, 1997). But by the 1970s, women still weren't being allowed to use the biggest telescopes.

Scott's Big Telescopes Story went as follows: For many years, women were not allowed to use the 100-inch telescope at Mount Wilson Observatory and the 200-inch telescope at Palomar Observatory in their own names, but had to sign up for observing time under men's names. These were the world's largest telescopes. The 100-inch was the largest between 1917 and 1948 (Mount Wilson Observatory, 2014), and the 200-inch was the largest effective telescope between 1948 and 1993 (Caltech, 2014). Scott alleged that, as of 1970, no woman had yet been allowed to use the 100-inch (Scott, E.L., 1970a), and as of 1976, no woman was allowed to use the 200-inch (Scott, E.L., 1976). The story, as it was repeated, was simply that women were not allowed to use the big telescopes. Stories, when passed down from one person to another, often lose their specificity and accuracy, but at the same time may become more compelling in the simplified form. Earlier, the great astronomer George Ellery Hale advanced the vision that big telescopes would be critical toward advancing our understanding of the fundamental nature of the universe (Hale, 1928). It is no wonder the Big Telescopes Story was both intriguing and emblematic of the problems faced by professional women.

Even compelling stories as told firsthand with specificity may be political or controversial. Scott seemed to sense this in 1970 when she wrote a detailed letter telling the Big Telescopes Story to colleagues who were working with her on the status of academic women at Berkeley and hesitated before sending it (Scott, E.L., 1970b). But as the years went by, Scott became bolder about telling the story. Then in 1974, the Hale Observatories Director H.W. Babcock challenged the story, saying that women had in fact used the telescopes at his observatories (Spiegel, 1974). But he apparently was not telling the whole story. What he was not revealing was that women had used telescopes at his observatories, but only the smaller telescopes, not the 200-inch. Scott had previously checked the observing records. Scott's professional network extended to the major women astronomers, and she also checked her facts with her network (Scott, E.L., 1974). Scott had a PhD in astronomy. She was a statistician and an empiricist. Her Big Telescopes Story was backed by the data. It was authentic.

Scott told the Big Telescopes Story to many professional or student groups when she talked about the problems of women in science. The effectiveness of the story in touching the hearts and minds of listeners is evident in both the number of people who repeated the story to others and in the story's lasting effect on the people who heard it. Indeed, when I was a graduate student at Berkeley, a number of my student colleagues repeated this story to me, and it has remained with me for many, many years. There was something about the concreteness of the object of the telescope... something about women being cut out of being part of the discovery of the universe.... It was clear that this was just not right and something needed to be done about it ... change for women in science needed to happen.... There needed to be a future world where women could use the biggest telescopes and be a full part of the discovery of the universe!

Scott became known for the Big Telescopes Story. It was a story that clearly made waves toward her vision of equality for women in academe. I soon recognized that storytelling would be important to the success of my own leadership.

At an earlier point in my career, I held the position of vice president for academic affairs for the Kansas Board of Regents. I had oversight responsibility for academic affairs matters for a state system of 36 postsecondary institutions that included universities, community colleges, and technical colleges/schools. As a statistician, reason backed by facts and figures was one of my comfort zones. But to be an effective academic affairs leader, I had to bridge many divides: between the nine regents, who were appointed by the governor and were laypeople, and the world of academe; between the system office and the campuses; and between the system office and the state legislature, in which many were without college educations. To bridge these huge divides, I needed to have another comfort zone, one that included storytelling. I needed to tell stories to the regents about the cultural context of colleges and universities in order to advance their understanding of how their institutions worked and inspire them to take action that would be efficacious. I needed to tell stories at the campuses about the needs of the system to advance the system vision and inspire the campuses to take action to support that vision. I needed to tell stories to the legislators to advance their understanding of needed legislative changes that related to postsecondary education and inspire them to take action toward those changes. I made sure my testimony included facts and figures, but I also made sure it included stories about how the vision or changes would benefit the constituents of the regents, campuses, and legislators. It was the stories, backed by the facts and figures, that captured their attention and had the greatest chance of bringing them to my point of view.

But it is not just organizational leaders who inspire action through storytelling. Project statisticians can also use storytelling to advance their influence on a project team. Consider an example that I have constructed from my own experience.

An interdisciplinary team of researchers has come together to determine which design should be used to conclude whether a treatment, in comparison with standard care, improves certain patient outcomes. The project statistician wants to influence the team to do a randomized trial and knows she/he needs to advocate for randomization. The other team members want to use a historical cohort design because they are uncomfortable with a design where not all patients get the new treatment, and they are worried about being able to get twice the number of current patients, which is what they would need over a historical cohort study, to consent to a trial. The project statistician is outnumbered. The statistician could provide a brief explanation that randomization would ensure every patient had equal opportunity to be in the treatment versus the standard care group, and therefore the two groups could be expected to be comparable relative to factors not explicitly controlled for in the study. Often, in my experience, only very statistically enlightened teams are moved by this kind of explanation. Instead, the statistician could tell a story, something like this: "I once was recruited to work on a research project that had already started, and it was too late for me to influence the type of research design. The design that was being used was a historical cohort design. The researchers put a lot of effort into that study, and a lot of money was spent. They really wanted to be able to come to a conclusion that was valid and reliable. But as the study went on and the analysis took place, they came to find that their treatment and standard care groups weren't very comparable, and that because of the limitations of the logistic regression methods, they couldn't use all of the many variables that they had at hand to adjust for these differences. At the end of the study they wished they could start over and do the randomized trial, because they could have stated their results with a much higher degree of confidence. I think in our study, we don't want to be in that unfortunate situation." Often, in my experience, these types of stories can move teams to make better decisions. Statisticians should consider the power of storytelling in addition to the power of facts and figures.

(4) **Successful leaders are relationship-builders. Statisticians should generously share their intellectual wealth as a way of exercising influence.**

A leader needs to form strong connections with people in order to be successful. A leader needs to be an initiator. Scott understood it was her responsibility as a leader to be the initiator and nurturer of relationships among those who could help her achieve her vision. She understood the power of sharing her intellectual wealth. And she undertook this responsibility and exercised this power with gusto.

Consider as an example the first half of the year 1974. Scott had just completed two papers on salary equity using regression methods, one using Berkeley data, and the other using national data. The first was a white paper from late 1973, presumed to be written by Scott and perhaps others: "Women generally receive less: Inequalities in employment in salary, and in other benefits at the University of California, Berkeley" (Scott, E.L., 1973). The "Women" paper contained estimates of how much women faculty members at Berkeley were being underpaid relative to men, and it suggested methods for correcting women faculty members' salaries. The other was a conference paper published in 1973 in the American Statistical Association proceedings: "Application of multivariate regression to salary differences between men and women faculty." The "Application" paper contained estimates of how much women faculty members across the nation were being underpaid relative to men and concluded there was strong evidence for sex discrimination in faculty salaries (Darland, Dawkins, Lovasich, Scott, Sherman, and Whipple, 1973).

Scott initiated and nurtured relationships among those fighting for equality of women in academe through these research papers. She wrote letters to people, letters which were personal and often lengthy and which explained her position, and she attached her research papers to the letters. She took the time and made the effort.

Scott sent the Women paper to the Berkeley vice chancellor, gently but firmly criticizing the administration for not including details of salary inequities between men and women in the current draft affirmative plan, and requesting permission to send the paper to others on the campus. The vice chancellor agreed to add Scott's work to the affirmative action plan and include it in future official discussions about the plan with the federal government. This opened the door for a long and continuing dialog between Scott and the vice chancellor about the methods that were being used by the administration to measure underutilization of women and minorities, the poor quality of the data being used in the measurement, and the problems of trying to use administrative data for affirmative action research. It also opened the door for Scott to advise the vice chancellor that the affirmative action coordinator should report directly to the chancellor or vice chancellor if the position was to have any real impact. Scott also sent the Women paper to one of the state legislators with whom she had a previous relationship about nonsalary women's issues, explaining in detail that women were being very much underpaid in comparison with men.

Scott sent the Application paper to other people on campus who were working with her on women's issues or who would be involved with the issue, such as the chancellor's office and the university budget committee, urging all to give salary equity more attention while making sure she explained the limitations of the statistical analysis in proving discrimination. She sent the paper to student lobbyists and the media, including the campus student newspaper and the university faculty and staff bulletin, which turned around and reported the explosive results.

Scott also sent the Application paper to other researchers around the country who were working generally on women's issues or particularly on regression analyses to examine sex differences in salary, urging the latter to update their methods or their data. She even sent it to statisticians who she thought might be interested. She sent the paper to national women's rights activists. She sent the paper to the federal Office of Civil Rights, after establishing a connection with it the year before, and it reported seeing an immediate use for Scott's work in its contract compliance investigations. Most influentially, Scott sent the Application paper to two staff members at the AAUP, asking that it forward the paper to its Committee W on Women in the Academic Profession, and to its Committee Z on the Economic Condition of the Profession.

As a result of all of these new and continuing relationships, Scott became the person on campus that people went to for the last word when it came to affirmative action data. She established a national reputation as a leader in the statistical analysis of affirmative action data. And she became the author of the AAUP Higher Education Salary Evaluation Kit (Scott, E.L., 1977). This kit was used by hundreds of colleges and universities to flag individual women and minorities whose salaries appeared to need additional review for possible correction.

Scott was exceptionally generous in sharing her intellectual wealth. She built relationships that helped further her vision for the advancement of women as she did the sharing. My current historical biographical research on Scott reinforces for me the importance of sharing my own intellectual wealth.

Leaders need to constantly be on the lookout for ways to improve their leadership. I wrote a few articles in the past few years that had to do with advocacy issues for women and implicit bias, including the invisibility of women at the Joint Statistical Meetings (JSM) (Golbeck, 2012) and the underrepresentation of women among winners of awards for statistics scholarship and research (Golbeck and Molgaard, 2013). The first article was published in the *Amstat News* and showed that at the JSM in 2012 there were no women among the seven keynote speakers, three speakers with lunch, four introductory overview lecturers, or eight meet and mingle well-known statisticians. The second article was published in the Proceedings of the International Statistical Institute and showed that in the eleven-year period between 2001 and 2012, there were no women recipients of either the Deming Lecturer Award or the W. J. Dixon Award for Excellence in Statistical Consulting; there was only one woman recipient of the Gottfried E. Noether Senior Scholar Award and of the Samuel S. Wilks Memorial Award; and women were better represented among recipients of awards that were based on recent substantive evidence, such as peer-reviewed publications, rather than lifetime achievements. It is important that statisticians be made aware of these issues, and I should not assume statisticians will necessarily find these articles and read them. I should follow the example of Scott, and proactively and prospectively reach out to appropriate individuals, representatives, administrators, committees, and the like with my research on women's issues and use the research as a vehicle to build connections toward the advancement of women in statistics. Another way to reach out is by organizing and moderating panel discussions at professional statistics meetings. For example, Yulia Gel and I organized, and I moderated, an invited panel at the 2013 JSM in Montreal, Canada, titled "Educating Future Leaders in Statistics and Maximizing the Likelihood of Leadership: Perspectives from and on Women in Statistics," which was the initial impetus for the book you are reading. Statisticians should look for strategic opportunities to share their intellectual wealth and build relationships toward the achievement of diversity goals.

Elizabeth L. Scott's Advocacy

This section originally appeared as a blog post, "Elizabeth L. Scott's Advocacy," that I was invited by the American Statistical Association to write in celebration of the International Year of Statistics and the International Women's Day. I present it here as it originally appeared (Golbeck 2013):

Women professors in the United States used to be scarce. Women professors of statistics were especially rare. When I was a graduate student at the University of California, Berkeley in the late 1970s, there was only one woman among the tenure ladder statistics faculty members. She was Elizabeth "Betty" L. Scott (1917–1988).

Betty regularly chose problems to solve where she could make a difference in real-world situations. One of these problems was the status of academic women. Here are eight basic principles, based on my experience with her and reports of her collaborators, which she followed to make her advocacy effective.

1. *Collaborate toward positive action.*
   Betty collaborated with many diverse groups of individuals toward positive action for women. She fostered collaborative relationships through competence and kindness. She routinely asked colleagues how she could help.

2. *Make evidence-based recommendations.*
   Betty used both new data obtained by conducting surveys and interviews, and existing administrative data, to support her advocacy. She carefully considered and explained the strengths and limitations of evidence when presenting her recommendations. There was nothing arbitrary or capricious about Betty.

3. *Use statistical methods.*
   Betty turned data into information for her advocacy by using statistical methods. She was an expert in both mathematical and applied statistics, and she carefully chose her statistical tools, such as regression techniques with matched samples, to illuminate problems.

4. *Be precise and accurate.*
   Betty understood the importance of good measurement from her early research days in astronomy when she published measurements of newly discovered comets. By attending to precision and accuracy, she gained a reputation where her words on advocacy problems could be trusted.

5. *Use channels.*
   Betty was from a family with a military tradition. She knew how to use communication channels in a hierarchy. Because of her strengths, individuals at the top of the channels often chose to work directly with her rather than work indirectly with her down through the chain of command.

6. *Work hard and persevere.*
   Betty knew that advocacy was not easy. Her advocacy for women was in addition to her normal research and teaching. But it was so important she

burned the midnight oil. She knew there would be setbacks and sometimes became discouraged, but she was not to be deterred from her mission.

7. *Keep monitoring.*
Betty understood that the goals of advocacy work are not achieved overnight. She knew that values, attitudes and behaviors are hard to change, and that culture change requires vigilance. She carefully monitored the data for changes, both positive and negative.

8. *Involve men.*
Betty's vision was complete integration of women into the professional academy. She accordingly advocated for increasing male participation on women's initiatives. She understood the importance of including men in efforts to elevate the status of women.

The presidents of four statistical societies, national and international, offer an award in the name of Elizabeth L. Scott. In this International Year of Statistics, we celebrate those who, in the tradition of Betty, use statistics for social advocacy.

Conclusion

Elizabeth L. Scott was a statistics leader. She advanced the quality of research methods or outcomes in astronomy, atmospheric science, cancer research, and women's studies. She inspired others to leap toward a future where academic men and women are treated equally. She made sure her positive impact on both people and science lasted in her absence. She held the door wide open for me to make my leap into the world of statistics, she influenced many other generations of statisticians, and she will surely influence others yet to come.

Acknowledgments

For their generous financial support of the Elizabeth L. Scott projects, I am grateful to the Phoebe Waterman Haas Charitable Trust and the Philadelphia Foundation, especially David Haas. I would also like to thank Yulia Gel, Craig Molgaard, and Ingram Olkin for their helpful suggestions on this manuscript.

Bibliography

AAN (American Academy of Neurology). *Women in Leadership Program.* Accessed on 5/11/2014 at http://tools.aan.com/science/awards/?fuseaction=home.info&id=84.

AASCU (American Association of State Colleges and Universities). *Millennium Leadership Initiative: Preparing the Next Generation of Leaders.* Accessed on at 5/4/14 http://www.aascu.org/MLI/.

ACE (American Council on Education). *ACE Leadership Academy for Department Chairs.* Accessed on 5/11/2014 at http://www.acenet.edu/leadership/programs/Pages/Leadership-Academy-for-Dept-Chairs.aspx.

Army JROTC. *Principles and Leadership (U2C1L4).* Accessed on 5/14/2014 at http://www.dimondjrotc.org/Leadership/Chapter1/Chapter1Lesson4/U2C1L4A0_Text.pdf.

ASA (American Statistical Association). *Education: Caucus of Academic Representatives: Caucus Activities.* Accessed on 5/11/2014 at http://www.amstat.org/education/caucusactivities.cfm.

Caltech. *Caltech Astronomy: The 200-inch Hale Telescope.* Accessed on 5/11/2014 at http://www.astro.caltech.edu/palomar/hale.html.

CASE (Council for Advancement and Support of Education). *Conferences & Training: Development for Deans and Academic Leaders.* Accessed on 5/11/2014 at http://www.case.org/Conferences_and_Training/DALW14.html.

COPSS (Committee of Presidents of Statistical Societies). *The Elizabeth L. Scott Award, Operating Principles.* Accessed on 5/6/2014 at http://nisla05.niss.org/copss/OperatingProceduresScott.pdf.

Darland, M.G., Dawkins, S.M., Lovasich, J.L., Scott, E.L., Sherman, M.E., and Whipple, J.L. (1973). Application of multivariate regression to studies of salary differences between men and women faculty. In *Proceedings of the Social Statistics Section.* Washington, DC: American Statistical Association, pp. 120–132.

Golbeck, A.L. (1981). A probability mixture model of completed parity. *Demography* 18:645–658.

Golbeck, A.L. (2012). Where are the women in the JSM Registration Guide? *Amstat News* 421:16–17.

Golbeck, A.L. Elizabeth L. Scott's advocacy. *The International Year of Statistics.* Invited blog at http://www.statistics2013.org/files/2013/03/Statistician-Elizabeth-Scotts-Advocacy-FINAL.pdf.

Golbeck, A.L. *Equivalence: Elizabeth L. Scott at Berkeley.* Chapman and Hall/CRC Press, forthcoming.

Golbeck, A.L. and Molgaard, C.A. (2013). Professional awards in statistics: Chipping away at gender disparities in the USA. *Proceedings of the 59th World Statistics Congress*, Hong Kong, China, pp. 5155–5160.

George, E.H. (1928). The possibilities of large telescopes. *Harper's Magazine*, p. 639, April.

Harvard Graduate School of Education. *Institute for Educational Management | IEM*. Accessed on 5/11/2014 at http://www.gse.harvard.edu/ppe/programs/higher-education/portfolio/educational-management.html.

Harvard Kennedy School. *Executive Education: Mastering Negotiation: Building Agreements Across Boundaries*. Accessed on 5/11/2014 at: https://exed.hks.harvard.edu/Programs/mn/overview.aspx.

Kotter, J.P. Management is (still) not leadership. *Harvard Business Review Blog Network*. Accessed on 5/4/2014 at

http://blogs.hbr.org/2013/01/management-is-still-not-leadership/.

Lankford, J. (1997). *American Astronomy: Community, Careers, and Power, 1859-1940*. University of Chicago Press, p. 290–291 & 340.

Maxwell, J.C. (1998). *The 21 Irrefutable Laws of Leadership: Follow Them and People Will Follow You*. Nashville: Thomas Nelson Publishers.

Military Times. John Julius Waterman. *Hall of Valor*. Accessed on 5/4/2014 at http://projects.militarytimes.com/citations-medals-awards/recipient.php?recipientid=76862.

Mount Wilson Observatory. *The 100-inch Hooker Telescope*. Accessed on 5/11/2014 at http://www.mtwilson.edu/vir/100in.php.

Scott, E.L. Letter to National Center for Education Statistics Director Dr. Dorothy M. Gilford, 6/10/1970a.

Scott, E.L. Draft letter to Berkeley Anthropology Professor Elizabeth Colson, 8/26/1970b.

Scott, E.L. (1973). Women generally receive less: Inequalities in employment in salary, and in other benefits at the University of California, Berkeley. Unpublished manuscript, author assumed.

Scott, E.L. Letter to Business and Professional Women's Foundation Librarian Jeanne Spiegel, 4/29/1974.

Scott, E.L. Influences, challenges, and problems. Speech to unidentified audience, 3 double-spaced, typewritten pages. Around 5/1976.

Scott, E.L. (1977). *Higher Education Salary Evaluation Kit.* American Association of University Professors.

Scott, R.W. *Loxley R. Scott 1945, No. 15170, Class of 1945.* Accessed on 5/4/2014 at http://apps.westpointaog.org/Memorials/Article/15170/.

Spiegel, J. Letter to Elizabeth L. Scott, 6/16/1974.

United States Department of Veterans Affairs. *Scott, John W.* Accessed on 5/4/2014 at http://gravelocator.cem.va.gov/index.html.

United States Navy (1942). WWII U.S. Navy Aircraft Carrier Muster Rolls, 1938–1949, roll MIUSA2006_082862.

West Point. John Charles Waterman, No. 2916, Class of 1881, *Annual Report of the Association of Graduates*, pp. 146–148, June 10, 1939.

Yang, G.L. (1999). A conversation with Lucien Le Cam. *Statistical Science* 14(2):223–241.

Part II

Fresh Opportunities and New Challenges for the Statistics Discipline

5

Robust Leadership in Statistics

Jon R. Kettenring

Research Institute for Scientists Emeriti (RISE) Drew University

Introduction

This is an exciting period for the statistics discipline. Fresh opportunities and new challenges are easy to spot. To a large extent, they stem from the widespread emergence of "big data," as in astronomy and genomics, and the sweep of machine learning and data science across the traditional bow of statistics. Above the fray, it is easy to be optimistic about the future of statistics in the 21st century (see, e.g., Lohr, 2009; Kettenring, 2011; and Speed, 2014).

However, robust leadership will be essential to fulfill the promise. By this I mean strong leadership across the board, without unnecessary limitations, covering all corners of statistics. In particular, it means that there must be special attention to attracting fresh talent and to realizing the full potential of all who work in the field — especially women and minorities. It means reducing the friction from bias and circumstance that too often surfaces in the course of careers. Some may question whether such problems are still with us after so many years of concern. An honest assessment would show that they are, albeit to varying degrees and acknowledging the progress that has been made on several fronts. In keeping with the theme of this book, my comments will be limited to leadership generally and the role of women in particular.

One goal for this chapter is to talk about robust leadership from a personal perspective. I chose to do this, hoping that my experiences might provide something of value for others. I will add in what I have learned from pondering the possibilities, conversations with others, and perusing the literature. Of course, leadership does not occur in a vacuum. It is naturally bound up with the personal traits and talents of those who would lead, but the circumstances surrounding leadership opportunities are equally important for success. I will illustrate both aspects with a variety of examples.

My relevant experiences can be separated conveniently into three categories: 35 years of industrial employment in the telecommunications sector, more than ten years of excursions into the academic world after retirement from industry, and 35 years of volunteer activities with various professional organizations. These will be discussed in order before turning to other topics relevant to leadership. I will end by suggesting a six-pronged strategy for fully achieving robust leadership in statistics.

A Little Data

To set the mood, we can point to a number of continuing disparities for women in broad categories of science and engineering (S&E) by drawing on recently released data from the National Science Foundation (NSF; 2014):

- Women continue to be underrepresented in the S&E workforce — 28% in 2010 — although the situation has improved modestly in recent years and the number employed has grown sharply.

- The proportion of women in the computer and mathematical sciences is a relatively low 25%.

- From 1993 to 2010, the number of women in these two fields nearly doubled, but the proportion fell from 31% to 25% because many more men were working in these areas.

- During the same period, the proportion of doctorate holders who are women in these same fields grew from 16% to 20%.

- Overall, the gender disparity in computer and mathematical sciences is second only to engineering among the S&E categories used in the report.

The overall picture (and reading between the lines for statistics) is that, while more women may be working in S&E generally and in the computer and mathematical sciences in particular, the percentages are still relatively low or moving in the wrong direction. These imbalances place women at a demographic disadvantage when it comes to developing leaders for the future.

The situation is worse if one factors in the impact of any lingering, even if subtle, biases in the system. A stark example of such gender bias faced by women working in the sciences in academia appeared recently in Moss-Racusin et al. (2012). Using a randomized, double-blind study design, they found that gender biases exist among both male and female faculty members in the biological and physical sciences that can impact the hiring process. The biases extend to salaries and support, such as mentoring. In a separate study, Carrillo and Karr (2013) estimate that women's salaries are about 7% lower than men's, based on NSF's Survey of Doctoral Recipients data. While the Moss-Racusin et al. study involved fields other than statistics, and only faculty from research-intensive universities, I see no reason to presume that results would be much different in the statistics field or other employment settings. Another systemic concern in the academic sector is the tenure system. See Olkin (2014) for an analysis of data on tenured women in statistics faculty at PhD-granting universities.

Experiences in Industry at Bell Labs and Bellcore

From mid-1969 until the end of 2003, I was employed in the telecommunications sector, first at Bell Telephone Laboratories (until 1984), and then a spinoff company, Bellcore (created in 1984), which later was renamed Telcordia Technologies. Actually, I started my employment at Bell Labs with a summer job in 1968. This made it much easier for me to become a full-time employee after completing my graduate work the following year. Indeed, it was widely understood that opening the door with a summer job was the path of least resistance to the coveted position as a member of the technical staff. You avoided the customary formal thesis review, which could be painful, and the rigorous interview process that typically followed. The first post out of school is very important for one's career. I was lucky to have had this early opportunity to work with the best.

For additional background on statistics research at Bell Labs and its offspring, see Kettenring (2001) and (2012), and Denby and Landwehr (2013). In the 2012 paper, I give a personal characterization of the many advances in statistics by scores of men and women that occurred prior to the breakup of the Bell System in 1984. Much of the research fell under the general heading of data analysis. "Through years of experience with small- to large-sized datasets from a wide problem base, a hands-on 'let-the-data-speak-for-itself' philosophy of analysis evolved and was cemented into the culture. It profoundly influenced the consulting and research activities at the labs."

The broader story of Bell Labs research, primarily between the late 1930s and the mid 1970s, was recently described by Gertner (2012). While he asserts on the opening page of his book that the employee body "included the world's most brilliant (and eccentric) men and women," it was largely a man's world. Indeed, the book revolves around the lives of six male scientists and their contributions. Women are barely mentioned.

However, within the statistics community at Bell Labs and Bellcore, at least during my time there, the situation was not quite so bleak. For example, the Bell Labs February 1970 telephone directory lists about twice as many males as females on the rosters of the two male-headed statistics research departments in Murray Hill, New Jersey, where I worked. This ratio, it must be clarified, is a bit misleading, because a majority of the women were cast in supporting roles as so-called associate members of the technical staff. This situation gradually improved and, much later, women became heads of statistics research units at both Bell Labs (Diane Lambert) and Bellcore (Diane Duffy).

At the end of his book, Gertner asks whether the success of Bell Labs was due to the "thousands of engineers and scientists working together, or the few exemplars who towered above everyone else." Then he mentions several names provided by Robert Lucky, a former Bell Labs research executive and later head of research at Bellcore and Telcordia. Gertner quotes him as saying of these individuals, "They set the examples that permeated the whole place.

They created the fame and were what other people aspired to be. They were the leaders, even if they weren't high up in management. If you knew them, you knew Bell Labs." There were no women on the list (at least as it appears in the book). One wonders what might have been if there had been a larger pipeline of scientifically-trained women at the time and — at least as important — if they had been encouraged to advance as leaders in the male-dominated culture. What would have been the impact on the history of Bell Labs and its several spin-off research laboratories? What would we have beyond the laser, the transistor, the quality control chart. . . ?

Experiences in Academia, Primarily at Drew University

In 2004, I joined a group of retirees at Drew University, a small liberal arts school in northern New Jersey. The group is known as RISE, which is short-hand for the Charles A. Dana Research Institute for Scientists Emeriti. It has existed for about 33 years (www.drew.edu/rise). As of early 2014, there were 12 scientists who were members of RISE. Collectively, they have had enormous cumulative experience — 322 years — working in industry, mainly in the pharmaceutical and telecommunications sectors. Two are members of the National Academy of Sciences, parasitologist William Campbell and micro-biologist Arnold Demain. Our main responsibility is to mentor undergraduates (potential future leaders!) who want to engage in research. Disciplinary coverage is consistent with the university's offerings in the sciences: biology, chemistry, biochemistry, physics, and the mathematical and computer sciences (including statistics).

It is a regrettable fact that, of the 12 scientists, there is only one woman member, Barbara Petrack, a biochemist who specializes in neuroscience. In fact, she is the only woman who has ever been a researcher in RISE! This lopsided state of affairs reflects the demographic composition of industrial research organizations in the 1950–1990 period — too few women in too few places. So, we struggle to improve our mix. A promising indicator for the future: We continue to observe that women are in the majority among the science majors at the university. Moreover, they tend to be the top science students. To illustrate, we award prizes annually to the most accomplished ones who have worked with members of RISE. Since 2001, there have been 20 such awards, and 14 of them have gone to women. Typically, these top students pursue advanced degrees in the sciences or medicine after graduation.

The mentoring role is enormously satisfying for senior scientists, such as the members of RISE. Over the years of its existence, we have mentored more than 350 students and watched many of them mature into productive scientists and practitioners of medicine. Since most of us have been leaders in our fields, one way or another, we hope some of our leadership experience rubs off alongside the technical skills.

While most of the mentoring takes place in the lab, we also run a special science seminar for honors students. It introduces them to scientific research during their first year at Drew. One of the science professors is in charge of the class, and each student is paired up with a RISE mentor. The students learn how to find and read scientific papers and to communicate with others in writing and orally about their research projects, which are developed with the help of mentors. After a few introductory sessions, the students effectively take over the class and run it for the remainder of the semester. For more details about the seminar, see Madden (2011).

Experiences such as those offered by the RISE program can go a long way toward beefing up the pipeline of well-prepared students interested in scientific careers. It is reasonable to expect that part of the fallout will be a corresponding increase in the number of young scholars who will grow into leadership positions as their careers progress. Adding quality and quantity to the pipeline with early-onset mentoring programs seems to be a very good way to improve the pool of leadership candidates generally and for women in statistics in particular. (See Section 5.7 for more discussion.)

Experiences with Professional Statistical Organizations, Primarily ASA and NISS

In the late 1970s, there was a movement afoot to start an American Statistical Association (ASA) chapter in northern New Jersey. Several of us from Bell Labs got involved in various ways. I became the first program chair and the second president. That somehow qualified me to represent the chapter in what was at that time the ASA Council. One opportunity led to another, and I became the vice chair of the council in 1980 and the chair in 1981. There seems to be a useful lesson here: Gaining leadership experience at the local level, even with relatively mundane committees, tasks, and assignments, can prepare one for larger opportunities at the regional and national levels. Of course, none of this was obvious to me at the time. I had no plan whatsoever to get involved in what lay ahead for me.

Because of my roles on the council, I began to attend ASA board meetings. These were tedious affairs, I must say. The board seemed too large to move with any agility, to get anything done. Too often the result was to refer an item to some committee for further study.

I recall the late John Flueck reporting to the board on a proposal to initiate a continuing education (CE) program. The result appeared to be a call for yet another study. Flueck, not so calmly, raised his voice and asked in essence, "Can't we ever decide anything?" People were taken aback. Within a few minutes, the mood changed, and it was decided to proceed with the program. And that — at least from my recollection — was the beginning of CE at ASA. His zeal to bring CE to life and then to nurture it along was a high-impact contribution to the profession and a great example of strong leadership.

Appropriately, he became the first chair of CE. My boss, Ramanathan Gnanadesikan, was invited to present the second ASA CE short course in 1977, based on his new book on multivariate analysis. He invited me and another Bell Labs colleague to join him. This was a terrific learning experience for me, and one of many occasions that I benefitted from "boss sharing spotlight with underling." (For information about Gnanadesikan's career, see Kettenring, 2001.)

Early in my time at Bellcore, one of the editors of this book (I.O.) called my boss to see if I might be available to run the Committee of Presidents of Statistical Societies (COPSS) Visiting Lecturer Program in Statistics. This was a rather different opportunity to make a contribution to the statistics community. It happened only because someone out there took the initiative to give me the opportunity. It was also my first experience at networking with a large number of statistics professionals nationally, almost none of whom I had met previously. While most of the effort was aimed at providing lecturers for colleges, we experimented a bit with visits to high schools. This didn't get very far. As I think back on the progress made introducing statistics into the high school curriculum, our efforts were probably a bit naive and premature, being well before the advent of AP Statistics. Another valuable lesson for fledgling leaders: Good ideas need to be timely to be successful. I have seen this over and over in my work with ASA. The accreditation movement is the perfect example. When first proposed some 20 years ago, it received too little support from ASA members to get off the ground. In today's environment the need seems clearer, the ASA's PStat accreditation program has been launched, and it is growing steadily. (The program recognizes the training, experience, professional development, ethical standards, and communication skills of participating members.)

My involvement with ASA intensified in the 1990s, during which I served as a vice president (1990–1992) and president (1997). By that time, the board had been reduced in size. It was much easier to get things done and yet still have time for important longer-range activities, such as strategic planning. One thing I grew to appreciate is just how important the entire infrastructure of statistics is to the success of the profession. The ASA, for example, has many types of meetings, journals, chapters, committees, and sections that constitute most of its backbone. Each component has a need for effective leadership, and fortunately numerous women have stepped up to these opportunities. Many have become board members and officers of the association. For example, during my year as president, I had the privilege of working closely with Sallie Keller, who served in dynamic fashion as program chair for the 1997 Joint Statistical Meetings, a huge responsibility. Later she became ASA president herself in 2006.

Robert Rodriguez devoted his March 2012 ASA President's Corner column to this same point: developing statistical leadership through service to the ASA. The column is packed with solid advice on how to develop leadership skills through the range of opportunities for contributing to the profession

Table 5.1
Statistics on Women ASA Presidents

Number of Women Elected ASA President 1950–2015						
1950–1959	1960–1969	1970–1979	1980–1989	1990–1999	2000–2009	2010–2015
2	0	0	3	2	3	2

that are open to statisticians. The advice comes from "a panel of four highly accomplished statistical leaders," one of whom is the 2014 president, Nathaniel Schenker!

It is a challenge to quantify how well women have fared overall and tempting to read too much into simple counts. However, with that caution in mind, Table 5.1 shows a far-from-perfect proxy for progress, the frequency at which women have been elected president of ASA going back to 1950 and forward to (the president-elect for) 2015.

There are lots of factors that influence these numbers, e.g., the number of women in the profession and the association, and the number who have developed a record of leadership in other capacities.

About 18% of the presidents during this 66-year period were women. For the more recent period of 1980–2015, the number increases to nearly 28%, a noticeable upward trend. To put these percentages in perspective, about one third of the 2014 membership of ASA is female. If the number of women in the field (and ASA) can be increased, we can expect more will be available, qualified, and interested in taking on leadership positions. This line of reasoning suggests — again — that a very high priority for promoting leadership by women in statistics should be directed at attracting even more into the field. One way of accomplishing this over the long haul would be to further increase the exposure of high school students to statistics.

For some, ASA is just too big and bulky. The National Institute of Statistical Sciences (NISS) works on a much smaller scale with few staff and a much different agenda, with particular emphasis on cross-disciplinary and cross-sector research involving statistics. Examples of topics of past scientific interest to NISS include data confidentiality, data quality, evaluation of computer models, financial risk and bank regulation, highway travel time reliability, and EKG variability.

I first joined the NISS Board of Trustees in 1993 and served as its chairperson from 2000 to 2004. When I joined the board, I had the same reaction that I had years earlier at ASA: this board — with 41 members! — is too big for its own good. Fortunately, others agreed, and the situation was corrected in 2002. With a right-sized board of 21 members, there was a much higher chance to have a positive impact and provide productive leadership.

The birth of SAMSI, the Statistical and Applied Mathematical Sciences Institute, in 2002, which is housed in the NISS building in the Research Triangle Park, in North Carolina, has strengthened the hand of both organizations. SAMSI's mission is "to forge a synthesis of the statistical sciences to confront the very hardest and most important data- and model-driven scientific challenges."

Numerous women have contributed their leadership talents to NISS throughout its 24-year history. Two of the eight past chairs of the NISS board have been women, Janet Norwood and Susan Ellenberg, the current chair. Looking ahead, there is plenty of opportunity for more involvement!

The early history of NISS is recounted from various perspectives in an unfinished document (NISS 2014). It's a great story of vision and action by leaders of the statistics community, local universities and industry, and state-level politicians. While there were about a dozen people whose roles were absolutely critical to launching NISS, as the document makes clear, here I will just mention one of them, quoting Daniel Solomon: "If it weren't for the *energy, commitment and optimism* of Dan Horvitz, we would never have stuck with putting this proposal together and having NISS where it is today." I added the italics to emphasize the three leadership traits that he mentions about Horvitz.

NISS launched its highly successful postdoctoral program in 1993. Later, SAMSI formed its own program as well. Consequently, there has been a steady flow of up-and-coming statistical talent that is being exposed to a wide range of problems and senior researchers not only inside the building, but also throughout the Research Triangle area. This is also the perfect opportunity and place for high-impact mentoring. One might argue that such a working environment for a fresh PhD provides the perfect incubator for nurturing future leaders of statistics. As of 2014, there have been 70 postdoctoral students, including 24 women, who have been through the NISS program and 96 for SAMSI's, which includes about 20 women. Overall, then, about 27% of the participants in both programs combined have been women.

A characteristic of robust leadership is the ability to be flexible and to change directions as needed. Under the leadership of Jerome Sacks and Alan Karr, NISS had many early successes with its ambitious and timely inter-disciplinary mission, which stretched the span of statistics in new directions. It also had extensive support from the three local research universities at the corners of the Research Triangle. However, the funding model for the institute became inadequate, as initial core money from the Research Triangle Foundation dried up and hopes for national funding failed to come through. As the reality of the situation sank in, then-Board Chair John Bailar took timely action by launching several task forces to identify possible next steps, including one on long-range financial planning. An idea that emerged from this task force was to create an affiliates program of members from academia, business, and government. This was agreed to and launched in 2000. This new arrangement not only helped to steady the finances of NISS, but also brought different components of the statistics community in closer touch with the institute and its activities. Put another way, it fairly radically changed the infrastructure of NISS and tightened the bonds with its natural constituents. Today, the number of affiliates is about 60.

Recognition and Awards

One way of encouraging leadership by women is to publicize and honor their accomplishments with coveted recognitions and awards. (See Gray and Ghosh-Dastidar, 2010, for an assessment of how well this is working out.) A long-standing example of such recognition is the election as an ASA Fellow. The pace of women being chosen as ASA fellows has picked up in recent years as more women have joined the profession and association. During 2000–2005, 69 women were elected as Fellows, which equates to 22% of the total number awarded for that period. Over the recent four years, 2010–2013, the percentage edged up to 28% — a solid move in the right direction — but still an underrepresentation.

The ASA's Founders Award is for members who have "rendered distinguished service" and is strongly correlated with leadership. For the period 1989–2000, 11 of the 43 awards, or 26%, went to women. However, from 2001–2013, the number increased to 17 out of 43, an impressive bump to nearly 40%. (Numbers for these two ASA awards are updated from Palta, 2005 and 2010, and Ghosh-Dastidar, Craven, and Stangl, 2012.) NISS has a distinguished service award of its own: 9 of the 23 recipients to date have been women.

There are many awards exclusively for, or in support of, women scientists and a few excellent ones already in statistics. The Elizabeth L. Scott Award (for individuals who create opportunities for women in statistics), the Florence Nightingale David Award (for female statisticians who exemplify David's contributions), and the Gertrude M. Cox Scholarship (to encourage women to enter statistically related positions) come to mind. Examples from other fields include the Pearl Meister Greengard Prize (PMG), an international award in biomedical research, and the Louise Hay Award for contributions to mathematics education. With the increased participation and accomplishments of women in statistics, the current statistics awards could be expanded by adding another, even more ambitious prize, a head-turner focused on leadership and designed to make bold headlines.

The PMG Prize illustrates the possible. As described by Strauss (2013), this prize rewards women who have made remarkable biomedical breakthroughs and "steered their way through professional environments that did not always welcome them and confronted barriers that might seem shocking to subsequent generations of investigators." The award is presented annually at a special event by a distinguished woman from another field (e.g., Sandra Day O'Connor in 2004) and includes a $100,000 honorarium. A statistics version of such an honor could be based on the broad-based definition of leadership as defined in this book. It might be named the Florence Nightingale Leadership Prize in Statistics (see next section).

Role Models and Mentors

Most of us mortals have had our share of influential role models and mentors. I could list several who have helped me in significant ways throughout my career. Bonetta (2010) discusses their importance for female scientists generally and mentions a variety of programs that have been successful in that regard. Rubin (2014) explains how mentors from different disciplines have helped him throughout his career. Such experiences deserve to be replicated!

Speaking of role models, a 2013 exhibit at the Grolier Club in New York City (Smeltzer, Ruben, and Rose, 2013) put on display the astonishing achievements of 32 women in science and medicine selected from the 17th–20th centuries. Five women were included under the heading of mathematics, with the most prominent example from a statistics perspective being Florence Nightingale (1820–1910). The authors, in the preface to the catalogue, comment that they tried to capture "the often proscribed educational opportunities available for women," how they dealt with "limitations imposed by contemporary society and the academy" and what they encountered in "gender discrimination, sometimes on an individual basis and sometimes organizational and societal." *The New York Times* (Grady, 2013) stated that "The exhibition... makes it plain that [the women] are all the more extraordinary given the deeply entrenched biases they had to overcome."

Nightingale was the consummate data analyst. Moreover, the exhibition's catalogue asserts that "Her use of statistics seems to be the true beginning of evidence-based medicine and health care." She became the first elected female member of the Royal Statistical Society in 1858. Diamond and Stone (1981) provide fascinating detail on her education, upbringing, and professional work. She had "the ability to draw to herself. . . a group of talented and influential men" and "by diplomacy and delegation, [she] drew her associates into a striking force of great efficiency in the cause which they all shared: sanitary reform." They also describe "the drive, the administrative ability, and above all the charisma of her name" as contributors to her success. Another trait worth mentioning was her skill as a communicator. As Diamond and Stone put it, she was a "superb propagandist of her own causes." In so many ways, she is the foremost example of a robust leader!

Achieving Equal Opportunity and Robust Leadership

A recent editorial in *Science* magazine (McNutt, 2013) is a timely reminder of the ongoing challenges for equal treatment of women in the sciences. This may be most apparent for women in academia, but it also concerns those in industry and government, too. McNutt highlights the work of the Committee on Women in Science, Engineering, and Medicine for its efforts to "level the playing field for all women in STEM disciplines," which, of course, includes statistics. The friction caused by unequal treatment can only hold back the advancement of women in science and limit their possibilities for leadership roles.

We are fortunate to have had many individuals, men and women, serve as strong leaders in statistics. I mentioned a few examples, with hints of why they were successful. A long-term strategy for developing future leaders should be based on attracting even more of the highest quality professionals into the field and providing them with the best possible support for their development. We already have the knowledge of how to do this. Here is a simple plan:

- Invest heavily in statistical education in high schools.

- Expand summer employment and internship opportunities.

- Facilitate mentoring in school and on the job.

- Promote postdoctoral appointments as career enhancers.

- Ensure that significant mid-career milestones are achievable by women.

- Establish a distinguished leadership award for women in statistics.

At Drew it is rare to find a student who comes to the university with a passion for statistics. I ran into an exception recently. It happened because of an outstanding course the student took at her high school. The continued development and expansion of appropriately designed and well-taught high school courses should help to spark interest in and increase awareness of statistics as a field of study when students enter college.

Industry and government could play an important role by creating more summer jobs and undergraduate internships for students of statistics. ASA and NISS, with support of their members and affiliates, could help make this happen. We've learned that it takes longer for students to find the right major than we might have thought in the past (Mervis, 2014). Boosting the number of opportunities for students to try out statistics on the job, in the real world, should attract more young talent to the discipline.

Mentoring is a tried-and-true way of accelerating and enhancing careers. The relationships can vary across the spectrum of informal to formal. The key, in my experience, is that both the mentor and mentee need to be fully committed and involved to ensure a good outcome.

Postdoctoral appointments are still not the norm for new PhDs in statistics. Yet they can be marvelous broadening experiences for students who have just completed the often narrowing process of completing the doctoral thesis. In particular, these appointments should provide ways of sharpening critical communication skills of speaking, listening, and writing. Done with a vision in mind for the future, the postdoctoral experience can be especially useful preparation for a career filled with satisfying leadership opportunities.

The mid-career opportunities for women can be especially challenging. McNutt (2013) put it bluntly: "Data show that women don't advance professionally at the same rate as men." This may be a larger problem in academia, where women are often caught up in the rigidity of a tenure system just as their family obligations are peaking. While finalizing this chapter, I was happy to learn that my daughter received tenure from a large public university. She benefitted from a flexible system that helped her to balance professional and family responsibilities. It can be done. Models that work need more airing and sharing.

Awards are a way of drawing attention to highly successful individuals and publicizing their accomplishments. A high-bar award for women that honors major leadership successes can catch the attention of and spur on younger tyros. At the same time, we need to make sure that women are considered equally for all the major honors in statistics; see Golbeck and Molgaard (2013) for a thorough discussion of best practices in this regard.

This six-pronged strategy package should help to achieve what we are looking for: a stable pipeline of robust leaders in statistics for today and tomorrow.

Acknowledgment

Thanks to Alan Karr and Sally Morton for their important contributions.

Bibliography

Bonetta, L. (2010). Reaching gender equity in science: The importance of role models and mentors. AAAS/Science Business Office Feature, Focus on Careers, 889–895.

Carrillo, I. and Karr, A.F. (2013). Combining cohorts in longitudinal surveys. *Survey Methodology*, 39, 149–182.

Denby, L. and Landwehr, J. (2013). A conversation with Colin L. Mallows. *Int Stat Rev*, 81, 338–360.

Diamond, M. and Stone, M. (1981). Nightingale on Quetelet. *J R Stat Soc Ser A*, 144, 66–79.

Gertner, J. (2012). *The Idea Factory: Bell Labs and the Great Age of American Innovation.* New York: The Penguin Press.

Ghosh-Dastidar, B., Craven, P., and Stangl, D. (2012). Gender balance in ASA activities. *Amstat News*, Issue 421, 12–15.

Golbeck, A.L. and Molgaard, C.A. (2013). Professional awards in statistics: Chipping away at gender disparities in the USA. *Proc 59th ISI World Congress*, 5155–5160.

Grady, D. (2013, November 12). Honoring female pioneers in science. *The New York Times*, p. D1.

Gray, M. and Ghosh-Dastidar, B. (2010). Awards for women fall short. *Amstat News*, Issue 400, 25.

Kettenring, J.R. (2001). A conversation with Ramanathan Gnanadesikan. *Stat Sci*, 16, 295–309.

Kettenring, J.R. (2011). Rise of statistics in the 21st century. In *International Encyclopedia of Statistical Science*, Ed. M. Lovric, pp. 1234–1237. Berlin: Springer.

Kettenring, J.R. (2012). Statistics research at Bell Labs in the regulated monopoly era. *Int Stat Rev*, 80, 205–218.

Lohr, S. (2009, August 5). For today's graduate, just one word: Statistics. *The New York Times*, p. A1 and p. A3.

Madden, K. (2011). A unique interdisciplinary science research experience for first-year students. *J Coll Sci Teach*, 41, 32–37.

McCullough, L. (2011). Women's leadership in science, technology, engineering & mathematics: Barriers to participation. *Forum on Public Policy Online*, V2011, n2.

McNutt, M. (2013). Leveling the playing field. *Science*, 341, 317.

Mervis, J. (2014). Studies suggest two-way street for science majors. *Science*, 343, 125–126.

Moss-Racusin, C.A., Dovidio, J.F., Brescoll, V.L., Graham, M.J., and Handelsman, J. (2012). Science faculty's subtle gender biases favor male students. *Proc Nat Acad Sci USA*, 109, 16474–16479.

National Science Foundation (2014). Science and engineering labor force. Chapter 3 of *Science and Engineering indicators 2014*. http://www.nsf.gov/statistics/seind34/content/chapter-3/chapter-3.pdf (accessed March 10, 2014).

NISS (2014). The National Institute of Statistical Sciences history. Unpublished paper.

Olkin, I. (2014). Where have all the tenured women gone? *Amstat News*, Issue 439, 31–32.

Palta, M. (2005). Election of women ASA fellows 1914-2005. http://www.amstat.org/committees/cowis/pdfs/fellowsreport.pdf (accessed February 17, 2014).

Palta, M. (2010). Women in science still overlooked. *Amstat News*, Issue 400, 21–24.

Rodriguez, R. (2012). Statistical leadership: Developing leaders through ASA service. *Amstat News*, Issue 417, 3.

Rubin, D. (2014). The importance of mentors. In *Past, Present, and Future of Statistical Science*, Eds. Lin, X., Genest, C., Banks, D.L., Molenberghs, G., Scott, D.W., Wang, J-L, pp. 605-613. New York: Chapman & Hall/CRC.

Smeltzer, R.K., Ruben, R.J., and Rose, P. (2013). *Extraordinary Women in Science & Medicine: Four Centuries of Achievement*. New York: The Grolier Club.

Speed, T. (2014). Trilobites and us. *Amstat News*, Issue 439, 9.

Starbuck, R. (2012). The ASA fellow award—revisited. *Amstat News*, Issue 422, 8–10.

Strauss, E. (2013). Introduction. The Pearl Meister Greengard Prize Brochure, Rockefeller University, 3.

6

Leading Significant Change in Official Statistics: A Woman's Place Is in the Counting House!

Cynthia Z.F. Clark*

Setting the Stage for a Change in Official Statistics

As it has worked out historically, the embedded system of checks and balances between the executive, legislative, and judicial branches of the US Government strongly maintains the status quo. It is exceedingly difficult to make program or organizational changes within government. This may entail improving either the methodology or processes to gain approval through the funding mechanism. Similarly, organizational changes that create or change internal structures require approval above the level of the individual agency located within an executive department. This approval must minimally come at the congressional committee level — and sometimes at the White House level. Less consequential changes (in terms of either cost or organizational structure) may occur within an organizational unit of a government agency — but only if they can be made within the budget or jurisdictional authority of that unit.

This chapter will address program and structural changes in several government agencies producing official statistics; the factors that were an impetus to make change; the characteristics that enabled changes to be implemented in the organizations; and the impact of the change on the organization in question. Its focus will be on change in statistical programs producing official statistics. Only in conclusion will I comment on the particular role — if any — that women have to play in promoting and facilitating those changes.

As a statistician, I have approached my role in official statistics as a facilitator of improvements in the statistical product. This may entail improving the methodology or processes used in creating that product; creating the human resource environment that produces that product; and obtaining funding necessary to improve the quality of product desired. My background includes work experience (at the staff, management, and senior leadership level) in

---

*Former Administrator, National Agricultural Statistics Service (2008 - 2014), the first female head of the agency.

four US statistical organizations and in the UK statistical system; it also includes national and international experience gained through interaction with colleagues in other major official statistical organizations in the US system and throughout the world.

Official statistics programs present issues that differ from those encountered in supervision of other types of government programs! Official statistics are often produced periodically. Data from the periodical release are often compared with data from the previous release to determine changes in the statistics due to changes in production, population, housing, etc. One often encounters resistance to the prospect of effecting change in any of the program's processes — design, collection, estimation, or other — as that may impact the data series. The impact may not necessarily be known, but it may affect the observed change in the statistics.

If there has been a change in the methodology producing the statistics, then the change properly ascribed to external factors and the change effected by the altered methodology are confounded. This issue may be addressed by producing parallel statistics for a given period — using both the previous method and the changed procedure — or by reproducing previously published data using the recently changed procedure. Sometimes, neither approach is feasible — particularly with programs that are released less frequently (such as a five- or ten-year census). At other times, the cost of either approach is prohibitive. In any case, guidance then needs to be provided to the user of statistics in comparing the two procedures.

Changes in methodology within an individual statistical program may have either a minor or major impact. Some changes can be made within existing budgets; others require additional funding. That funding might conceivably be obtained through rebudgeting within the given program — or within the agency. More substantive changes require additional congressional funding, which must be obtained through an extensive, almost two-year-long federal budgeting process. This chapter will address changes of both types.

Examples of Programs That Experienced Change

I will first provide some background on six specific programs that experienced change during my tenure. These come from my personal experience at the US Census Bureau (1983–1990, 1996–2004), the National Agricultural Statistics Service/USDA (1990–1996, 2008–2014), and at the UK Office for National Statistics (2004–2007). Such experience enables one to specify, more generally, the factors significant in leading and implementing a change.

- Research and Methodology for the 1987 Census of Agriculture at the Census Bureau

- Initiation of the Agriculture Resource Management Survey (ARMS) at the National Agricultural Statistics Service (NASS)

- Research Program on Survey Methodology for Establishment Surveys at the Census Bureau

- Quality Management Program at the Census Bureau

- UK Office for National Statistics (ONS) Modernisation Program

- National Agricultural Statistics Service (NASS) Transformation

Research and Methodology for the 1987 Census of Agriculture. Prior to 1997, the Census of Agriculture and several follow-on surveys were produced in the Agriculture Division in the Economic Fields Directorate of the Census Bureau. A unit within that division — known as Research and Methods — had historically been responsible for the sample design employed in surveys used to weight census counts of farm operations that did not respond; also to evaluate the coverage of the farm population enumerated by the census. Research and Methods also dealt with quality control of statistical processes and with sample design and estimation respecting follow-on surveys. The manager of this research unit had the vision that the unit could provide a wider range of statistical input to the census program. I was hired to implement this vision; within two years the research manager became the division chief and I took his former position as manager of the Research and Methods Unit.

The Agriculture Division routinely conducted a test of the upcoming census — generally a dry run of the collection and processing procedures. However, the test proposed for 1985 was of an entirely different character. It was conceived of as a factorial design comparing data collection materials, sequence timing, and questionnaire format. It meant to compare the traditional foldout version with a booklet format, and to perform classroom tests of respondents' cognition of the questions. It projected a reinterview of a sample of the respondent population — focusing on known data reporting issues and on the reaction of nonrespondents — obtaining their reason for not responding.

Initiation of the Agriculture Resource Management Survey (ARMS) at NASS. The National Agricultural Statistics Service (NASS) had conducted an annual Farm Cost and Returns Survey since the 1970s as a shared survey initiative with the Economic Research Service (ERS) and also, in the mid-1980s, NASS solely funded an annual Chemical Use Survey. There was some overlap between the two surveys in the survey population, and in the agriculture production and environmental data that was sought. The NASS goal was to collect aggregate statistics for economic and environmental data concepts; ERS was interested in linking data records from the two surveys relating to the farm household with farm financial, farm practice, and environmental information for econometric analysis. This would enable researchers to test hypotheses about relationships between a rich set of variables.

The ARMS program arose when NASS recognized that multiple surveys collected similar information, and then acted to reduce respondent burden occasioned by separate data collections. The program was envisioned as a three-phase survey that initially screened for the annual target population, had a second phase that collected the chemical use sample, and a third phase that collected economic, agricultural production, resource management, and environmental data on a sample of records from the chemical use sample, augmented by records from other agricultural populations of interest in that year.

Research Program on Survey Methodology for Establishment Surveys. Application of cognitive psychology to surveys had its inception in the 1970s, with its application to surveys of populations and households. Its application to establishment (a business enterprise) surveys had been limited — but the idea had arisen among survey methodologists that there was some potential in cognitive psychology. In 1998, I (as associate director for Research, Methodology, and Standards) was able to identify a recognized academic (a survey methodologist and statistician) who was willing to spend a sabbatical year at the Census Bureau working on survey methodology research of our choice relevant to establishment surveys. His only condition was that he have two researchers with whom to work. It was challenging to create two new positions appropriately placed within the organization — filled by appropriately skilled individuals — but in the process, support was gained from two individuals in management and leadership roles who contributed to the ultimate success of the research project.

Quality Management Program at the Census Bureau. The Census Bureau has long had a focus on the statistical methodology and survey collection procedures required to produce quality products. However, its procedures and standards were often unwritten or, if written, were issued under different executive authority. There was no consistent source for standards that were relevant across all its programs. In the early 1990s, a Census Quality Management (CQM) program was initiated by the then-Census Bureau director. Many quality circles were organized, and lots of ideas put forth to the bureau leadership. However, none were funded. The upshot was that the CQM effort left a bad taste at the bureau. Later, in the 1990s, I (as associate director for Research, Methodology, and Standards) felt the need to establish consistent standards as part of a broader Quality Management Program.

ONS Modernisation Program. The UK's Office for National Statistics (ONS) faced a problem common to official statistics organizations — its funding sources did not often appropriate funds for allocation to infrastructure and its modernization. The infrastructure for an official statistics office would include development of systems for collection, processing, and publication of data; acquisition of human resources that had a more diverse skill set and higher level of statistical and technology competence; and the procurement of technology relevant for these systems. The infrastructure for ONS was particularly challenged, as ONS had been formed in 1996 by merging three formerly independent statistical offices — the Central Statistical Office, the Office of Population Censuses and Surveys, and the Employment Department Statistics — all with separate locations, processing systems, and tools. The merging of these offices also presented the opportunity to reengineer programs that had components coming from several of these former offices.

The leader of the office, the National Statistician, was successful in getting the British Government to attend to this need (which was hampering the development and production of national statistics). ONS developed plans for an extensive modernization program; the Government agreed to the plan, and appropriated a large incremental sum of money, beginning in 2003/2004, expecting results beginning in 2006. I joined the office in 2004 as Executive Director for Methodology, participating in the overall direction of the program with particular focus on the statistical components — where my directorate played a major role.

National Agricultural Statistics Service Transformation. The National Agricultural Statistics Service (NASS) had been a very stable agency — producing agricultural statistics since its inception in 1957. It had, however, neither invested in research of a substantial enough breadth nor in technology to keep up with new developments in these fields. It had a strong culture promoting a decentralized structure for collection and review of agency data. This led to a lack of standardization in agency processes and, ultimately, to biases and variability in its data. It also lacked transparency in the description of its statistical methodology. As the new administrator in 2008, I felt the need to provide more transparency on agency processes and measures of the quality of the agency's data. I also foresaw a need for the agency to develop more efficient processes, given that the country had just entered into a major recession.

This led to a program that centralized and consolidated network servers and administration from 46 locations; standardized survey metadata and integrated survey data into easily accessible databases across all surveys; consolidated and generalized survey applications for the agency's diverse survey program; standardized processes by creating a central data collection and processing center and regional rather than state offices; centralized methodological support into one headquarters division; instituted an inclusive quality management program; and enhanced the research effort to enable the agency to use transparent estimating procedures. The end result held the promise of better-quality data, transparency of agency estimation methodology, and less costly production of agricultural statistics.

Factors Significant in Implementing a Change

Defining the vision for a new paradigm. The first step in improving a process or organization is to have a vision of what might be improved. In statistical processes, this generally occurs when someone becomes aware of a method or procedure that might improve a process — and envisions an opportunity to make that improvement. To develop a vision for change, it is not always necessary, initially, to have a plan as to how the improvement might be made. It is desirable to have some general notions of what needs to occur, but some communication of a vision may begin without a developed plan.

In the case of the 1987 Census of Agriculture Program, the research manager who hired me recognized that improvement in the census was needed and that he did not have the staff expertise to accomplish any substantive improvement. He became aware of an expert in survey methodology who had just retired and recruited her as a contractor; he also hired me as a mid-level statistical researcher and directed us to develop research that could improve the 1987 census. Additionally, he supported building a research team by hiring more statisticians than were previously in the division, and individuals with different skills. He communicated that he was seeking a new vision by taking the initiative to set up a research team where one had not previously existed.

Consider the National Agricultural Statistics Service (NASS) transformation: From the first, I had a goal to provide more transparency in NASS data release processes when I assumed the position of Administrator of the agency. I began communicating that goal in my first few days on the job — and cited an Office of Management and Budget Statistical Policy Directive requirement. I did not have a plan as to how to make that happen at the time.

Communicating the vision and motivating people. As the vision evolves and becomes more defined, it is necessary to discuss and present that vision. Most often, initial discussions begin with trusted colleagues; as the vision becomes more defined, it is desirable to expand the reach of discussion. The author of the vision needs to be able to motivate others — first, organizational leaders and managers, and then their staff — to engage with the vision. This is a necessity if the change is to be pervasive within an organization. Such was the prevailing condition in connection with the ONS modernization and the NASS transformations.

Less pervasive was the change in establishment survey methodology that I initiated as Associate Director for Research, Methodology, and Standards. The initiative began with the formation of a research team. This was immediately followed by the creation of a research steering committee convened by the Assistant Director for Economic Fields. This ensured that there was both buy-in and direction for the research from two of the Census Bureau directorates. Thus, there was very senior involvement initially that motivated staff to participate.

Creating a team that is committed to the changed vision. Most improvements in statistical programs, whether they be within a division, at a program level, or at an organizational level, require team efforts. One of the first steps in a change that leads to a new vision for that entity is to gain acceptance of the vision by others on the team. In some cases, this may happen initially, in other cases, it happens gradually — as the plans for the change develop and are communicated within the organization. When one contemplates a transformative change within an organization, the change agent may gain important support by engaging an executive coach. The coach is a hands-on practitioner of organizational theory, as taught (but not necessarily practiced) in a generic "School of Management." The coach stands outside the organization's structure, and can thus counsel managers in a nonthreatening way. The coach's services can provide an avenue for managers to discuss and consider alternative approaches to achievement of the vision.

In order to develop consistent standards, I began working with a staff member to compile all existing standards issued by various organizational units within the Census Bureau. Those compiled standards were organized in a manner consistent with categorizations in the metadata template to identify gaps in the current standards. They were also reviewed to determine a format for new standards. Once a format and conceptual structure for the standards was established, the Research and Methodology Council (the senior research and methodology managers in the Census Bureau) agreed to pursue an area in need of a standard to establish a prototype. The Council asked to be the first line of approval. Once a final product was agreed on it would be presented to the Associate Directors for all the program directorates of the bureau. The key to the success of this initiative was gaining agreement with the Associate Directors for the program directorates that this process would be followed for the development of individual standards. They would have the opportunity to review and to sign off, but then would have to take the responsibility for ensuring that the standards were adhered to within their program scope of responsibility. This was to ensure that the standards would be consistently followed for all Census Bureau programs.

Initially, creating a senior team that was committed to a new National Agricultural Statistics Service (NASS) vision was very challenging. The new vision was based on a totally different organizational paradigm — with responsibility for many statistical functions passing to a more centralized structure. Such centralization facilitated a higher degree of standardization — needed to reduce both bias and variability in the data and to realize cost efficiencies. The centralization occurred in stages, not all defined at the outset. Several of the senior team members who had spent their entire careers with the agency had difficulty grasping the perspectives of the changed vision and altered agency mission — and ultimately chose to leave the organization. This allowed the administrator to select, as their replacements, individuals who would champion the proposed changes in the organization. As Administrator of NASS during this phase of development of the change vision, I found a trustworthy executive coach invaluable.

Finding support and leadership for the vision at all levels of an organization affected by the change. In order for a change or improvement in a statistical program to occur, support is required at several levels within the affected organization. The program manager would certainly need to be supportive. Sometimes, the program manager initiates the change, but the initiator might be a member of the program manager's staff. All of the program staff would need to be supportive — as an official statistics program often involves analysts, statisticians, programmers, and public affairs. The funding for the program often is determined outside the program manager's responsibility — sometimes at the division or directorate level, but often at the agency level.

At the Office of National Statistics (ONS), support for the vision was engendered at all levels of the organization. The initial plan for change had four components: to modernize and upgrade the agency's technical infrastructure; to integrate generalized statistical tools into the technical environment; to re-engineer thirteen statistical areas (including national accounts, labor market, population, life events, business register, prices, surveys, regional ethnicity, industry, health, and care); and to transform access to ONS statistics via the web. Leaders, managers, and staff were engaged, and the program grew well beyond the initial bounds to include modernizing corporate services and registration as well as projects within the purview of the directors — with each seeking to make use of funding available to the Office overall (Penneck, 2009). The plans were neither well developed nor coordinated within the organization. The leaders had the vision, but the scale of the contemplated program practically dictated that there would be great difficulty in developing detailed plans and implementing them within the organization. The implementation plan must be understood and accepted by those in the trenches who can actually effect results. It becomes crucial for the change agent to identify — and trust — those who can do this.

Implementing the vision. The first step in making a change in a program or process at a statistical agency is to identify a champion for the change. It is the responsibility of the champion to identify others to be part of the change process and to promote the importance of the change. Generally, one of the first steps is for the champion to develop a plan. If it is a change in program or process, then the plan should include a description of the current program/process, also of the new program/process, and of the impact of the change that the new process will entail. If the change is major, it is useful to prepare a business case that looks at all aspects of the changed program or process — including its interactions with other agency programs; its resource requirements; and its funding requirements — and at the potential risks and barriers to implementation of the change. The plan should include steps to mitigate risks and to address identified barriers.

In the National Agricultural Statistics Service transformation, the process for an agency plan began with the senior managers/leaders of the agency reviewing proposals to become more cost-efficient. On the basis of their discussion, five projects were identified as efficiency initiatives. A senior leader was asked to champion each project and to develop a business case for the project. I had already begun to work with the Research Director to develop a plan to enhance the agency research program, which the director led and championed. A subsequent meeting was held with the next level of managers so that they could have input into the development of the business cases. Staff members were informed about these projects.

Once a plan or business case has been prepared, it is useful to obtain comments from the group involved in leading the change and then from those affected by the change. Discussions will be required to assess staff resource and initial funding requirements. On account of those requirements, adjustments in the plan will often be required. If the change involves a large organization, or unit within the organization, then involvement and commitment is needed at all levels of leadership, management, and supervision — to ensure that all staff are on board. Individual dissenters can influence others in the organization to oppose the changes. If their concerns are not addressed, opposition will continue to occur and confusion will exist respecting the direction the organization means to take.

If the change requires a substantial increase in funding, then an appeal may have to be made through the complex budget process to gain funding support. This means that support will need to be garnered in the organization within the department in which the agency resides; at the department; at the Office of Management and Budget; and, ultimately, with the congressional authorizing committee for the department. This later process can be quite time consuming — at least 18 months from the time when a proposal is made until funds are appropriated (and often longer when congressional budgets are not consummated within the agreed upon schedule). There is a similar process in the UK, equally complex and time consuming.

If there is support within an organization for a substantial change, steps may be taken, on either a temporary or permanent basis, to avoid the lengthy budget authorization process. The best strategy is to find funds within the control of the change agent rather than rely on requests for extraneous added resources. This was the situation in all of the examples, with the exception of the UK modernization program. If there is commitment within an agency for a major change, budgetary measures might be taken in the short run to transfer funds between programs; to reduce costs of travel, training, conferences, and supplies; or to reduce hiring. However, these produce only short-term savings. In the US budgetary system, any savings in one year must be expended that same year. Often, research initiatives can be accommodated within the agency budgetary process; this is not, however, the case for major changes in survey or estimating programs.

The NASS business cases were reviewed several months later to determine whether each was feasible and whether the agency could expect to be able to fund each project. In the interim after the progress since the projects were conceived, the agency unexpectedly received funding for a major survey that had been suspended and also for a new program, thus increasing the agency budget by almost 7%. An assumption was made that the agency budget would remain at the increased level for the following fiscal years - 2010, 2011, and 2012 - with the exception of the incremental increase for the periodic census of agriculture. The decision was also made by the administrator and the Chief Operating Officer to be very prudent with budget expenditures — limiting relocation of staff, travel, some hardware and supplies. Costs related to the efficiencies were prioritized and allocated annually. For example, the videoconferencing equipment was procured with FY 2010 funds by reducing the training budget for that year; the cost of providing a thin client network using Citrix across all offices was budgeted and the project phased in during FY 2011; the cost of leasing and building out a centralized facility for data collection and processing was paid for out of FY 2011 and FY 2012 funds and savings achieved from the two initiatives previously implemented. Costs of developing new technology and systems and enhancing the research program were spread across four to five years.

No matter what the funding situation is within the organization, there is often a need for additional staff to assist with the transformation. Often, those staff resources may need to be reassigned from elsewhere in the organization undergoing a change in organizational priorities. Sometimes, staff with the needed expertise cannot be found within the organization. Thus, either new staff must be hired, or existing staff retrained. In the short run, it can eventuate that neither is possible.

The UK Office for National Statistics (ONS) was overambitious, thinking that many things could be accomplished simultaneously. This put the office in a position of overstretching its management, overhiring staff, and expending large sums on contractors to perform information management and technology services for the office (as there was no adequately knowledgeable internal staff to deploy for this important aspect of the project). Projects were defined on too large a scale, rather than in manageable phases. By 2005–2006, the office was in a major financial crisis, requiring downsizing of staff and a rescoping of the activities in the modernization program.

Once there is agreement with the plan, it is advisable to use project management tools to ensure that the project can proceed to completion. Those tools require oversight and management at several levels. If the change within a program or organization is extensive, there is a need for staff involvement, management, leadership, and oversight. As the improvement or change process proceeds (whether it be research or program development or organizational change), there will arise suggestions for changes in the plan. It is useful to have a formal process that assesses supplemental proposals for changes — often referred to as a change control process. This process would then provide recommendations respecting whether or not to change the path.

Both ONS and NASS engaged in formal project management. The approach differed in that ONS had project managers who moved between subject areas, whereas NASS project managers were selected within their subject matter expertise and provided with formal project management training. The ONS approach did not allow the project managers to contribute substantively to the project design. The NASS approach allowed the project manager to bring to the task both the tools of project management and significant subject matter or statistical expertise. NASS engaged in a formal training program in project management, selecting for the training those who were otherwise slated to be project managers.

Communicating the improvement or change within the organi-

zation. The role of communication is often underrated in an organization. Most organizations have formal mechanisms through which staff members are kept informed about activities within their part of the organization and also across the broader organization. When an improvement or change in a process or program is being proposed, it is even more important that communication relating to the change occur across the organization. The communication needs continually to reiterate the change vision. It is difficult to know how much to communicate and when to communicate it. Yet, as a change moves forward, it is important to ensure that information regarding the change be communicated. It must be understood how the change will impact different staff members in the organization. It is desirable that communication be an integral part of the project manager's task.

Different patterns of communication occurred within the noted examples. For the census of agriculture, communication was entirely within the Agriculture Division. It occurred through ongoing channels and also through project team meetings. For the Agricultural Resource Management Survey, design, survey, and communication took place within the interagency project team. When the survey was launched a broader based training occurred — first with headquarters staff of the two agencies and then many sessions with those in the field who would be doing the data collection. The success of the new survey was hampered by the fact that NASS did not have a strong investment in economic or environmental data and did not see such data as important to its mission. That view gradually changed as there developed a stronger realization of the importance of this data — gained, in part, as knowledge of the results of econometric analyses was obtained.

For the establishment survey methodology research, communication was primarily done within the project teams and the Economic Directorate. Several research seminars were held to communicate the research findings. The seminars discussing this research were the best attended of any research seminars given at the Census Bureau — with standing room only in the auditorium. The research agenda expanded from the initial project, leading to the publication of numerous papers in refereed journals. The impact of the project was to establish a unit that has maintained an ongoing program of research in establishment methodology. The Research, Methodology, and Standards Directorate proposed a plan to link quality management and methodology standards with the quality component of project management. At the time, there was a heavy emphasis throughout the organization on project management training. This strategy worked well for communicating knowledge about the importance of standards and quality in project management. It also provided a broader understanding of the goals of creating quality standards. Then, as an individual quality standard was developed, training was provided to the relevant staff.

At NASS and ONS, many forums were used to communicate the status of the initiatives throughout the process. These included, initially, management and staff meetings and monthly progress reports. Since staff members were in several locations, videoconferencing technology was an extremely useful tool. External communication was also very important to ensure that support was obtained when working with external stakeholders, and to ensure that the organization's priorities were understood when conflicting departmental or ministerial initiatives were proposed.

Periodically reviewing progress on the change. For any change process to be effective, the leaders of the implementation of the change must periodically review the status of the change's implementation. The timing of this review differs for each improvement or change, depending on many varied factors. However, it is important that several levels of review occur in relation to any substantive change.

Midway through the planned three-year program of implementing the efficiency and research initiatives at NASS, the federal budget was in a crisis situation. The leaders of NASS recognized that the agency would have to take the major step of resizing its state offices. An additional plan was developed to effectively create regional offices (with a two-person state structure kept in place only to facilitate agricultural survey collaboration with individual states and for state data collection). The NASS plan for its transformation risked being stalled — not just for the short term but much longer — as the agency was soon in the midst of its large five-year census program. Because I and the senior executives of the agency had communicated the agency transformation plan rather well within the department, the departmental leadership agreed that the agency might suspend or reduce programs rather than delay its efficiency initiatives. The NASS agency plan to resize its offices was the most traumatic of all the initiatives because it entailed moving many staff to different physical locations. Although it had been the agency culture for many years that staff had to move to have the opportunity for more senior positions, some staff members chose to stay in their current positions where they wanted to live. The trauma relating to the disruption was minimized because the new structure provided numerous opportunities for promotion — and many staff members had the opportunity to select where they wanted to relocate. However, some did retire early or leave the agency. This loss of staff put the agency in the position of needing to hire to replace. This effort was engaged in very selectively, so as to not over-hire when the agency was still expecting to realize staff reductions through the implementation of technology and research initiatives.

Final Analysis of the Change Process

All the programs mentioned achieved some degree of success:

The plan for the 1985 Census of Agriculture test was significantly different than previous census tests, in that it used a formal experimental design. Analysts in the division were not supportive of all the procedures being tested in the experimental design but were more willing to consent, as it was only a test. The results from the test, however, produced the evidence to implement different data collection procedures in the 1987 census. In comparison with previous censuses, these procedures increased by 17% the early response rate to the census, and by 7% the final response rate. The new procedures shortened the data collection period — thus reducing costs — and likely improved the data quality, given that the data recall period was shortened. It also provided information that a booklet form received the same response as the traditional foldout census form. However, there was not enough confidence in that result within the management of the agricultural analysts in the division to implement the booklet for the 1987 census. It did lay the groundwork for implementation in a later census. The research agenda raised the expectations for the Research and Methods staff in the division and demonstrated that their work could benefit the census program in more ways than had previously been assumed.

The initial introduction of the Agricultural Resource Management Survey was followed by several years in which the survey did not receive the attention needed for it to gain momentum within NASS. This may have been due to the lack of a champion within NASS — and to a failure to recognize how ARMS might further agency programs. The results of the ensuing econometric research (and a more recent focus on environmental issues within society as a whole) have enhanced the value placed on the survey data by agency managers and staff.

The initial establishment survey methodology research was very effective because of the involvement of senior managers and leaders at the Census Bureau; because of an academic who was committed to improving official establishment survey data broadly using new methodology; and because of the recruitment of well-qualified staff to work on the project. The assistant director for Economic Fields took this as an opportunity to learn about issues of interest to him, and then used his management resources to engage others in the research process. He continued to support the new field of research even after the conclusion of the project.

In the case of the Quality Management Program, the senior methodology managers were committed to a new vision. They knew that an initial effort had to start small because of the previous bad CQM experience. In this case, no focused communication effort was thought useful. Rather, the change occurred through a managed process, initially small, which was, however, backed by the managers of the statistical processes for the bureau programs. It was, nevertheless, necessary to obtain support from the program associate directors at the Census Bureau because the final standard was issued under all of their signatures. That entailed an extensive discussion and review process as each new standard was developed. The final result was a set of current, documented, consistent statistical program standards at the Census Bureau.

In my opinion, the initial mistake at ONS was the failure to set priorities within the modernization program vision. The vision was very expansive, and there did not seem to be any understanding that a long time would be required to realize its goals. Projects were defined on too large a scale, rather than in manageable phases. There was no realistic assessment as to whether staff had the skills required to execute the program. The modernization program should have been conceived as a longer-term initiative with funding coming over a multiyear period. Rather, a large sum of money was received over a three-year period — an inadequate period of time to accomplish or even make significant progress on any of the goals. The office was overambitious, thinking that many things could be accomplished at once. This put the Office in a position of overhiring staff and expending large sums on contractors to perform information management and technology services for the office. By 2005–2006 the office was in a major financial crisis, requiring downsizing of staff and re-scoping the activities in the modernization program. There were some program accomplishments, e.g., reengineering of the national accounts system, modernized monthly and annual survey of employment, hours and earnings, and the integrated business register and employment survey. The program accomplishments did not, however, live up to the vision presented to stakeholders.

The NASS implementation experienced challenges throughout the process, but did achieve its goals (though not on the original time schedule). Challenges were both internal and external — initially from staff not committed to the changes — and subsequently from budget restrictions and the extraneous agendas of elected politicians as the schedule slipped into an election year. The agency was fortunate in having a senior individual to lead the design (or reengineering) of over twenty technology and processing systems who understood the information technology, statistical, and programmatic aspects of the business. This individual was willing to address difficult issues and work with people to achieve the goals of the individual systems (Nealon and Gleaton, 2013). After the designs were in place, a subsequent manager worked successfully to coordinate and integrate the different systems. Penneck (2009) provides a discussion of efforts by several national statistical offices to redesign their statistical systems, commenting on the role of statistical leadership, the need for methodologists and technologists to work together, and the requirement for strong governance.

One major challenge that the agency faced was conducting the 2012 Census of Agriculture while still completing the regional restructuring (not all the needed managers and staff were in place), and doing all this during a governmentwide shutdown at a crucial review point. The agency rose to meet the ultimate challenge of delivering on a predetermined schedule, and was even successful in implementing innovative research to adjust the census for non-response and under-coverage (using a capture-recapture estimator) on a compromised time schedule. The ultimate success of the transformational change derived from the commitment of the entire NASS staff to the mission of the agency, realizing that a new vision was required to meet the next generation of challenges facing the agency.

A Woman's Place Is In the Counting House!

Do statistical data have a gender? Obviously not! Female statisticians and economists have played important leadership roles in official statistics, both internationally and in the United States. Indeed, official statistics has been more open to women than other avenues in the profession have been.

This is not to say that there is now an adequate pipeline for advancement of women in official statistical agencies. At NASS in 2014, for example, there were roughly equal numbers of male and female professional staff (including statisticians, economists, computer specialists, and analysts in the subject matter fields of the agency); what is lacking, overall, are women in government with advanced degrees who have developed leadership skills.

The required skills are:

- a breadth of knowledge — both general and specific – concerning the subject of the agency and its programs;

- an ability to think strategically;

- an ability to work well with people;

- facility in oral and written communication;

- a developed understanding of how one works within a government bureaucracy.

Some exceptional women have emerged as leaders in both the US and the UK. Female Commissioners have consistently led the Bureau of Labor Statistics since the late 1970s and have been Administrators at the Economic Research Service since 1995. There have been two female Census Bureau Directors in the past twenty-five years; so also several Associate Directors — and the current US Chief Statistician. Additionally, the two most recent UK National Statisticians were female. Most of these women earned their graduate degrees in economics or demography rather than statistics. There is an opportunity for women with graduate degrees in statistics to aspire to these positions.

It should be evident that no woman can rightly claim a unique gender-related capacity to manage significant change, but no organization aspiring to effect insightful self-improvement can afford to deprive itself of the services of individuals of either gender who have honed the skills required for effective leadership.

At the Salt Lake Tabernacle, on 18 July 1869, the enlightened leader of a beleaguered people — facing a stark need to capitalize on every available capability — had this to say:

> [W]e have sisters here who, if they had the privilege of studying, would make just as good mathematicians or accountants as any man; and we think they ought to have the privilege to study

these branches of knowledge that they may develop the powers with which they are endowed. We believe that women...should stand behind the counter, study law or physic, or become good book-keepers and be able to do the business in any counting house....

Discourses of Brigham Young [Journal of Discourses 13:61], selected and arranged by John A. Widtsoe, Salt Lake City, Utah: Deseret Book, 1925 edition, p. 335.

Even today, some others would say: "A woman's place is in the home."

A woman's place is also in the counting house!

Acknowledgments

The author thanks colleagues who contributed to the examples discussed in this paper and provided comments on a draft — Tom Mesenbourg, former Acting Director, US Census Bureau; Charles P. Pautler, Jr. and Cheryl Landman, senior executives, US Census Bureau; Stephen Penneck, former Director General, UK Office for National Statistics; Michael Hidiroglou, Research Director, Statistics Canada; Mary Bohman, Administrator, Economic Research Service; Joseph Reilly, Administrator, National Agricultural Statistics Service.

Bibliography

Nealon, J.P. and Gleaton, E. (2013). Consolidation and Standardization of Survey Operations at a Decentralized Federal Statistical Agency, Journal of Official Statistics, 29, No. 1.

Penneck, S., The Office for National Statistics (ONS) Statistical Modernisation Programme: What went right? What went wrong? Proceedings of Modernisation of Statistical Production, Helsinki, Finland, 2009. (available at URL= http://www.scb/se/Grupp/Produkter_Tjanster/Kurser/ModernisationWorkshop/final_papers/D_1_management_Penneck_final.pdf).

Penneck, S. (2013). Discussion (of Nealon and other papers in issue), Journal of Official Statistics, 29, No. 1.

7

What Makes a Leader?

Lynne Billard

University of Georgia

Introduction

This article reviews the literature on leadership: types of leadership (transformational, transactional, laissez-faire, etc.), and actions that define effective leadership. Since no leadership can occur in a vacuum, this review includes how to respect and treat followers, i.e., those under the supervision of the leader. First, truly effective leaders put the core interests of their organization ahead of any personal goals. While much of the vast body of research has its roots in industry and government, most of the principles carry over to the academic world. Therefore, this article adapts those principles to academia. However, while a nonacademic setting may function with more of a vertical leadership pattern, a unique feature of the academic world is its horizontal leadership structure, wherein full professors tend to be the governing body, and peer review and its referents dominate. Unfortunately, but not surprisingly, until recently there have been very few studies on the roles of women as leaders. A section on the specific characteristics of leadership by women is included. Though few in number, and though leadership is still more difficult for women than for men, most women display transformational leadership patterns.

Leadership can be a most elusive concept. History has shown many examples of leaders who have assiduously worked for the betterment of society and the environment in which they exercised their skills. Equally, there are numerous historical figures whose leadership has been devastating to those around them. Some of these "devastating" leaders were even seen as positive influences at the time, with the passages of time revealing such figures in entirely different colors. Indeed, an effective leader is basically someone who was able to effect change — for good or for evil — by successfully engaging those around them to support their agendas.

Important historical leaders tend to have exercised their art on a national or international stage. It is easy to think of major political leaders here. However, leaders can occur at all levels of society and all stages of a person's life and career, including in the field of education. What about the school principal who engendered such a love of learning and such a high standard of achievement that four of the graduating class of fifteen students received four of the only twelve nationwide university fellowships? Or, what about the parents who worked tirelessly to raise funds to equip their schools with much needed school supplies and opportunities? These are certainly leaders.

While there were important studies on leadership before the 1980s (e.g., Kerr and Jermier, 1978), it was the definitive work of Bass (1985) that generated a large, rich, and diverse literature on what makes an effective leader: What are the leader's defining characteristics? What are the differing styles of leaders? What roles do emotions play? and so on. There are some key differences in the modus operandi of a successful academic leader, but many of the basic principles apply or can be adapted to the academic world. Indeed, over an academic career, it is easy to identify all the major leadership types.

The present work will draw upon this literature as it describes the roles of leadership, particularly in the academic environment, though as in the literature the terminology will flow loosely back and forth between organizations, industry, government, and academia. Further, the studies mostly assumed that the leadership cadre is exclusively male. However, in recent years, some research is emerging about women as leaders. Certainly, today, women are slowly being appointed primarily to the lower levels of academic leadership and to leadership roles in associations. There are some notable exceptions of women appointed as presidents or chancellors of colleges and universities. A Google search lists sixty-one current and former such appointments, since Hanna Gray became the president of the University of Chicago in 1978, ten years prior to the second and third appointees in 1988; forty-five (74%) of these were appointed in 2004 or later, and twenty-one (34%) since 2010; see Appendix. Some thoughts that are especially applicable to women are addressed in a final section.

I have served academia both as a department head and associate to the dean, and I have worked and served in various leadership roles, including presidents of professional statistical societies. More importantly, over the years, I have observed many leaders both in academia and professional associations, both nationally and internationally — both inside and outside the statistics profession. This chapter addresses principles that make this leadership effective (not all of which, incidentally, did I realize when serving in these roles myself — learning from observing others is a great leveler!). Many of my illustrations are drawn from my own observations and experiences, including those of other leaders, though specific details of most events have been altered so that no one event or person involved can be identified; their only purpose is to illustrate some principle.

Types of Leadership

In any setting, there are both followers and leaders. In the often existing hierarchical structure, a leader at one level is a follower at a higher level. Of course, leadership cannot occur in isolation; it has to have its followers. Bresnen (1995) asserts that "effective leadership (or the lack of it) is regarded as being crucial to overall organizational success." He is speaking of the construction industry; but surely this applies to all sectors of the workforce, industry, government, and academia. While the literature is extensive, it essentially defines three broad categories of leadership — transformational, charismatic, and transactional — though Yukl (2006) opines that "influence is the essence of leadership." The literature investigates how leaders lead, how they inspire their followers to meet, or even exceed, the goals of their organizations. A detailed catalogue of the characteristics of competent leaders can be found in Bolden and Gosling (2006).

Transactional leadership occurs when there is a form of contract between the leader and her followers, as when an employee is hired to undertake a specific task, a traditional leadership style that has been especially applicable in industry. It involves rewards for attaining the contractual goals, but also interventions, maybe even threats, when those goals are not met. Transactional leaders are quick to take credit when their program performs well, but are equally often quick to blame their followers when they themselves are criticized; see Bligh, Kohles, and Pilai (2011). Bass (1990) shows how this style is not very effective, and in fact counterproductive in the long run, particularly when the transactional leader passively employs "management by exception," i.e., only intervenes when problems have become severe. This passivity is in contrast to "active management by exception" where attention is paid only when standards are not met, as described by Eagly, Johannesen-Schmidt, and van Engen (2003). Either way, the follower is left with an unpleasant feeling and can eventually feel disloyal to her institution. Such leaders can forget that transactions must depend on the interests of all followers. For, as Bass and Steidlmeier (1999) report, when "only the interests of the strongest faction dominate, more factional conflict will emerge with less tolerance for minority views." Since a minority view may well be the correct view, this intolerance can be destructive. Also, this inadvertently or otherwise implicitly fosters deeper and wider divisions. They further say that rival and opposing interests contrast with the sharing of a common vision and purpose propagated by transformational leadership. Bass (1990) sums up transactional leadership as "a prescription for mediocrity."

Transformational leadership is the ideal, but it is also harder — it is an art and a science (Bass, 1990). Prime and foremost, the leader places the good of the company, or institution, ahead of her own self-interests (which always take a back seat for such leaders). This leader facilitates the work of her followers, articulates what has to be done and how, works energetically, tolerates stress, is hard working, etc. — she is a cheerleader of sorts. This role involves frequent engagement with the followers in the trenches; she cannot live in splendid isolation if she is to succeed. Nor does she only listen to followers who think like herself. She recognizes that diverse opinions, especially contrary views, are invaluable when studying issues; they help expand the mind, expand horizons, and help to entertain new ways of doing things, especially when changes are on the agenda. It behooves us to recall President Lincoln's deliberate appointment of men of diverse views to his cabinet (Goodwin, 2006).

In addition, the transformational form of leadership is grounded in high moral and ethical standards and based on trust (Bass and Steidlmeier, 1999). Followers feel safer, more included, and more valued than under transactional leadership. They understand that their opinions cannot always prevail, but their opinions are heard, and they will be more loyal and more supportive of their leader when final decisions are made. There is a collective inclusiveness. For those of us in the statistical world, we are obliged (!) to embrace Bass's (1990) assertion that candidates for higher management levels be assessed according to an "intellectual stimulation — general, creative, or *mathematical*" (emphasis added). This is particularly applicable to statisticians, since we are used to tackling problems using logical reasoning. Indeed, although I led a statistics department, it was often clear that creative problem solving was needed whether the department was English, or history, or statistics.

An extensive review of the moral character of transformational leaders is in Bass and Steidlmeier (1999). Personal integrity is necessary. Bass (1998) talks about the need to avoid "stretching the truth or going beyond the evidence" (something that should be easy for statisticians!). For those of us in academia, meeting with higher-level administrators with our requests requires "doing your homework"; we should not expect to receive favors (positions, funds, support, etc.) just because we asked, but because the needs can be backed up with sound facts, evidence, data. For example, in my early years as head, the demand for course enrollments was accelerating well beyond our capacity to meet them. To justify requests for additional faculty lines, equivalent full time (EFT) numbers generated by the department enrollments were calculated routinely, and compared with the actual number of faculty lines. These EFT calculations along with those available from other college departments were given to the dean where/when appropriate. Since the imbalance was stark, it was therefore easy for the dean to give the department new positions.

This preparation of a leader's leader should also include discussions — often disguised as seemingly idle chatter — alerting the dean to future requests that might be submitted eventually. Ideally, this involvement with the now-well-informed dean will generate a suggestion that the chair submit a formal proposal on whatever issue is at stake. The chair should be ready to do so. Such a dean is highly unlikely to refuse that chair, since the groundwork has been laid.

The inclusive nature of transformational leadership is clearly the ideal for academia. Among other assets, Bass and Steidlmeier (1999) summarize this engagement of followers as uplifting, and a sharing of mutual success. In academia, we well know that when one faculty member is successful and honored, it brings accolades to the entire department, i.e., the maxim "a rising tide lifts all ships" is at work. Likewise, when everyone shares in the successes of the entire program, the leader does in fact receive credit. In contrast, the head who is frightened that credit will go to someone else and so is reluctant to share or engage all faculty, will ultimately fail.

There can be *pseudo-transformational leaders* – those who from afar may seem to be transformational but are not, as they put their own interests first rather than those of the collective good. According to Bass and Steidlmeier (1999), these leaders may at first be able to motivate their followers but eventually segue into "encourag[ing] (a) 'we-they' competitiveness, ... (and) are more likely to foment envy, greed, hate, and conflict rather than altruism, harmony, and cooperation." There is a sense of "playing favorites". Such leadership eventually results in destructive outcomes - for all. Some, but not all, charismatic leaders fall under this same paradigm. Bass and Steidlmeier also provide a nice comparison and contrast of authentic and pseudo-transformational leaders.

Another form of leader is the *laissez-faire leader.* This person is a type of transactional leader, but is one who abdicates her responsibilities by avoiding making necessary decisions when needed (Bass, 1990; Avolio, 1999; Eagly and Carli, 2003). Other leaders are viewed as being *charismatic*, though this charisma can be used for good or not so good (see, e.g., Yukl, 2006). Socialized charismatic leadership is based on the collective interests of the department, while personalized charismatic leadership is based on the self-interests of the leader and her own self-aggrandizement; see House and Howell (1992) and Popper (2002). Other studies suggest that the contemporary world needs transformational leaders as replacements for the transactional leader that may have been the norm in some industrial settings (e.g., Cascio, 1995).

In recent years, studies have emerged on *unethical leadership* and its devastating impact. We have already seen that an implicit assumption of a good leader is her high level of integrity. Unfortunately, situations exist where unethical behavior flourishes. An example is when appointing someone who is ineligible or unqualified for a position, yet is selected anyway. When confronted, the unethical person will often reply along the lines of, "We want to win" (where "win" can be variously defined by circumstances). When further confronted with the ineligibility of the candidate, this person will say, "No one will know." The ethical leader should remind the unethical person that it is a question of doing what is right, not whether they might be caught; and should then proceed to divorce herself from the situation if her concerns are dismissed. How she might do this will depend on the situation. Sometimes she may be the lone voice speaking up at a faculty meeting, so that at least minimally the faculty minutes might reflect that sentiment, and her integrity is intact. Brown and Treviño (2006) takes the reader through that unethical quagmire and contrasts this with ethical leadership, a hallmark of the transformational leader. Followers of unethical leaders also become unethical (role models can be powerful indicators of any enterprise, be those good or ill). In academia, some wonder why upper administrators are sometimes slow or even fail to respond to seemingly unhealthy situations. On those occasions, there is more of a laissez-faire approach borne of a not wanting to micro-manage.

Treviño, Butterfield, and McCabe (1998) found that leaders who are focused on their own self-interest are most strongly associated with unethical behavior. In any environment, leaders will inevitably face situations where decisions have the potential to cause harm. As Brown and Treviño (2006) conclude, the ethical leader will find the ethically correct response, whereas unethical leaders will make decisions that can bring "significant harm to others."

Bringing about institutional change is, according to Yukl (2006), one of the most difficult and most important responsibilities of a leader. Some of the reasons for these difficulties are obvious, such as the entrenched values of those who in effect do not want to change. Some simply feel threatened by what a new cultural norm may entail. In these situations, transformational leadership is essential. Effective and permanent change, rather than superficial changes, can only occur when the followers/faculty/employees assume "ownership" of the transformed department/university/company. There needs to be a shared vision of the new goals, but reaching all followers to understand that vision is not always easy. The leader must be open, be accessible, and maintain a sense of humility; she must be ready to modify ideas, and certainly not be threatened by the effectiveness of others — indeed, the goal of the greater good of all must be held high and firm, and it must be shared. More than on other occasions, the leader effecting change must regard the good of the organization as more important than her own ambitions. Nevertheless, though she must encourage open and collegial discussions and advice, when the time comes, she has to be decisive, to make the tough decisions. She has to be "tough but fair" (see also Bresnen, 1995). This firm action is essential if the department is to advance. In my opinion, this ability — and willingness — to make the tough decisions, is the most important stratagem a leader can make — one that is seen the most favorably by upper administrators.

Further, when contemplating how to proceed when confronting change, it is important to establish priorities as to how these changes are to be implemented. Substantial changes mean simply there are many aspects to be tackled simultaneously. However, the leader and her program, her followers, are not working in a vacuum. Thus, while the program may have its own top priority, a particular goal may not be immediately attainable in the wider context. This is what Pye (2005) refers to as "sensemaking in action." It is the leader's responsibility to be aware of what can be achieved in the immediate context. There is a constant juggling and reordering of priorities as they together keep their eyes on the ultimate goals. Everyone must understand that deep change cannot occur overnight, that there will be frustrations along the way, and hard work all the way. Although it is always important to have the followers collectively on board in making decisions, it is more relevant in situations of change. Followers need to "buy in" to these changes, feel they have ownership of them; otherwise, eventually they will undermine whatever progress may have been achieved.

In their study of transformational leaders, Alimo-Metcalfe and Alban-Metcalfe (2005) remind us that it is important to value individuals as individuals, to network, to enable the followers by empowering employees to develop their potential. In the academic context, this translates into encouraging and mentoring junior faculty. Further, it is well known that not all new junior faculty will be able to develop research programs sufficient to earn tenure at their current institutions. In these cases, a good leader — a transformational leader — will still help the young faculty member to reach her/his potential, including advice so as to prepare for employment elsewhere. For example, if the individual excels in the classroom, then encouragement to look for positions in a teaching institution or liberal arts college is an option. Another option is to encourage such faculty to teach courses in new areas, essentially preparing them for positions in industry. We are well set up in statistics with many applied statistics offerings that would serve this purpose.

For someone already tenured but whose research program is struggling, an appropriate adjustment to the teaching load would be in order. Such reassignment of tasks is routine in the nonacademic environment; indeed, Bass (1990) suggests that a good transformational leader will assign (or reassign) tasks as necessary. This is what Alimo-Metcalfe and Alban-Metcalfe (2005) refer to as treating people with dignity, recognizing and valuing their individual differences, and building on their strengths. Also, not all faculty, not all followers in any context, have equal abilities. It is important to treat everyone fairly, however. Of course, "fair" is not necessarily "equable." For example, giving all faculty the same salary raise is not fair to those who accomplished more. However, all have their own differing strengths. The key is to find that strength for each person, to develop ways to work with him/her, and thence to build on it for the productive good of all.

According to Yukl (2006), there are five types of power, specifically — reward, coercive, legitimate, expert, and referent. Importantly, power and authority are two different entities. Any leader has to be cognizant of the pitfalls of corrupting power. These can be mitigated in an academic setting when rotating chairs/heads are the norm, such as when a limit of one five-year term or two three-year terms is imposed, though leaders can be rotated back into a leadership role after a time out of these positions. Such term limits are valuable, anyhow, for academics, since appointment to an administrative role often takes the individual away from her or his research, and so is not developing her or his own value to the institution. Fortunately, any corruptive influence in academia should be minimized by the shared leadership roles played by senior faculty.

Within academia itself, the most important forum is the faculty meeting. In theory, all faculty are present, and in theory all can participate in collective discussions about departmental issues. Of course, an effective head will engage in frequent one-on-one discussions with all faculty; but her talk should not be limited to a select few. Thus, the faculty meeting is the forum which brings the issues to a more formal level of discussion. Viewpoints from the full professors are essential if only because of their greater experiences — to ignore them is the proverbial "kiss of death"; but the views of the younger faculty are also important, because it is their department's future at stake, and they may have fresh ideas based on how things were done in their former institutions. Indeed, Kerr and Jermier (1978) suggest that to ignore segments of faculty is both insulting and time-wasting. They distinguish an academic environment as a "horizontal" instead of a "vertical" leadership structure, with its peer review and external referents being key components of the academy. A good head/chair will usually have some form of advisory committee made up of some or all full professors. This helps protect against decisions that might have inadvertently circumvented the rules or procedures, and the like.

Of course, engagement of the entire faculty does not necessarily mean all faculty are actively involved in all aspects of academic life. Committees can be set up to focus on different issues. In some cases, the head merely charges the members by delegating responsibilities, though providing some input as to where useful background information may be found might be important. In other cases, the chair may be an active participant in committee deliberations. In all cases, the head should define the issues, set parameters, and provide time frames. Then, committees bring their (necessarily preliminary) ideas and proposals to the full faculty for discussion. If need be, this process is iterated. Kerr and Jermier (1978) refer to this process as providing structuring information.

With this committee structure, progress can be slow, especially in situations of change. In some sense, it might be suggested that the leader (new head, say) should simply impose the changes autocratically. However, this is not likely to bring your followers "on board" (or most of them; some will never comply). Remember, it is important to treat the followers/faculty with dignity and respect. For example, in my own department (then computer science and statistics), it was necessary to revamp completely the requirements for all six degrees and all course offerings (thirty-six new courses added and content of old ones revised) where initially resistance ruled. Lots of committees were set up — some to look at say a particular course content, some to evaluate certain degree requirements, and so on, i.e., small pieces of the big picture. Of course, here, a "piece" can be but one step of many involved on some particular aspect (e.g., revising comprehensive examination requirements). Faculty were given the option to focus on computer science or statistics issues — their choice. Instead of taking a single giant step to achieve the desired goals, we zig-zagged along, taking small steps, each one becoming acceptable to faculty members as we progressed ever so slowly but assuredly and determinedly toward the goals. It took four years, but by then substantial changes had been effected and were acceptable to the department as a whole. Because of their active participation, faculty "owned" them. To illustrate the scope, breadth, and depth of these changes, at the beginning, the most advanced course in the statistics doctoral program was a course based on Hogg and Craig (1965), while at the end, the first course used Bickel and Doksum (1977). One or two courses were left untouched, such was their political value to some. Not all battles have to be fought!

Generally, email "discussions" are not effective, perhaps even dangerous; this includes contemporary evolutions such as Twitter and texting. Emails are a great medium for transferring information (equivalently, the modern-day "Drop-box" or its variants). However, emails ignore the synergism of faculty meetings. They cannot transmit the body language and facial expressions that can make face-to-face discussions so much more revealing. As a trivial example, you start to say that "A=B," and you hear someone correct you, "No, A=C"; so you make an immediate correction, and proceed. Had this been an email, the writer would have gone down the wrong path, making arguments to match the premise, etc., perhaps even creating an unfortunate hard copy trail that comes back to haunt you later. Valuable time has been lost. In a faculty forum, a comment by one member may stimulate others to think along new lines and come up with ideas that no single person working alone would have produced.

In many situations, it is important to document conversations and events precisely to keep a proper paper trail. It may sound far-fetched, but it behooves the writer to remember that any written text could appear in court ten years later, and so it is important not to embarrass the university by its contents, no matter how egregious the incident that elicited it. Of course, write the letter that must be written, but however hard it might be, emotions must be set aside. Indeed, there is a growing body of literature on "emotional" leadership. (See, e.g., George, 2000, for an extensive coverage, and Caruso, Mayer, and Salovey, 2002, for an overview of this topic.)

When appointed to the head/chair position, it is imperative that budgetary promises be incorporated into the department's budgets along with the usual written commitments before accepting the job. Furthermore, any faculty lines resulting from losses — retirements/resignations/terminations/death — should be added to those promised prior to appointment, with a written understanding that the timing will be flexible to make these appointments, and that temporary ones in the interim are approved.

Following long-time heads can be problematic. Two newly appointed heads (who happened to be women) successfully navigated this potential cataclysm by engaging their predecessors in some meaningful way, different for each, depending on their situation and backgrounds. In each case, they followed founding chairs who might have legitimately felt resentful of their replacements. All it takes, really, is to give that predecessor a prominent role, make him feel he is still a valued member, treat him with dignity. In one case, the predecessor even wrote to the institution's president applauding his successor. This is leadership, maybe of a different kind, but leadership nevertheless.

For example, in my own case, my predecessor would be asked who (in the upper administration) was responsible for something. Over coffee, he would fill me in with relevant details. He became one of my greatest supporters. Initially, coffee was every day, but in time it became increasingly less frequent. One day he came in with the keys to his office; he was permanently leaving town and offered me all the books he had left there. Instead, the books were deposited in the newly established departmental library except for one — a long-out-of-print text I had been trying to find for years; this, for me, was an extravagant reward for endless enjoyable cups of coffee.

Although this advice also applies to both men and women, the potential for trauma is lessened in the case of men, since men have traditionally been followed by men. In contrast, new appointees to leadership positions who proceed to "clean the slate" overnight rarely succeed ultimately; they are not transformational leaders, though some may display the charismatic style. This modus operandi has the unfortunate effect of essentially removing the experienced followers who could have helped them achieve their goals.

Women as Leaders

Beware of the appointment as a woman "because" of the underlying premise that — as a woman — she can be manipulated. This also applies to ostensibly good "support" from others in the community. My own appointment as head is a case in point. The department had been split into two opposing factions, regardless of issue. All other interviewees (all men) had received the obligatory split vote of approval, but surprisingly there was a unanimous vote of approval for my appointment. Not unexpectedly, then, the administration moved hard to appoint me. It was only later that I realized that both factions thought that "as a woman" they could control me (and hence, by inference, they were convinced I would support their agenda for the department's future). Since the administration had charged me to "change everything," this was potentially an extremely difficult position. It was imperative that true transformational leadership had to be exercised abundantly. The only "change" either faction desired was that their ideas be imposed on everyone, rather than the across-the-board changes sought by the university. It was important to be strong, to maintain the goals required for the department, and not to be diverted into unproductive endeavors.

In her book on women of influence, in speaking of sponsorship and mentorship, Doogue (2014) concluded that most of the women leaders (from inside and outside academia) interviewed accepted that it was crucially important "to bring people along." Typically, men are naturally mentored without any formal mechanism from their male colleagues — senior faculty and major professors — as they go about their daily work, by drawing attention to some important conference that the junior person should attend, asking how the publication of his research is progressing, and so on; all these conversations are subtle ways of ensuring the person's career advances. Men are likely to receive such information on the proverbial "golf course" — translation, "women are not included."

A key role as a formal mentoring program in the statistical profession was played through the National Science Foundation (with some support also from the Office of Navy Research later) sponsored "Pathways to the Future"

Workshops designed for young women and traditionally disadvantaged faculty. Begun in 1988 and continuing until 2004, these were held prior to an associated statistical meeting (usually the annual Joint Statistical Meetings, though the first was held before a stand-alone Institute of Mathematical Statistics meeting in Fort Collins). The grant funds supported young women academics in their first five years since graduation. A handful of senior women also attended to assist with the many discussions that became a part of the formal presentations. The two-day workshops would start with a major keynote talk that presented data showing the progress or lack thereof for women in the academic statistical world. Elizabeth Scott was the keynote speaker for the first Pathways, and she had data! Almost universally, the junior participants who just "knew" they were going to be successful had their eyes opened wide by these data and so became very attentive throughout the rest of the workshop presentations. For them, the underlying theme had suddenly been transformed to strategies to ensure their own survival in this real world. Subsequent talks focused on such issues — e.g., how to and the importance of publishing their research, how to read and respond to referees' reports of their submitted manuscripts, grantsmanship, how to give talks/presentations of their work, how to handle teaching obligations in and outside the classroom, how to manage their time, how to be an integral part of their department but without being taken advantage of with inequitable assignments (commonly imposed on women), the importance of attending and presenting their work at professional meetings, and how to maximize the benefits of these efforts. All these presentations evolved into discussions. Often, a young person might speak up with a version of "that couldn't/doesn't happen," to be quickly brought to reality by a senior woman's "Oh, yes, it does," eliciting yet more discussion. The final set of presentations consisted of each junior attendee giving a short summary of her own area of research.

As an aside, Scott died a few months after the first Pathways. That her Pathways keynote address was her last presentation is, in a way, a fitting tribute to her influential leadership role over the years in furthering justice for women in the workplace, regardless of discipline. The staff at Berkeley was delighted to give me copies of her materials that formed the basis of that presentation. Those papers, along with others cited therein, became the foundation for subsequent work on measuring women's progress in academia; see, e.g., Billard (1989,1994). Following Scott's example, future opening keynote addresses focused on these data. Attendees from early workshops have since become important leaders to the statistical field. Of course, not everyone is capable of being a leader, per se; however, all people, women included, should have the opportunities to excel to their potential, and that is what the Pathways Workshops and mentoring in general should achieve.

An opportunity for women faculty to exercise local leadership is to embark on a salary study with her institution. Although I was not part of the group of women faculty who sought such a study from our university president, I was engaged to work on this, first at a university wide level, and then for two of the colleges. At the university level, a preliminary study was done using the regression methods developed for the American Association of University Professors (AAUP) by Scott (1977). Partly because, at the time, salary compression and inversion existed for all faculty, it was important to untangle the gender inequity from the compression and inversion component; i.e., to calculate and report on the gender inequity only. Thus, an adjustment to the Scott AAUP regression method was adopted; see Billard, Cooper, and Kaluba (1991). Following that initial analysis, detailed studies were undertaken at the college level. For this, it was now important to have a committee including men (there was a former male department head on the committee) to conduct this study. In addition to the salary regressions, department heads were asked to supply a full curriculum vita for all their faculty — men and women, they were told the committee would be compiling comparative information about productivity measures, they were asked to provide input as to their perceptions of the equity of their female faculty salaries, and they were told the regression analysis would be part of the input. Interestingly, but consistent with the literature, the heads' inputs universally assured that no woman needed a salary adjustment. Individual faculty were also invited to meet with the committee to discuss their own salary profile. As an aside, adjustments made were, on average, consistent with those made elsewhere. It is important, however to be aware that subsequently attempts will be made to undermine or to discredit these efforts, regardless of the facts that such inequities are known to exist nationwide (and, unfortunately, known to return in about a decade). Billard and Kafadar (2015) show the deficits between men's and women's salaries have in fact, widened over the years 1972–2013.

Eagly and Carli (2003) suggest that women are currently making their way to the top, even though men still dominate the leadership population. They feel this movement is partly due to a changing culture, to changing views of what makes a good leader, especially those views which try to embrace teamwork and collaboration, which are adopted more often by women (see, Rosener, 1990, 1995; Eagly and Johnson, 1990). Despite the potential gains, however, Eagly and Carli (2003) and Eagly and Karau (2002) describe extensively the prejudices and discriminations that prevail against women. There are incongruities between societal expectations from women and those from leaders. Women are held to higher levels of competence than men. Ford (2006) concludes that progress for women up the administrative ranks is still slower for women than for men; men are more likely to apply for higher level positions (Why not "have a go?" they ask themselves), whereas women hold back wanting to be sure they can "do it" before allowing themselves to be considered. An academic analogue here includes the promotion process through the professorial ranks.

There are parallels with what is called the "Goldberg paradigm" originating from the Goldberg (1968) study showing that academic articles perceived to be written by women are evaluated as being of lesser quality than when perceived to be written by men; and that women also rate the women authors lower than they rated men authors, though the disparity is not as great as it was for men evaluators. Paludi and Bauer (1983) observed that this downplaying of women's abilities and contributions still persisted fifteen years later. These disparities in perception still carry over to all aspects of the working world, including leadership roles.

By drawing upon the Goldberg effect, Eagly and Carli (2003) go on to document how women can be unfairly viewed by their colleagues. If they express opposing views, or appear to be dominant, and are seen to be assertive, they will encounter more dislike and rejection than do men exhibiting the same opinions. Social isolation can occur. All this makes it harder for women to be seen as candidates for leadership, as they are subsequently seen as being less able to lead. This is particularly true still today for male-dominated environments, which, according to Eagly and Carli (2003), "can be difficult for women." Yet, they do suggest that women who do survive in such environments tend to be "very competent" — given the barriers against which they had to fight.

Powell, Butterfield, and Parent (2002), among others, provide evidence that management ability is perceived to be associated with being male. In a meta-analysis of forty-five studies of transformational, transactional, and laissez-faire leadership, Eagly, Johannesen-Schmidt, and van Engen (2003) concluded that women were more likely to be transformational leaders than were men, which is encouraging for aspirants, given the preponderance of evidence that transformational leaders are more effective than are other leadership styles. This was also a finding of Alimo-Metcalfe and Alban-Metcalfe (2005) though these authors also noted that prior to the early 1990s few studies included women as subjects.

Conclusion

This chapter has attempted to review some of the extensive literature on leadership, with illustrations particularly relevant for the academic setting wound in as these roads were traveled. A key question not answered is whether or not women are truly having an effect on academic leadership. Anecdotal evidence — one only has to look around the campus — suggests there are more women being appointed as department heads/chairs. However, progress is painfully slow, since, in the statistical world, first Gertrude Cox (1941–49) at North Carolina State University, Elizabeth Scott (1968–73) at the University of California at Berkeley, F. N. David (1971–77) at the University of California at Riverside, and Lynne Billard (1980–89) at the University of Georgia, were appointed to head departments of statistics. As an aside, these appointments reflect three different scenarios, viz., coming into a well-established, top ranked program (Scott), starting a brand-new program (Cox, David), which is considerably easier than trying to build a program by changing what was there (Billard).

Perceptions of progress can be deceptive, however. In a study of the rates of promotion to tenure of men compared to women, Billard and Kafadar (2015) found that the differentials between the percentage of men and of women being tenured were essentially unchanged from 1972 to 2013. That is, the data do not support the prevailing perception that things were now "equal" for men and women. There are certainly more women being appointed to faculty positions; but what percentage will eventually reach leadership positions is unknown. At any rate, as has been demonstrated from the literature, women who are appointed are held to higher standards than men. The encouraging news is that most women leaders tend to engage under the transformational leadership paradigm whose style is viewed as being more effective than other leadership styles.

Perhaps my appreciation of this literature explains in part my own experiences — good and bad.

Bibliography

Alimo-Metcalfe, B. and Alban-Metcalfe, J. (2005). Leadership: Time for a new direction. *Leadership* 1, 51–71.

Avolio, B.J. (1999). *Full Leadership Development: Building the Vital Forces in Organizations.* Sage: Thousand Oaks CA.

Bass, B.M. (1985). *Leadership and Performance Beyond Expectations.* Free Press: New York.

Bass, B.M. (1990). From transactional to transformational leadership: Learning to share the vision. *Organizational Dynamics* 18, 19–31.

Bass, B.M. (1998). The ethics of transformational leadership. In *Ethics, The Heart of Leadership* (ed. J. Ciulla). Praeger: Westport.

Bass, B.M. and Steidlmeier, P. (1999). Ethics, character, and authentic transformational leadership behavior. *Leadership Quarterly* 10, 181–217.

Bickel P.J. and Doksum, K.A. (1977). *Mathematical Statistics.* Holden-Day: Oakland.

Billard, L. (1989). The past, present and future of women in academia. *American Statistical Association Sesquicentennial Invited Paper Sessions* (eds. M.H. Gail and N.L. Johnson), 645–656.

Billard, L. (1994). Twenty years later: Is there parity for women in academia? *NEA Higher Education Journal* 10, 115–144.

Billard, L., Cooper. T.R., and Kaluba, J.A. (1991). A statistical remedy to gender and race based salary inequities. University of Georgia Technical report.

Billard, L. and Kafadar, K. (2015). Women in statistics: Scientific contributions versus rewards. In *Advancing Women in Science* (eds. W. Pearson Jr., L.M. Frehill and C. L. McNeely). Springer, in press.

Bligh, M.C., Kohles, J., and Pilai, R. (2011). Romancing leadership: Past, present, and future. *The Leadership Quarterly* 22, 1058–1077.

Bolden, R. and Gosling, J. (2006). Leadership competencies: Time to change the tune? *Leadership* 2, 147–163.

Bresnen, M.J. (1995). All things to all people? Perceptions, attributions, and constructions of leadership. *Leadership Quarterly* 6, 495–513.

Brown, M.E. and Treviño, L. K. (2006). Ethical leadership: A review and future directions. *The Leadership Quarterly* 17, 595–616.

Caruso, D.R., Mayer, J.D., and Salovey, P. (2002). Emotional intelligence and emotional leadership. In *Multiple Intelligences and Leadership* (eds. R.E. Riggio, S.E. Murphy, and F.J. Pirozzolo). Lawrence Erlbaum Associates Publishers: New Jersey.

Cascio, W. (1995). Whither industrial and organizational psychology in a changing world of work? *American Psychologist* 50, 928–939.

Doogue, G. (2014). *The Climb: Conversations with Australian Women in Power.* Text Publishing: Melbourne.

Eagly, A.H. and Carli, L.L. (2003). The female leadership advantage: An evaluation of the evidence. *The Leadership Quarterly* 14, 807–834.

Eagly, A.H., Johannesen-Schmidt, M.C., and van Engen, M.L. (2003). Transformational, transactional, and laissez-faire leadership styles: A meta-analysis comparing women and men. *Psychological Bulletin* 129, 569–591.

Eagly, A.H. and Johnson, B.T. (1990). Gender and leadership style: A meta-analysis. *Psychological Bulletin* 108, 233–256.

Eagly, A.H. and Karau, S.J. (2002). Role congruity theory of prejudice toward female leaders. *Psychological Review* 109, 573–598.

Fiedler, F.E. (1964). A contingency model of leadership effectiveness. *Advances in Experimental Social Psychology* 1, 149–190.

Ford, J. (2006). Discourses of leadership: Gender, identity and contradiction in a UK public sector organization. *Leadership* 2, 77–99.

George, J.M. (2000). Emotions and leadership: The role of emotional intelligence. *Human Relations* 53, 1027–1055.

Goldberg, P. (1968). Are women prejudiced against women? *Transaction* 5, 316–322.

Goodwin, D.K. (2006). *Team of Rivals: The Political Genius of Abraham Lincoln.* Simon and Schuster: New York.

Hogg, R.V. and Craig, A.T. (1965). *Introduction to Mathematical Statistics.* Macmillan: New York.

House, R.J. and Howell, J.M. (1992). Personality and charismatic leadership. *Leadership Quarterly* 3, 81–108.

Howell, J.M. and Avolio, B.J. (1992). The ethics of charismatic leadership: Submission or liberation? *Academy of Management Perspectives* 6, 43–54.

Kerr, S. and Jermier, J.M. (1978). Substitutes for leadership: Their meaning and measurement. *Organizational Behavior and Human Performance* 22, 378–403.

Paludi, M.A. and Bauer, W.D. (1983). Goldberg revisited: What's in an author's name? *Sex Roles* 9, 387–390.

Popper, M. (2002). Narcissism and attachment patterns of personalized and socialized charismatic leaders. *Journal of Social and Personal Relationships* 19, 797–809.

Powell, G.N., Butterfield, D.A. and Parent, J.D. (2002). Gender and managerial stereotypes: Have the times changed? *Journal of Management* 28, 177–193.

Pye, A. (2005). Leadership and organizing: Sensemaking in action. *Leadership* 1, 31–50.

Rosener, J.B. (1990). Ways women lead. *Harvard Business Review.* November–December, 119–125.

Rosener, J.B. (1995). *America's Competitive Secret: Utilizing Women as Management Strategy.* Oxford University Press: New York.

Scott, E.L. (1977). *Higher Education Salary Evaluation Kit.* American Association of University Professors, Washington, DC, 55 pages.

Treviño, L.K., Butterfield, K.D., and McCabe, D.M. (1998). The ethical context in organizations: Influences on employee attitudes and behaviors. *Business Ethics Quarterly* 8, 447–476.

Yukl, G. (2006). *Leadership in Organizations* (6th ed.). Prentice Hall: New Jersey.

Appendix — Women Presidents or Chancellors

A Google search for "women university presidents" produced a Wikipedia page of women presidents or chancellors of co-ed colleges and universities, listed in Table A. This list has been slightly edited to reflect end of appointment, where applicable. These are noninterim appointments.

Also, the disciplines of these accomplished women have been added. Two-thirds of these fields are in the social sciences (especially political science) and education (usually educational administration). The sciences are dominated by biology-related fields. Three women had undergraduate degrees in mathematics, but all moved to other fields for graduate work; a fourth studied computer science as an undergraduate and has a mathematics doctorate, but then moved back to computer science. A few women studied law as a second discipline. The details are listed in Table A where *present* means 2014:

Table A: Women University Presidents

Hanna Holborn Gray	1978–1993	U. Chicago	history
Diana Natalico	1988–present	U. Texas El Paso	modern languages
Donna Shalaha	1988–1993	U. Wisconsin Madison	political science
	2001–present	U. Miami	
Nanerl O. Keohane	1993–2004	Duke U.	political science
Judith Rodin	1994–2004	U. Pennsylvania	psychology
Mary Sue Coleman	1992–2002	U. Iowa	biochemistry
	2002–2014	U. Michigan	
Deborah F. Stanley	1997–present	State U. of New York Oswego	English law
Molly Corbett Broad	1997–2006	U. North Carolina System	economics
Shirley Ann Jackson	1999–present	Rensselaer Polytechnic Institute	physics
Pamela Brooks Gann	1999–2013	Claremont McKenna College	mathematics law
Judy Genshaft	2000–present	U. South Florida	social work
Shirley C. Raines	2001–2013	U. Memphis	education
Ruth Simmons	2001–2012	Brown U.	comparative literature
Shirley M. Tilghman	2001–2013	Princeton U.	biology
Karen A. Holbrook	2002–2007	Ohio State U.	zoology
Kay Norton	2002–present	U. Northern Colorado	English law
Nancy Cantor	2004–2013	Syracuse U.	psychology
	2014–present	Rutgers U.	
Marye Anne Fox	2004–2012	U. California San Diego	chemistry
Jo Ann M. Gora	2004–2014	Ball State U.	sociology
Amy Gutmann	2004–present	U. Pennsylvania	political science
Susan Hockfield	2004–2012	Massachusetts Institute of Technology	neuroscience
Lou Anna Simon	2005–present	Michigan State U.	mathematics education
Ann Weaver Hart	2006–2012	Temple U.	history education
	2012–present	U. Arizona	
Maria Klawe	2006–present	Harvey Mudd College	computer science mathematics
Drew Gilpin Faust	2007–present	Harvard U.	history
Sally Mason	2007–present	U. Iowa	biology
Barbara Snyder	2007–present	Case Western Reserve U.	sociology law

France Córdova	2007–2012	Purdue U.	astrophysics
Virginia Hinshaw	2007–2012	U. Hawaii Manoa	microbiology
Judith A. Bense	2008–present	U. West Florida	anthropology
Linda P. Brady	2008–present	U. North Carolina Greensboro	political science
Carolyn Martin	2008–2011	U. Wisconsin Madison	German
	2011–present	Amherst College	
Elsa A. Murano	2008–2009	Texas A&M University	microbiology
Renu Khator	2008–present	U. Houston	political science
Pamela Trotman Reid	2008–2015	U. Saint Joseph	psychology
Paula Allen–Meares	2009–present	U. Illinois Chicago	social work
Bernadette Gray–Little	2009–present	U. Kansas	psychology
M.R.C. Greenwood	2009–present	U. Hawaii System	biology
Linda P.B. Katehi	2009–present	U. California Davis	engineer
Nancy L. Zimpher	2009–present	State U. New York System	education
Judy Bonner	2010–present	U. Alabama at Tuscaloosa	nutrition
Barbara Couture	2010–2012	New Mexico State U. Las Cruces	English
Waded Cruzado–Salas	2010–present	Montana State U.	Spanish
Dana L. Gibson	2010–present	Sam Houston State U.	business
Mary Jane Saunders	2010–present	Florida Atlantic U.	biology
Teresa A. Sullivan	2010–2012	U. Virginia	sociology
Sandra K. Woodley	2010–present	U. Louisiana System	business
Mary Ellen Mazey	2011–present	Bowling Green State U.	geography
Carolyn W. Meyers	2011–present	Jackson State U.	engineer
Phyllis M. Wise	2011–present	U. Illinois Urbana–Champaign	biology
Susan Herbst	2011–present	U. Connecticut	political science
Jane C. Conoley	2012–present	U. California Riverside	psychology
Christina Paxson	2012–present	Brown U.	economics
Cheryl B. Schrader	2012–present	Missouri U. of Science and Technology	engineer
Rebecca M. Blank	2013–present	U. Wisconsin Madison	economics
Sheri Noren Everts	2014–present	Appalachian State U.	education
Carol Folt	2013–present	U. North Carolina Chapel Hill	biology
Glenda Baskin Glover	2013–present	Tennessee State U.	mathematics business
Melody Rose	2013–2014	Oregon U. System	political science
	2014–present	Marylhurst U.	
Gwendolyn Boyd	2014–present	Alabama State U.	engineer
Elmira Mangum	2014–present	Florida A&M U.	management

8

To Be or Not to Be Bold: Effective Leadership in Various Individual, Cultural, and Organizational Contexts

Kelly H. Zou

Pfizer Inc

Introduction

In the current era of globalization and proliferation of so-called "netizens," we discuss additional new challenges for leaders to ponder. This chapter is organized as follows: We first review the phenomenon of the glass ceiling that many women face, which is related to my own personal experience in academia and industry; the universal qualities, namely, confidence, preparation, and self-assessment, are elucidated and then discussed in greater detail.

We focus on individual contexts, including personality, heredity or learned qualities. Besides individual contexts, cultural influences from my own experience as an Asian-American are analyzed. In particular, the analogy of facing a bamboo ceiling is discussed. Biases and barriers in such contexts are elaborated on. Furthermore, organizational contexts, including leadership tracks, gender parity, and various functional roles in the ever more complex relationships while having flatter organizational structures, are reviewed. Finally, to many of the readers who are "Gen X-ers," new challenges with required skills, such as possessing cross-functional skills, developing virtual communication methods, and mastering social media and networking opportunities, are presented.

Finally, further remarks are given at the end. I hope that readers will gain some insights on complex leadership topics through my own personal lenses and experiences in academia and industry across various countries in over three continents.

The Glass Ceiling

Women have traditionally faced the so-called glass ceiling, both outside or within the board room. The Federal Glass Ceiling Commission (1995) defined the glass ceiling as "the unseen, yet unbreachable barrier that keeps minorities and women from rising to the upper rungs of the corporate ladder, regardless of their qualifications or achievements."

The same report urged that, "Breaking the glass ceiling is an economic priority that this nation can no longer afford to ignore. It is an economic imperative driven by recent dramatic shifts in three areas that are fundamental to business success: (1) changes in the demographics of the labor force; (2) changes in the demographics of national consumer markets; and (3) the rapid globalization of the marketplace."

There has been good progress toward achieving gender parity in education, and yet it is still a moving target throughout the world. For example, in 2001, 91 developing countries and 34 industrialized countries had as many girls as boys in schools, according to the United Nations International Children's Emergency Fund (2005). In terms of senior leadership opportunities, DiversityInc (2014) reports that there are merely 23 (i.e., 4.6%) women chief executive officers (CEOs) among Fortune 500 companies.

In our statistical profession, women are ever more present as accomplished college students, post-graduate researchers, and attendees of the annual Joint Statistical Meetings (JSM). Meanwhile, it is a concern that a much lower proportion of women emerge as leaders in the statistical profession.

Golbeck (2012) observed that during the 2011 JSM conference, "Only 2 of 12 (16.7%) keynote speakers were women. None (0%) of the three speakers with lunch was a woman. One of the four (25%) introductory overview lecturers was a woman, and only one of the five (20%) meet and mingle well-known statisticians was a woman."

Furthermore, Olkin (2013) pointed out concerning those prestigious lectureships during the 2013 conference: "The JSM meeting in Montreal made it abundantly clear that women are under-represented in named lectures. Of the four named lectures, the seven medallion lectures, and the two invited lectures, none were women."

President Summers' Hypotheses

Here is a little background information about my own interest in gender parity both in academia and industry. About ten years ago, the former Harvard President Lawrence (Larry) H. Summers drew a firestorm (e.g., detailed in Bombardieri, 2005) in his presentation at the National Bureau of Economic Research Conference on Diversifying the Science & Engineering Workforce on January 14, 2005. According to Summers, the following would attribute toward the scarcity of women in science and engineering.

In his own words and the transcript released by Harvard, Summers (2005) had three hypotheses in mind to explain the substantial disparities in so-called high-end scientific professions: "The first is what I call the high-powered job hypothesis. The second is what I would call different availability of aptitude at the high end, and the third is what I would call different socialization and patterns of discrimination in a search. And in my own view, their importance probably ranks in exactly the order that I just described."

Afterward, three university presidents, John L. Henessy of Stanford University, Susan Hockfield of Massachusetts Institute of Technology, and Shirley M. Tighman of Princeton University, jointly voiced their opinions and responses to the above hypotheses. These presidents (2005) rightfully commented: "Extensive research on the abilities and representation of males and females in science and mathematics has identified the need to address important cultural and societal factors.... Colleges and universities must develop a culture, as well as specific policies, that enable women with children to strike a sustainable balance between workplace and home. Of course, achieving such a balance is a challenge in many highly demanding careers. As a society we must develop methods for assessing productivity and potential that take into account the long-term potential of an individual and encourage greater harmony between the cycle of work and the cycle of life — so that both women and men may better excel in the careers of their choice."

My Personal Experience

Following President Summers' comments that struck a chord of many in academia, I decided to apply to and subsequently got invited to join the Faculty Subcommittee of the Joint Committee on the Status of Women (JCSW) at the Harvard Medical School and Harvard School of Dental Medicine. One of the early tasks of our committee was to systematically analyze the outcomes of grant applications of women faculty members. There must be a great deal of sensitivity about how to handle situations within cultural and organizational contexts, which are common in the statistical profession. I will provide further details on results from this study during the remainder of this chapter.

During my swift career transition from academia to industry, as well as my residences spanning across three different continents, i.e., Asia, North America, and Europe, I have been fortunate enough to witness some universal qualities of women, as well as of men, to achieve successes.

After coming to the US as a native of Shanghai, China, I received my education in Hawaii, and New York, with additional post-doctoral training in Massachusetts, respectively. I remained on its faculty and was promoted from instructorship to associate professorship at Harvard Medical School within ten years. I had the opportunity to serve as the principal investigator on several multidisciplinary teams, including two R01 grants funded by the National Institutes of Health (NIH) in the United States. One grant focused on bioinformatics in neurosurgical planning, and the other focused on hierarchical modeling in medical imaging. Furthermore, I led an R01 grant on innovative sample size design methods, as well as serving as the principal investigator of the Biostatistical Validation Core, on a P01 grant.

After ten years at Harvard in Boston, Massachusetts, I utilized my doctoral thesis work in signal detection, processing, and prediction and worked as an associate director of rates in the financial industry at Barclays Capital at its headquarter in London, England.

Since then, I took on a new role and am currently a senior director and statistics lead, with a specialization in real-world "big data" analysis in the Statistical Center for Outcomes, Real-World and Aggregate Data at Pfizer Inc in New York, NY, back in the US.

In the next section, before diving deeper into individual, cultural, and organizational aspects of leadership qualities, I first discuss universal success qualities that a leader should possess, regardless of the special contexts. I provide some insights and shed some light from my own experience as an immigrant and an expat during my education and career. These views may be subject to my own unique background as an Asian-American, with an upbringing in Asia, immigration to North America, and living as an expat in Europe.

Universal Successful Qualities

Before I elaborate on such subtle or not-so-subtle contexts, there are some universal qualities of success that could propel a woman to be ready for and to excel at leadership. Sandberg (2013) wrote in her widely read book, *Lean In: Women, Work, and the Will to Lead*, that "if I had to embrace a definition of success, it would be that success is making the best choices we can . . . and accepting them."

To be specific about the necessary keys to successes, I have turned to Miller and Miller (2011), who suggested in their book, *A Woman's Guide to Successful Negotiating: How to Convince, Collaborate, and Create Your Way to Agreement*, three conditions for successful negotiations, namely confidence, preparation, and willingness to walk away.

In the following subsections, I expand and adapt these qualities into confidence, preparation, and self-assessment, as elaborated below. Although I would rely on literature within the last decade to support my personal views, I identify myself as a statistician, since I embarked on my doctoral education over two decades ago.

The principle of understanding oneself is based on active classroom and extracurricular participation, dedicated practice in interpersonal and non-interpersonal skills, as well as self-assessment through internal and external feedback.

Confidence

A timely article published by Kay and Shipman (2014) has suggested that women tend to be less confident than men. For instance, when applying for positions, "Women applied for a promotion only when they met 100 percent of the qualifications. Men applied when they met 50 percent." This is why Sandberg (2013) advised women to avoid "I'm not ready to do that" and to instead roll up our sleeves and say, "I want to do that — and I'll learn by doing it!"

Similarly, a lack of confidence was reflected by the reality associated with grant applications in academia. As mentioned, I joined the JCSW at Harvard, and as part of the Grant Parity working group, I participated in and coauthored an article by Wasbren et al. (2008). This grant parity research was supported by the Center for Public Leadership, John F. Kennedy School of Government, Harvard University, and by the JCSW of the Harvard Medical School and the Harvard School of Dental Medicine. The authors analyzed more than 6,300 grant applications submitted by 2,480 faculty members within a three-year period, all of whom were throughout eight Harvard-affiliated institutions.

Wasbren et al. (2008) found in the grant parity research that I participated in that, compared with men, female investigators submitted fewer proposals (on average, 2.3 submissions for women versus 2.7 submissions for men), asked for shorter project duration periods (on average, 3.1 years of duration for women versus 3.4 years for men), and requested smaller dollar amounts (on average, \$115,325 budget for women versus \$150,000 for men), according to recent research. In short, women applied for smaller grants and asked for less funding.

According to the most recent data on grants funded by the NIH, its Deputy Director for Extramural Research, Sally J. Rockey (2014), wrote that "about 30% of NIH-funded research grants are held by female PIs and that this has not changed substantially in the past few years." See Data by Gender from the NIH (2014).

Preparation

It goes without saying that many of us have been hard workers from the time when we were kindergartners. Steve Jobs once said that, "Your time is limited, so don't waste it living someone else's life. Don't be trapped by dogma — which is living with the results of other people's thinking. Don't let the noise of other's opinions drown out your own inner voice. And most important, have the courage to follow your heart and intuition. They somehow already know what you truly want to become. Everything else is secondary." See, e.g., Prive (2013).

After educational preparation, there are also ways to further one's statistical knowledge and training. For example, a useful preparation for junior and mid-career statisticians is to seek accreditation. The American Statistical Association (ASA, 2014) started offering two levels of accreditation: Accredited Professional Statistician (PStat®) and Graduate Statistician (GStat®).

Similarly, the Institute for the Operations Research and the Management Sciences (INFORMS, 2014) is a sister society of the ASA. It offers a path to Certified Analytics Professional (CAP®) through a general analytics certification examination. Attaining the CAP® credential requires passing this exam, adhering to the CAP® code of ethics, and meeting eligibility requirements for experience and education.

Knowing that with preparation through hard work, dedication, and an inner voice, if an opportunity appears on one's horizon, it takes bravery to have the mentality of "carpe diem et carpe noctem," translated as "to seize the day and to seize the night."

Perhaps there may never be enough time to master the next ideal job, the next big role, or the next key assignment at 100 percent ahead of time until one makes an attempt.

Self-Assessment

According to the guide to effective negotiations, written by Miller and Miller (2011), besides confidence and preparation, being able to walk away is also a valuable strategy. Here, we expand this characteristic as self-assessment of the situation before making a decision.

Young children, especially when I grew up in Asia, we were taught to be diligent and tenacious. Although not giving up is a quality that many entrepreneurs possess, we also need to evaluate the pros and cons of the existing environment and potential new opportunities. For those who need to maintain a work-life balance and be aspired to success and lead in our profession, however, the self-assessments would allow us to be able to walk away from micro-managing all aspects of our lives.

Below, I focus on individual contexts. However, it is apparent that we are all unique individuals, and therefore, some of these "generalizations" or "typecasts" must be taken with caution. It is particularly alarming if one is in a stereotypical category because I believe that leaders come in different forms, and leadership qualities are associated with charisma rather than a rigid set of rules or norms.

Individual Contexts

In the individual contexts, I will focus on the influences of personality, heredity, and gender-specific roles.

Personality Types

The Myers Briggs Type Indicator® instrument classifies personality types into 16 categories (see Table 8.1, adapted from The Myers & Briggs Foundation, 2003a).

- Extroversion (E) versus Introversion (I): Extroverts are action-oriented who enjoy a wide array of experiences, while introverts prefer to ponder before acting and prefer in-depth knowledge.

Table 8.1

Sixteen Personality Types Based on the Myers-Briggs Type Indicator®
Instrument, along with General Tendencies in Personalities

	Sensing to Intuitive →				
Judging to Perceiving to Judging ↓	ISTJ ISTP ESTP ESTJ	ISFJ ISFP ESFP ESFJ	INFJ INFP ENFP ENFJ	INTJ INTP ENTP ENTJ	Introvert to Extrovert ↓
	Thinking to Feeling to Thinking →				

- Sensing (S) versus Intuition (N): The sensing types tend to be realistic and practical, while the intuitive ones are creative and imaginative.

- Thinking (T) versus Feeling (F): Thinkers are logical and evaluative, while the feeling types are subjective and empathetic.

- Judgment (J) versus Perception (P): The judges like preplanning and evaluations, while the perception types enjoy open-ended discussions and flexible arrangements.

Lawrence and Martin (2001) wrote that "since type provides a framework for understanding individual differences, and provides a dynamic model of individual development, it has found wide application in the many functions that compose an organization." The reliability and validation of the instrument have been evaluated (see The Myers & Briggs Foundation, 2003b).

As leaders and followers, knowing your own personality types and understanding the associated strengths may be helpful. Various ways are available such as the Myers-Briggs type indicator or the 72-question version that is free online (see The Myers & Briggs Foundation, 2003c; and Humanmetrics, 2014). The reliability and validity of the instruments have not been fully determined as yet.

Although it may be controversial how well this instrument performs in terms of its reliability and validity, I have found that taking such a test, which I had the opportunity to do while at Harvard and via its online short form, to be empowering. This showed me the type of leadership style that I should be aware of. On the other hand, one should be careful not to over interpret and narrowly put my own communication skills into a specific type. The reason is that there are still many factors and contexts to consider, and being one type of a communicator may not work for a variety of scenarios.

For example, as a "judging type," I found myself to be quite suitable as a deputy editor, theme editor or associate editor for scientific and statistical journals. In fact, I was honored to be chosen by the editors of the medical journal, *Radiology*, as a reviewer with special distinction, in the top 5% of all reviewers over several years.

Nevertheless, I should emphasize to inspiring leaders to be very careful not to overplan or jump to conclusions. As an "extrovert," I should keep an open mind when colleagues or coinvestigators present their views or solutions to a particular challenge, especially those who might have creatively thought "outside the box."

Nature Versus Nurture

Are leaders born or trained? In addition to the 16 personality types, it may be tempting to think that traits, charisma, creativity, or entrepreneurship may or may not be innate. Let's take an example by singling out the personality in Table 8.1 that represents the combination of extroverted, intuitive, thinking, and judging type, i.e., ENTJ. This type thrives on complex organizational challenges, abundance of information, and making multiple decisions.

Research conducted by Johnson et al. (1998) has shown that many leadership tendencies are genetic and inherited. The authors studied twins using the Multifactor Leadership Questionnaire, the Leadership Ability Evaluation, and the Adjective Checklist. They concluded that there was a large overlap in the underlying genes responsible for leadership dimensions. Specifically, close to half of the variance in transactional leadership can be explained by additive heritability, while greater than half of the variance in transformational leadership may be explained by dominance heritability. Transactional leadership focuses on supervision, organization, and group performance, while transformational leadership enhances the motivation, morale, and individual performance of followers. More broadly, in Bouchard and McGue (2004), there was evidence indicating that reliably measured individual psychological differences are moderately to substantially heritable. Therefore, leaders tend to run within the same family.

When interacting with a statistician who typically is an introvert like my father, a mathematician, or an extrovert like my mother, a clinician, I would prepare a short list of specific questions in topics of interest before bringing up in-depth discussions. Finally, I expand their imaginations and creativity through follow-up questions. Such communication would see introverts come alive by accomplishing a task with their collaborative participation and by giving them additional room for input and enthusiasm.

On the other hand, my personal opinion is that I strongly believe that leadership qualities can be nurtured and enhanced by increasing one's cultural and organizational sensitivities. I elaborate on this theme in the next part of this chapter. Such effort also echoes the lean-in concept that Sandberg (2013) has advised to women.

Successful leadership involves much harder work than examining the employees' psychological traits. It also involves behaviors that can be learned. I recall that at an event hosted by the International Chinese Statistical Association, Xiao-Li Meng, now a Harvard dean, told the audience about his experience of having selective graduate students undergo a coaching program. After that experience, the communication skills were improved remarkably. Meng (2013) also talked about the usefulness of such coaching during a witty alumni panel speech.

There are personal strategies that a junior or mid-career professional should consider. The first is to have the "Own-It" and "straight talk" culture that is prevalent throughout my company. By raising one's hand and volunteering for assignments, this action may signal the "can do" attitude.

I believe that women are generally better at communication that they learn from mentors and peers on how to improve performances, and become take-charge people by joining in the leaning-in circles. More importantly, they learn how to influence and interrupt in an effective and constructive manner.

Mars versus Venus

In John Gray's (2008) book *Why Mars and Venus Collide? Improving Relationships by Understanding How Men and Women Cope Differently with Stress*, the author wrote: "A man's brain is single-focused, while a woman's brain tends to multitask." Using image analysis of brain wave connectivity, there is evidence that supports that the brains of both sexes tend to be wired differently.

More recently, Ingalhalikar (2014) has suggested that there is difference in how brains are wired. For example, the structure of the female brain generally facilitates socialization and communication through analytical and intuitive processing modes. In contrast, the male brain gears toward perception and coordination action.

I have found such research fascinating mainly because a large portion of my work at Harvard dealt with functional magnetic imaging (e.g., Zou, 2005), as well as medical image analysis and post-processing statistical research. Some of my colleagues have examined brain connectivity through diffusion weight imaging (see, e.g., a summary in a review article that I coauthored with Talos et al., 2006). Much more sophisticated statistical research needs to be done in such cutting edge of medical imaging.

In the era when communication skills are one of the keys to success, perhaps women could take advantage of this ability as part of the leadership style. In particular, a "full spectrum" mother is much more descriptive and colorful, according to Sheehan (2013), who received the best speaker award titled "You Take The Cake" during the M2Moms annual conference that year.

Now I present my personal views on cultural contexts, particularly because I am an Asian-American who has become a naturalized US citizen. Hence, the views represented are in my own belief system. Due to the sensitivity of the topics covered, I will provide my own interpretations of such contexts.

Cultural Contexts

As an Asian-American and a naturalized US citizen, I discuss several cultural contexts in the form of a bamboo ceiling hindering career advancement, implicit biases Asian-Americans face, and language barriers that many immigrants face.

The Bamboo Ceiling

As a Chinese native and a naturalized American, I will speak my mind. Based on a recent report following a comprehensive survey conducted by the Pew Research Center (2013), Asian-Americans have become a racial group that has achieved the highest income and is the fastest growing in the US Generally, Asians are satisfied with their lives, financial status, and where this country is heading.

Asians had above-average household incomes. In 2010, the median household income for Asian-Americans was $66,000, compared with $49,800 for the entire United States population. At statistical conferences, there is visibly no shortage of Asian authors, presenters, and attendees. I have been regularly attending activities sponsored by the Global Asian Alliance within our institution. A disheartening topic that gets raised often is why Asian-Americans rarely reach the board room.

The glass ceiling that women face has also extended to Asians, particularly, for Asian women who aspire to reach further in their career. There is even a custom-made term called the bamboo ceiling (see, e.g., Hyun, 2006). As noted by Fisher (2005), 44% of Asian-Americans were college graduates, in comparison with only 27% for the entire country.

Unfortunately, "only 1% of corporate directors are Asian. Even in Silicon Valley, where about 30% of tech professionals or their forebears hail from Pacific Rim countries, Asian Americans account for only 12.5% of managers; 80% of tech bosses are Caucasian." According to DiversityInc (2014), of the Fortune 500 companies, there are 9 (i.e., 1.8%) Asian CEOs, 8 (i.e., 1.6%) Latino CEOs, and only 6 Black (i.e., 1.2%) CEOs.

A series of experiments conducted by Berdahl and Min (2012) have found the following: "The dominant East Asian employee was relatively disliked as a coworker compared to the nondominant East Asian employee, the nondominant White employee, and the dominant White employee. Even in this majority Asian sample, people preferred a White coworker over an Asian coworker if that coworker had a dominant personality. There were no differences in this preference based on the rater's own racial background, gender, or the gender of the potential coworker." This phenomenon may not be racial. It could be that dominant employees are disliked overall. However, if Asians become dominant, then they may be disliked as a group.

Implicit Bias

Implicit or subconscious bias could set in regardless of how well you prepare or achieve. Trix and Psenka (2003) analyzed more than 300 recommendation letters for successful medical faculty applicants. For females, the letters tended to be shorter, with descriptions such as "hardworking" and "diligent," besides twice as many doubt-raisers. In comparison, males were more likely to receive praises such as "brilliant" and "superb." Then, how much preparation is truly needed when women feel ready to make the next career move or advancement?

"Project Implicit" is a collaborative and concerted effort between three researchers, Greenwald, Banaji, and Nosek (1998). Their Implicit Association Test can be taken online to demonstrate whether there is a subconscious bias via computerized analysis. However, if such a bias exists, it does not necessarily mean prejudice or discrimination in any conscious actions of the test-takers. It is worth finding out how powerful and prevalent such biases are, without our full awareness.

I have first-handedly experienced implicit assumptions even before, as well as after, I came to America. I had grown up using my Chinese name, "Hong," which was given to me after my grandfather had a dream of a rainbow before I was born. I will explain below.

In Chinese, unless one sees the corresponding character, there can be several distinct characters with the same pronunciation. To complicate the matter, there are four tones within the "Hanyu Pingyin" system, the official phonetic system mapping Mandarin pronunciation of Chinese characters into the Latin alphabet.

In the original pingyin, the four tone marks are Tone 1 = macron; Tone 2 = acute accent; Tone 3 = caron; Tone 4 = grave accent. My particular letter has Tone 2. Whenever I made phone calls and gave my name, the phone receptionist or clerk always wrote down another hong, which means "red." Since red is the most popular and lucky color in China, empirically, my given name was assumed to be that letter. These two characters have some similarities in appearances, making them quite confusing even to native Chinese speakers.

To make the matter worse, my last name, or surname, is somewhat rare, which is spelled "Zou" rather than the commonly seen character "Zhou," with Tone 1. The pronunciations have subtle differences, and in my hometown, Shanghai, they sound identical. Thus, I had to explain endlessly that these should be written entirely differently. Once clarified, the next assumption is that "ou" sound would be pronounced as in the word "our."

After arriving in China, the assumption of going with "Zhou" occurs just as frequently. On the other hand, I have discovered a different assumption with respect to "Hong." Over the phone, the first few words I heard would be "Mr. Zou" or "Mr. Hong" and apparently, my given name was mistaken as a male name. At that time, I was an undergraduate student, and even at the Office of the Registrar, I had to explain that my name was a female's name. Finally, I decided to officially change and amend my name in the court system here in America by including a feminine Irish-sounding name, while keeping my Chinese name as my middle name, and my last name intact.

Since then, I have never been mistaken as a male by someone who only looks at my name in advance. Interestingly, the next assumption in the United Kingdom, where I also lived, is that "Zou" would be pronounced as "zoo." This tells you how much even one's name would say about the person before even any face-to-face interaction.

Well, in terms of stereotypical Asians, the "tiger mom" image and "cultural superiority" imposed by Chua (2011a; 2014) certainly does not help but leave the impression that the Asian workers lack social skills, are robotic, and have high-pressure parents who push them to the extreme. When *The Wall Street Journal* ran an essay by Chua (2011b) to explain the superiority of Chinese mothers' influence, it generated heated debates and interests in well over 8800 comments.

Chua declared: "Western parents try to respect their children's individuality, encouraging them to pursue their true passions, supporting their choices, and providing positive reinforcement and a nurturing environment. By contrast, the Chinese believe that the best way to protect their children is by preparing them for the future, letting them see what they're capable of, and arming them with skills, work habits and inner confidence that no one can ever take away."

It is indeed quite appalling to read because I believe that true passion, positive reinforcement, and a nurturing environment are a combination of qualities an excellent leader can display as a visionary. By depriving the children of this type of confidence and vision, they may grow up being part of the silent bunch.

Mehta (2014) recently questioned Chua's evidence by pointing out the issues of selection bias and networking opportunities. The author commented as follows: "The groups Chua and Rubenfeld and the other new racialists typically pick out as success stories are almost without fail examples of self-selection. Forty-two percent of Indians in the U.S. ages 25 and older have a postgraduate degree. But only about 20% of those they've left behind in the motherland even graduate from high school, and 26% of the population is illiterate.…

"Further, the authors pay almost no attention to the role of networking, which accounts for so much of the success of groups like Jews, Cubans and Indians. Part of the reason so many immigrant groups thrive is that when they arrive in the U.S., they already have an uncle who runs a store and cousins who are tutors, doctors or lawyers who can help them negotiate the new country."

With a combination of the above factors, Asians may still appear to be foreign and silent, and thus it is more difficult to ascent to the top of Corporate America. Chinese statisticians have made outstanding contributions to the statistical profession. Chen and Olkin (2012) wrote "Pao-Lu Hsu, Luo-Keng Hua and Shiing-Shen Chern. These three began the development of modern mathematics in China: Hsu in probability and statistics, Hua in number theory, and Chern in differential geometry."

Language Barriers

During my undergraduate education in America, I took a summer class to earn a few credits. It was a Communication and Speech class, which turned out to be one of the most beneficial and brilliant courses for my professional life down the road. The instructor had all of our sessions video taped with our permission. That was well before the YouTube era.

One of the tasks was to demonstrate step by step how to complete a project like a cooking show on television. I chose to show how to make scented candles with a gradually layered color spectrum. A seemingly simple project was not so easy to perfect in front of the audience, to time it right, and to maintain the eye contact to invoke a rapport with the audience. My classmates demonstrated how to make sushi, how to make an all-natural car air freshener using lemon and spice, etc. We also learned how to debate impromptu, how to teach a seminar, as well as how to give and receive awards. I thought of how useful these skills were when I presented travel and poster awards to recipients as the cochair of the Awards Committee at the inaugural Women in Statistics conference in 2014.

The lesson that remains vividly in my mind until this day was not to lean against the podium. Instead, I would briefly scan the document at hand, while periodically looking at the audience in their eyes. This would exude confidence, expertise, and interactions, making a memorable impression as a speaker. This training has turned out to be valuable down the road because it prepared me for my thesis defense, seminar talks, and podium presentations. Above all, it has given me the confidence to be at a podium and to speak in public. In a professional life, there are still other opportunities to practice one's linguistic and speech skills, such as to be a part of Toastmasters International (2014). Non-native speakers must practice communication skills and overcome language barriers.

Another lesson was to write clearly and make sure to use spell and grammatical checks. While typing my doctoral thesis using the scientific editor, LaTex, it was common to write in a text file and ignore spelling and grammar. Thankfully, my advisor was particularly thoughtful in an old-school way. Nowadays, it is quite easy to write and submit online following several options for spell checking.

It is understandable that navigating through cultural gaps is difficult enough, and it is more challenging to emerge as an effective leader in organizational contexts because a hierarchical or vertical career ladder may not be as evident as before. In the meantime, there is the pressure of "publish or perish" frequently seen in the academic environment. To many junior or mid-career statisticians, different roles such as advisor, mentor, manager, and supervisor may have blurred lines. In the next section, these topics are discussed.

Organizational Contexts

I now go over organizational contexts. On a modern organizational level, the career path may no longer be a straightforward vertical or upward movement, but rather lateral. In the academic setting where "soft money" is key, grant funding parity becomes an important topic. Furthermore, several common roles emerging in the organizations are reviewed.

Vertical Ladder Versus Lateral Path

During the 2012 JSM conference, Robert N. Rodriguez, president of the American Statistical Association, said in his presidential address: "Our future strength over the next 25 years will be serving anyone who uses statistical methods to solve problems. These include problems that draw on new types of data in science, business, and policy making." Rodriguez further emphasized that "Regardless of whom we serve, whether they call themselves statisticians, scientists, or business analysts, we must convey the importance of statistical contributions. We must have the ability to explain, demonstrate, and prove that importance to students, to the media — and most of all — to society" (See Rodriguez, 2013).

I have been working as a statistician in the industry for a number of years. Organizationally, there has been a tendency to have a "flat structure" rather than a straightforward vertical one. Traditionally, career advancement means promotions and climbing through the ranks or up the ladder. Nowadays, however, a wide breadth of job functions, at equivalent authority, similar income levels, or career successes, is the new norm. Such horizontal movements are via a lateral path. Of course, in a somewhat zigzag pattern, the career trajectory may also be simultaneously lateral and vertical (see Ashkenas, 2012).

Due to the cross-functional nature of many project teams, it is important to make and recognize contributions toward the overall team success as a process leader or an owner. Women generally have excellent multitasking and communication skills (see, e.g., Ingalhalikar, 2014), and it would be rewarding for organizations to recognize and encourage lateral movements and team contributions. Nevertheless, it is also important to cultivate excellence and expertise, so that there would not just be a "Jack (or Jill) of all trades" without in-depth knowledge.

Grant Funding Parity

For academicians, it is well-known that one needs to "publish or perish." Thus, grant funding for conducting cutting-edge research in a successful career has become an important benchmark, as well as for providing critical support for developing ideas that result in quality peer-reviewed publications.

As a member of the Faculty Taskforce on the JCSW, I coauthored an article by Waisbren et al. (2008), as noted before. The committee analyzed research grants between 2001 and 2003 at eight participating sites affiliated with Harvard Medical School and Harvard School of Dental Medicine. Of these, approximately 60% of the applicants and submissions from four institutions had academic rank information.

The good news was that women at higher academic ranks ultimately achieved parity in terms of grant funding success and the awarded amounts. Stratified by academic rank, there was no statistically significant difference between the proportions of successful grants submitted by women versus men. Such success stories gravitated toward high ranks, reflected by increased success rates.

Apart from success rates, unfortunately, women tended to receive smaller grants in terms of amounts awarded. The differences in median amounts awarded were statistically significant between male and female principal investigators within the category of instructor, as well as within the category of associate professor.

Leaders in academic institutions need to be aware that many beginner female researchers have a so-called confidence gap, as Kay and Shipman (2014) noted. Providing support to relatively junior faculty members would improve their confidence and help their negotiation skills.

Advisor, Manager, Supervisor, Mentor, and Sponsor

Scanning across the above roles, they may sound like a mouthful and with some overlap. Indeed, a leader nowadays is likely to wear multiple hats. In both my doctoral thesis planning and my post-doctoral training, I was fortunate enough to have two visionary advisors who pointed toward my research topics of interest for years to come.

In the corporate world, there is typically the structure of line management with an immediate manager or supervisor. Along the reporting line, there are higher-level general managers, senior leadership team, and ultimately, the head of the division, department, school, or corporation. The organizational structure still contains a vertical path, as well as interrelated cross-functional branches to members within the same division or between different divisions in various collaborative functional areas.

Professional societies and large companies encourage and embrace mentorship, in the hopes that it will empower the workers and ultimately improve their performances and lead to senior management. The employees seek out mentors through a "matching" process. The sessions may last for several months to years, and meeting one on one at regular intervals would be helpful. This relationship may somewhat resemble the advisor–advisee relationship in graduate school, but a mentor does not have supervising responsibility to the mentee.

Why are women still not promoted fast enough, despite being mentored? A major reason, based on Ibarra, Carter, and Silva's (2010) findings, is the following: "All mentoring is not created equal, we discovered. There is a special kind of relationship called sponsorship — in which the mentor goes beyond giving feedback and advice and uses her or his influence with senior executives to advocate for the mentee. Our interviews and surveys alike suggest that high-potential women are overmentored and undersponsored relative to their male peers — and that they are not advancing in their organizations. Furthermore, without sponsorship, women not only are less likely than men to be appointed to top roles but may also be more reluctant to go for them."

For aspiring leaders who have passed the stage of having an advisor and who are currently having a manager or supervisor, it would still be valuable to reach out and seek mentorship. Although not sought but earned, a sponsor, typically a senior executive, will speak on employees' behalf for a more successful career.

As demonstrated and discussed previously, there are many opportunities for leaders of diversity to shine in the statistical careers. In an era when we face globalization, leaders must exhibit agility and adaptability. In the next section, I present new challenges facing the next generation leaders, as well as practicing statisticians.

Challenges Facing the Next Generation

Because it is highly challenging to navigate in the new social era, it is imperative that next generation leaders must possess several skill sets, such as cross-functional, virtual, social media, and networking skills. In this section, I will elaborate on these emerging topics.

Broad Cross-Functional Skills

Former US President Dwight Eisenhower once said: "In preparing for battle I have always found that plans are useless but planning is indispensable" (see, e.g., Ronco and Ronco, 2005). Often seen on large, multisite collaborative studies, projects, and grants nowadays, today's research teams often incorporate a cross-functional and sometimes a cross-institutional paradigm.

For example, there were several grants that I led as a principal investigator while I was an associate professor at Harvard Medical School. One R01 grant-funded research team included coinvestigators from the Department of Neurosurgery, the Decision Systems Group, the Surgical Planning Laboratory, and Medical Physics.

Another R01 grant-funded team included coinvestigators and consultants from the Departments of Radiology from two different institutions, the Department of Health Care Policy, the Decision Systems Group, the Computational Radiological Laboratory, and the Surgical Planning Laboratory.

The new challenge arising from multidisciplinary teams is that the success in completing tasks may not require layers of hierarchical management. Instead, there are parallel and sequential processes that require different expertise and skills, different time zones spanning across various domestic and international regions, and specialty knowledge, as well as general understanding of different functions. Thus, in order to complete the task, it would be useful and imperative to have a wide knowledge base, versatility, and agility.

More importantly, the leaders need to be transparent on the expectations, be good at process management, and be able to align different stakeholders who have overall goals, as well as individual interests.

To foster collaborative spirits, effective communication, meeting planning, continuing education, webinars, and team building are useful tools. An inspiring collection of statistical leaders, who have excelled in a wide array of topics and collaborations, are featured in the book *Past, Present, and Future of Statistical Science*, edited by Lin, Genest, and Banks et al. (2014).

Specific guidelines, steps, and checklists for effective partnerships between research organizations and groups are detailed in Ronco and Ronco (2005). Participating in local chapters, sections, webinars, and women's leadership networking opportunities may provide connection in areas of interest (e.g., science, technology, engineering, and mathematics, in short, STEM in the US, and by including medicine, STEMM in the UK).

Virtual Communication

Through The Collaborative on Academic Careers in Higher Education (or COACHE), Helms (2010) has shed some insight on the type of communication that Generation X (i.e., those born from 1964 to 1980) have chosen in their lives and careers. In addition, Trower (2010) has observed that a new generation of academicians, particularly women, may choose to wait and stay put, rather than immediately pursuing tenure. Therefore, it is imperative to provide helpful support for professional development and career mentorship. Even among older academicians relative to those among Generation X, including the so-called traditionalists (i.e., those born before 1946), older boomers (i.e., those born from 1946 to 1955), and younger boomers (i.e., those born from 1956 to 1963), work–life balance has always been an important factor.

Nevertheless, because Generation X workers tend to be busy in many directions with competing needs, achieving this balance is not a simple task. More and more due to possibility of virtual communication, flexible work arrangements, and limited budgets for professional interactions, there are "lone" academicians in the statistical fields who may interact more with nonstatisticians (e.g., in medical, engineering, outcomes research, and marketing).

It is worth noting that face-to-face meetings still have their merits. For example, Yahoo's controversial internal memo stated the following reason: "To become the absolute best place to work, communication and collaboration will be important, so we need to be working side-by-side. That is why it is critical that we all be present in our offices. Some of the best decisions and insights come from hallway and cafeteria discussions, meeting new people, and impromptu team meetings. Speed and quality are often sacrificed when we work from home. We need to be one Yahoo!, and that starts with physically being together." (see, e.g., Swisher, 2013).

Social Media and Networking

Since the birth of social media in the last decade, more institutions and businesses have interfaced with such new communication channels, as analyzed and summarized by O'Malley (2013). There are emerging tools that are frequently used, such as Wikipedia for searching of information, the members' discussion boards hosted by the American Statistical Association, Facebook, LinkedIn, Google+, Twitter, WeChat, etc., for communication, and YouTube for clips on authors' monographs, not to mention various apps on the iPad, iPod, iPhone, Android, and Blackberry platforms. In business, sophisticated analysis of real-world "big data" has been increasingly more and more extensive (see, e.g., Thomas, 2013).

For the next generation leaders, it is important to shape up their digital communication strategy and establish policies on the use of such platforms. Furthermore, it also requires a different way of effective dissemination of information that may be more visual, instantaneous, compelling, and engaging.

Thus, it is of importance to sharpen one's abilities on speech, technology, and information exchange in this new virtual infrastructure with its fewer physical boundaries. In Deiser and Newton (2013), the six specific social-media skills that leaders require are listed, which include creating compelling content, leveraging dissemination dynamics, managing communication overflow, driving strategic social-media utilization, creating an enabling organizational infrastructure, and staying ahead of the curve.

Next, I have a few remarks on the golden era of statistics when there rains "data." How we seize the opportunity due to the abundance of different types of data, new analytic problems, untapped areas for statistical expertise, and own the profession as women leaders is not simple but is exactly why we can make our marks in our statistical community. I believe that there is hope that women can make statistics an important component in our societies.

Closing Remarks

The Wall Street Journal has just listed mathematicians, tenured professors, and statisticians, which may not be mutually exclusive, as the top three jobs in 2014 (see Auriemma, 2014). In his whimsical writing, Harvard's Dean and Professor Xiao-Li Meng (2011) predicted the rapid growth of quantitative fields, such as statistics, several years ago. With more and more degree candidates in statistics and related fields, women can reach high and set our sights even higher.

In my opinion, practice makes perfect, and these can be aspired to and be learned. I disagree that leadership fully relies on so-called charisma. An analogy is the storyline in the movie, *The King's Speech*, directed by Hooper (2010), which was about King George VI of the United Kingdom. He ascended to the throne as the head of the kingdom, which motivated and required him to seek speech therapies in order to lead effectively.

The mastery of the speech, as a metaphor for leadership in general, would take years of dedication, practice, unwaveringness, and perseverance. These are some of the traits that we, as women and many of whom are great mothers and/or daughters, can be extremely good at, depending on cultural and organization contexts.

One of the founding fathers of modern Chinese literature, writer Lu Xun (from 1881 to 1936), once famously wrote (e.g., quoted in Spence, 1999): "Hope cannot be said to exist, nor can it be said not to exist. It is just like roads across the earth. For actually the earth had no roads to begin with, but when many people pass one way, a road is made."

What Lu Xun wrote about at the time was his old home, and yet his words have echoed so deeply in my mind because it could be the hope that women may break the glass ceiling, and Asian women, especially, may ultimately break their bamboo ceiling, in a figure-of-speech way of viewing leadership. As women, we collectively owe it to ourselves to maximize our potential.

Acknowledgment

I would like to thank the Caucus for Women in Statistics, particularly Professor Amanda L. Golbeck, University of Montana, and Professor Yulia R. Gel, University of Waterloo, who organized a panel discussion during the Joint Statistical Meetings in Montreal, Canada, in 2013. I am particularly grateful to Professor Emeritus, Ingram Olkin, Stanford University, who has championed the careers and leaderships of women in statistics.

In addition, my doctoral thesis advisor, Professor W. Jackson Hall, University of Rochester, my post-doctoral advieser, Professor Sharon-Lise Normand, Harvard Medical School and Harvard School of Public Health, and my former colleague, Professor Ferenc A. Jolesz, Harvard Medical School, as well as many others with whom I have crossed paths in various capacities, have instilled in me the belief that, as a woman, one can reach out to the stars and make a difference in the world.

In particular, I would like to acknowledge the Fudan Fuzhong Overseas Foundation, which is associated with my beloved High School Affiliated to Fudan University in my hometown, Shanghai, China. The school has provided me with a stimulating environment to maximize the students' potential, including leadership qualities. The students' creativity and talents are discovered and encouraged through research, discussions, performances, and many other opportunities. Above all, it has fostered a nurturing environment for students to gain self-confidence, which would lead to their long-term career excellence.

Finally, the views expressed in this chapter are solely my own and thus do not necessarily reflect the opinions of the institution at which I am employed.

Bibliography

American Statistical Association (2014). Accreditation. Alexandria, VA. http://www.amstat.org/accreditation/index.cfm (accessed on June 23, 2014).

Ashkenas, R. (2012). Your career needs to be horizontal. *Harvard Business Review*. Harvard Business Publishing: Boston, MA. http://blogs.hbr.org/2012/03/your-career-needs-to-be-horizo/ (accessed on June 23, 2014).

Auriemma, A. (2014). The 10 best jobs of 2014. *The Wall Street Journal*. New York, NY. http://blogs.wsj.com/atwork/2014/04/15/best-jobs-of-2014-congratulations-mathematicians/ (accessed on June 23, 2014).

Berdahl, J.L. and Min, J.A. (2012). Prescriptive stereotypes and workplace consequences for East Asians in North America. *Cultural Diversity and Ethnic Minority Psychology*. 18: 141–152.

Bombardieri, M. (2005). Summers' remarks on women draw fire. *Boston Globe*, Boston, MA. http://boston.com/news/education/higher/articles/2005/01/17/summers_remarks_on_women_draw_fire (accessed on June 23, 2014).

Bouchard, T.J. and Jr., McGue, M. (2003). Genetic and environmental influences on human psychological differences. *Journal of Neurobiology*. 54:4–45.

Chen, D. and Olkn, I. (2012). Pao-Lu Hsu (Xu, Bao-lu): The grandparent of probability and statistics in China. *Statistical Science*. 27:434–445. http://arxiv.org/pdf/1210.1031.pdf (accessed on June 23, 2014).

Chua, A. (2011a). *Battle Hymn of the Tiger Mother*. Reprint Edition. The Penguin Press: New York, NY.

Chua, A. (2011b). Why Chinese mothers are superior. *The Wall Street Journal*: New York, NY.

Chua, A. (2014). *The Triple Package: How Three Unlikely Traits Explain the Rise and Fall of Cultural Groups in America*. The Penguin Press: New York, NY.

Deiser, R. and Newton, S. (2013). Six social-media skills every leader needs. *McKinsey Quarterly*. McKinsey & Company: New York, NY. http://www.mckinsey.com/insights/high_tech_telecoms_internet/six_social-media_skills_every_leader_needs (accessed on June 23, 2014).

DiversityInc. (2014). Where's the Diversity in Fortune 500 CEOs? DiversityInc Media, LLC: Princeton, NJ. http://www.diversityinc.com/ diversity-facts/wheres-the-diversity-in-fortune-500-ceos/ (accessed on June 23, 2014).

Federal Glass Ceiling Commission (1995). *Solid Investments: Making Full Use of the Nation's Human Capital.* Department of Labor, Washington, DC. http://www.dol.gov/oasam/programs/ history/reich/reports/ceiling2.pdf (accessed on June 23, 2014).

Fisher, A. (2005). Piercing the 'bamboo ceiling.' *CNN, Fortune & Money.* CNNMoney: New York, NY. http://money.cnn.com/2005/08/08/ news/economy/annie/fortune_annie080805/ (accessed on June 23, 2014).

Gray, J. (2008). Why *Mars and Venus Collide? Improving Relationships by Understanding How Men and Women Cope Differently with Stress.* Reprint Ed. HarperCollins Publishers: New York, NY.

Greenwald, A.G., Banaji, M.R., and Nosek, B.A. (1998). ProjectImplicit. https://www.projectimplicit.net/index.html (accessed on June 23, 2014).

Golbeck, A. (2012). Where are the women in the JSM registration guide? *Amstat News.* http://magazine.amstat.org/blog/ 2012/07/01/statviewguide/ (accessed on June 23, 2014).

Helms, R.M. (2010). New challenges, new priorities: The experience of generation X faculty: A study for the collaborative on academic careers in higher education. Harvard Graduate School of Education, Cambridge, MA. http://camp.rice.edu/uploadedFiles/CAMP/ Non-Rice_Resources/COACHE%20Gen%20X.pdf (accessed on June 23, 2014).

Henessy, J., Hockfield, S., and Tilghman, S. (2005). Women in math, engineering and science: Drawing on our country's entire talent pool. *MIT News.* Massachusetts Institute of Technology: Cambridge, MA. http://web.mit.edu/newsoffice/2005/hockfield-presidents.html (accessed on June 23, 2014).

Hooper, T. (2010). *The King's Speech.* British Board of Film Classification. UK Film Council: London, UK. http://www.bbfc.co.uk/ releases/kings-speech-film (accessed on June 23, 2014).

Humanmetrics (2014). Jung Typology Test[TM]. http://www. humanmetrics.com/cgi-win/jtypes2.asp (accessed on June 23, 2014).

Hyun, J. (2006). *Breaking the Bamboo Ceiling: Career Strategies for Asians.* Reprint Ed. HarperCollins Publishers Inc: New York, NY.

Ibarra, H., Carter, N.M., and Silva, C. (2010). Why men still get more promotions than women. *Harvard Business Review*. Harvard Business Publishing: Boston, MA. http://hbr.org/2010/09/why-men-still-get-more-promotions-than-women/ar/1 (accessed on June 23, 2014).

Ingalhalikar, M., Smith, A., Parker, D., Satterthwaite, T.D., Elliott, M.A., Ruparel, K., Hakonarson, H., Gur, R.E., Gur, R.C., and Verma, R. (2014). Sex differences in the structural connectome of the human brain. *Proceedings of the National Academy of Sciences of the United States of America*. 111:823–828.

Institute for Operations Research and the Management Sciences (2014). Certified Analytics Professional. Catonsville, MD. https://www.informs.org/Certification-Continuing-Ed/Analytics-Certification (accessed on June 23, 2014).

Johnson, A.M., Vernon, P.A., McCarthy, J.M., Molson, M., Harris, J.A., and Jang, K.L. (1998). Nature vs nurture: Are leaders born or made? A behavior genetic investigation of leadership style. *Twin Research*. 1:216–223.

Kay, K. and Shipman, C. (2014). The confidence gap. *The Atlantic*. Atlantic Media: Washington, DC. http://www.theatlantic.com/features/archive/2014/04/the-confidence-gap/359815/ (accessed on June 23, 2014).

Lawrence, G. and Martin, C.R. (2001). *Building People, Building Programs: A Practitioner's Guide for Introducing the MBTI to Individuals and Organizations*. Center for Applications of Psychological Type: Gainsville, FL.

Lin, X., Genest, C., Banks, D.L., Molenberghs, G., Scott, D.W., and Wang, J.-L. Eds. (2014). *Past, Present, and Future of Statistical Science*. Chapman and Hall/CRC: Boca Raton, FL.

Mehta, S. (2014). The "Tiger Mom" Superiority Complex. *Time*. Time Inc: New York, NY. http://content.time.com/time/magazine/article/0,9171,2163555-6,00.html (accessed June 23, 2014).

Meng, X.-L. (2011). Statistics: Your chance for happiness (or misery). *The Harvard Undergraduate Research Journal*, 2:21–26.

Meng, X.-L. (2013). One Harvard Faculty and Alumni Panel — Xiao-Li Meng. YouTube. A subsidiary of Google Inc., San Bruno, CA. https://www.youtube.com/watch?v=XjFRgA1WziU (accessed on June 23, 2014).

Miller, L.E. and Miller, J. (2011). *A Woman's Guide to Successful Negotiation*. 2nd Ed. McGraw-Hill: New York, NY.

National Institutes of Health (NIH). (2014). Data by Gender. Research Grant Investigators: Representation of women, by mechanism. Bethesda, MD. http://report.nih.gov/NIHDatabook/Charts/Default.aspx?sid=0&index=1&catId=15&chartId=169 (accessed on June 23, 2014).

Olkin, I. (2014). Where have all the tenured women gone? *Amstat News.* January 1. http://magazine.amstat.org/blog/2014/01/01/tenured-women (accessed on June 23, 2014).

O'Malley, A.J. (2013). The analysis of social network data: An exciting frontier for statisticians. *Statistics in Medicine.* 32:539–555.

Pew Research Center. (2013). The Rise of Asian Americans. Pew Research Center: Washington, D.C. http://www.pewsocialtrends.org/files/2013/04/Asian-Americans-new-full-report-04-2013.pdf (accessed on June 23, 2014).

Prive, T. (2013). Top 32 quotes every entrepreneur should live by. *Forbes.* Forbes, Inc.: New York, NY. http://www.forbes.com/sites/tanyaprive/2013/05/02/top-32-quotes-every-entrepreneur-should-live-by (accessed on June 23, 2014).

Rockey, S. (2014). FY2013 By The Numbers: Research Applications, Funding, and Awards. *Rock talk: Helping connect you with the NIH perspective.* Bethesda, MD. http://nexus.od.nih.gov/all/2014/01/10/fy2013-by-the-numbers/ (accessed on June 23, 2014).

Rodriguez, R.N. (2013). Building the big tent for statistics. *Journal of the American Statistical Association.* 108: 1–6.

Ronco, W. and Ronco, J.S. (2005). *The Partnering Solution.* The Career Press, Inc.: Pompton Plains, NJ.

Summers, L.H. (2005). Remarks at NBER conference on diversifying the science & engineering workforce. The Office of the President, Harvard University, Cambridge, MA. http://www.harvard.edu/president/speeches/summers_2005/nber.php (accessed on June 23, 2014).

Sandberg, S. (2013). *Women, Work, and the Will to Lead.* Alfred A. Knopt: New York, NY.

Sheehan, K. (2013). Today's Full Spectrum Mom. 9th Annual M2Moms: The Marketing to Moms Conference. Chicago, IL. http://www.m2moms.com/news_12_16_13.php (accessed on June 23, 2014).

Spence, J.D. (1999). *The Search for Modern China,* 2nd Ed. W.W. Norton & Company: New York, NY.

Swisher, K. (2013). "Physically together": Here's the internal Yahoo no-work-from-home memo for remote workers and maybe more. *All Things Digital*. Dow Jones: New York, NY. http://allthingsd.com/20130222/physically-together-heres-the-internal-yahoo-no-work-from-home-memo-which-extends-beyond-remote-workers/ (accessed on June 23, 2014).

Talos, I.F., Mian, A.Z., Zou, K.H., Hsu, L., Goldberg-Zimring, D., Haker, S., Bhagwat, J.G., and Mulkern, R.V. (2006). Magnetic resonance and the human brain: Anatomy, function and metabolism. *Cellular and Molecular Life Sciences*. 63:1106–24.

The Myers & Briggs Foundation (2003a). MBTI®Basics. http://www.myersbriggs.org/my-mbti-personality-type/mbti-basics/ (accessed on June 23, 2014).

The Myers & Briggs Foundation (2003b). MBTI®Basics. http://www.myersbriggs.org/my-mbti-personality-type/mbti-basics/reliability-and-validity.asp (accessed on June 23, 2014).

The Myers & Briggs Foundation (2003c). Ways to take the Myers-Briggs Type Indicator® (MBTI®) instrument. http://www.myersbriggs.org/frequently-asked-questions/ways-to-take-the-mbti/ (accessed on June 23, 2014).

Thomas, K. (2013). Pills tracked from doctor to patient to aid drug marketing. The *New York Times*: New York, NY. http://www.nytimes.com/2013/05/17/business/a-data-trove-now-guides-drug-company-pitches.html?_r=0 (accessed on June 23, 2014).

Thompson, K. (2014). Author, N. Va. native Helen Wan on the 'bamboo ceiling'. *The Washington Post*: Washington, DC http://www.washingtonpost.com/lifestyle/magazine/author-n-va-native-helen-wan-on-the-bamboo-ceiling/2014/02/12/89cc0b76-5151-11e3-9e2c-e1d01116fd98_story.html (accessed on June 23, 2014).

Toastmasters International. (2014). Rancho Santa Margarita, CA. http://www.toastmasters.org/Members/MembersFunctionalCategories/AboutTI.aspx (accessed on June 23, 2014).

Trix, F. and Psenka, C. (2003). Exploring the color of glass: Letters of recommendation for female and male medical faculty. *Discourse & Society*. 14:191–220.

Trower, C.A. (2010). A new generation of faculty: Similar core values in a different world. *Peer Review* 12:3. Association of American Colleges and Universities. https://www.aacu.org/peerreview/pr-su10/pr_su10_NewGen.cfm (accessed on June 23, 2014).

United Nations International Children's Emergency Fund. (2005). Gender parity: A moving target. *Gender Parity and Primary Education*. 2. http://www.unicef.org/progressforchildren/2005n2/gender.php (accessed on June 23, 2014).

Villarica, H. (2012). Study of the day: There's a 'bamboo ceiling' for would-be Asian leaders. *The Atlantic*. Atlantic Media: Washington, DC. http://www.theatlantic.com/health/archive/2012/05/study-of-the-day-theres-a-bamboo-ceiling-for-would-be-asian-leaders/257135/ (accessed on June 23, 2014).

Zou, K.H., Greve, D.N., Wang, M., Pieper, S.D., Warfield, S.K., White, N.S., Manandhar, S., Brown, G.G., Vangel, M.G., Kikinis, R., and Wells, W.M. 3rd; FIRST BIRN Research Group. 2005. Reproducibility of functional MR imaging: Preliminary results of prospective multi-institutional study performed by Biomedical Informatics Research Network. *Radiology*. 237:781–9.

Part III

Project Leadership

9

Statistical Challenges in Leading Large-Scale Collaborations: Does Gender Play a Role?

Bhramar Mukherjee and Yun Li

University of Michigan

Introduction

The landscape of leadership for women has transformed over the last century, but leadership opportunities for women are still far from being equal to men (Kellerman and Rhode, 2007). In statistics and biostatistics departments/divisions across the world, the situation is no better. This article focuses on the issue of underrepresentation of women statisticians leading large-scale collaborative projects (primarily in biomedicine and public health), the challenges associated with taking on such roles, and what can be done to address those challenges. It is a general theoretical belief that any workplace will ideally benefit from a woman's values, needs, and life experiences being reflected in decision-making positions, but it is a fact that decision-making bodies are mostly dominated by men (O'Connor, 2007). The style of successful leadership is often found to be different for men and women. While a self-asserting, threatening style may work well for a male leader, women using an inclusive and supportive style are often perceived as more effective leaders, and this type of empathic response may be biologically more natural for women (Singer et al., 2006). Do these general stereotypical conjectures or assertions about gender-specific roles translate to the field of statistics? In particular, our focus is on large-scale collaborations that involve partnership of statisticians and nonstatisticians to tackle a question of scientific/societal relevance. Communicating and convincing the nonstatistician colleague itself can pose certain basic challenges irrespective of the gender of the statistician on a project, but in this article we explore if there is an "effect modification" by gender: namely, do the challenges and the solutions that appear to work for lead statisticians in large collaborative teams comprising of diverse scientific expertise vary by gender?

We would like to begin with an apology and a confession. As we embarked on writing this article, unlike the articles we both are used to writing for our methodological and collaborative publications in peer-reviewed scientific journals, we quickly realized that it is incredibly hard to write a nontechnical article on an issue we experience and confront daily, but have no formal training to write about! The more we plunge through TED talks, books, and articles on the web about leadership of women, we come to understand that the specific leadership issues for statisticians working on large collaborative projects are quite different than the general context, though there are some latent commonalities. Thus, in pursuit of empirical data, we started talking to colleagues in and beyond our institutions on some common themes, trying to gain a better understanding of the topic we were trying to write about. We also wanted to explore whether a collective pattern emerges from the sporadic recollections and individual experiences of peer women colleagues working as lead statisticians in the forefront of biomedical research. Much of this article is based on our conversations and personal experiences, rather than a thorough review of the literature in the general field of female leadership. We ask the reader to take this article in that spirit and pardon our lack of formalism. The structure of the article is as follows: We will start with a section on getting started and staying engaged in collaborations, followed by discussion on three topics: rising to leadership roles, best practices for statistical leadership in project teams, and challenges for female statisticians as leaders in a scientific team. We will then conclude with some thoughts on fostering the next generation of women leaders in statistical science, in particular in collaborative teams.

Getting Started with Collaborations and Staying Involved

Getting started in large-scale collaboration is sometimes a matter of a series of coincidences, or sometimes a pursued effort. There are several routes that are often found to be useful strategies to initiate a collaboration: (1) let senior members/head of your group or your assigned mentors know about the types of collaborations you are interested in; (2) attend seminars by clinical or public health researchers in your institution, and approach them if their research sparks your interest; (3) contact other department heads or individuals whose research areas are appealing to you, either directly or through your mentors, and be proactive; and (4) affiliate yourself with existing research centers associated with a disease area or a theme of interest to you (for example: comprehensive cancer center, center for global health, and the like that may exist at your institution). When you are new to an organization, it is a good idea to make yourself known by introducing your own research by giving seminars and tutorials locally and by availing opportunities to converse with scientists from other disciplines. In an academic department, to offer a special topics or seminar course is a proven strategy to attract an audience toward your work.

Once you have a foot in the door and an initial opportunity to collaborate, you have to engage in the science of your team to stay involved. One typically has to spend an extensive amount of time in the beginning to learn the language and subject-matter jargons used by the collaborative team. Understanding the science and communicating effectively is critical to establishing a respectable entity in a collaborative team. When you get involved, you are unaware of the possibilities that a project may hold; however, it is important to listen, improvise, and create at each consultation. Often you will have tight deadlines, and delivering on time is an asset for a successful collaborating statistician. In many project teams, statisticians are already held in high esteem, and what they bring to the table are well-recognized as key elements for a successful study. However, in some collaborative situations, the challenge is harder, where the statistician is viewed as a service or support provider rather than an independent thinker and a cocontributor. Being a woman statistician in such an environment could often enhance these challenges, as women tend to be less assertive, at least in the beginning phase of a collaboration. The principal investigator (PI) leading the science part may often dominate the discussion and even claim boldly something to the effect of "if there is anything really going on in my study, I do not need a statistician to tell me." However, she/he DOES need a statistician. In the modern scientific era, with massive amounts of heterogeneous data coming from, for example, next generation sequencing, magnetic resonance imaging, or electronic health records, it is impossible to just know "what is going on." Classically trained intuitions and visualizations often fail in these high-dimensional settings with sparse data. Even for the much simpler setting where dimensionality is not an issue, there are plenty of examples where confounding or selection bias has produced misleading results without a careful analysis and inference (Bosco et al., 2010).

In this latter situation, where the culture in the team is not conducive for a statistician to have a strong voice, one has to be patient and prove one's worth without being frustrated and agitated by the situation. The only way to earn trust is through hard work that is *consistent and persistent*. It often helps to get involved with collaborations that are aligned with your own methodological research or can serve as motivations for new methodological research. It allows you to develop new methods as opposed to just providing run-of-the-mill analyses. This can enhance the impact and respect for the statistician as an independent thinker and creator. We illustrate this point via a small story from a shared experience. Dr. X (say), a young female assistant professor of biostatistics, approached Dr. E (with E standing for ego), a very senior scientist at the same university with a never-ending list of degrees from ivy-league institutions. Dr. X was interested in the study that Dr. E had launched and published on, as the inferential set-up matched with her own research interest. For the first few months, Dr. X simply sat in Dr. E's lab meetings, listening through the work of the team members, understanding a small fraction of it, filtering through the seemingly impenetrable barrier of jargons. Dr. E neither gave her a specific project to work on, nor offered any financial support; a friendly nod was all they exchanged each week. During this time, Dr. X wrote a new methodological paper for answering one of the questions Dr. E was interested in, used his data to illustrate the method, and sent a relatively complete and mature draft for Dr. E's review. Dr. E was pleasantly surprised that the silent spectator in the lab had latently absorbed what the actual clinical question was, and came up with a new method that did better justice to the data than the analysis he was doing in his lab. Thus, they started talking about science and analysis tools. Five years later, they had 20 papers together, and Dr. X was in charge of a large study that Dr. E launched as a principal investigator. This is an example where perseverance had paid off. Women are often better at creating such long-term relationships even when immediate success or positive response via the direct initial contact did not materialize. Quietly leaning in is also an effective option in life (Kahnweiler, 2013).

Rising to Leadership Roles in Collaborative Projects

Based on a long-term relationship with the research team and individual career trajectory, statisticians are sometimes offered or asked to assume leadership roles in large scientific projects. These roles could be diverse in nature; for example, it could be as the director of a data coordinating center, director of a statistics/biostatistics core in a large center or program project grant, or senior and lead statistician on a large study in a consortium. Women often turn away from taking responsibility of large-scale analysis. There are examples of statisticians being the principal investigators for large clinical and epidemiological studies, but almost none of these examples involve female leaders. Many women work on instrumental analysis for these large and impactful studies but often do not claim lead authorship or do not get the external recognition they deserve. This may be simply related to there being fewer women at senior-level positions (a fact that we hope is going to change in the coming decade). There may be several other reasons underlying this under representation of women statisticians as leaders in large studies, including, but not limited to, reluctance to lead a large, diverse team composed of scientists at various levels, not just of statisticians, taking care of routine paper work involving reporting to the regulatory/funding agency; managing a substantial budget; limited flexibility to travel as a leader on large consortium; fear of losing time from methodological research, and possibly a tendency to stay away from high publicity and limelight. Whereas on the one hand, you should not pursue this path merely to set an example for other women, or for changing the culture; on the other hand you should not be afraid to take on a leadership role if it is compatible with your personal goals and milestones. It is important to internally define the "purpose" while taking on leadership roles in large teams. Before embarking on the journey, decide why you want to be a leader and at which phase/time point of your career and life you would like to become a leader. Being a leader may mean that you will have less time dedicated to your statistical expertise or technical skills but more time for management, or will potentially have less time for your personal life. It is important to understand your desire, your goal, and the potential trade-offs in achieving the goal before undertaking a huge responsibility.

It obviously takes an incredible amount of stamina to lead large analysis projects. When one assumes such a role, one has to be prepared for the fact that much of the work is not technically challenging or stimulating, but managing logistics and people, attending many meetings, preparing progress reports for the external advisory committee or funding agencies, and delegating and distributing work to others. The motivation and incentive for taking on such a leadership role, as opposed to settling down comfortably with individual research projects, has to be clear from the point of reaching your own personal objectives and aspirations. It could be for the science that one grows to care about, it could be for bringing in funding resources and infrastructure to the home unit, it could be for establishing leadership and administrative capabilities for future career aspirations. Once the purpose is clear, it is easier to devise the leadership style and choices that cater toward that purpose.

Leadership becomes easier if during the time spent in a collaboration you invest effort to build relationships with the entire study group. In addition to your key collaborators, building relationships with analysts, data managers, programmers, junior investigators, and post-doctoral and doctoral trainees makes your decision to assume a future lead role in the group easier. It is also good practice to hold tutorials for your entire group to educate them in statistics or hold journal clubs. If they expect you to understand their science, you can expect your team to try to understand what you do. These types of inclusive activities help with recognizing the importance of statistics.

Showcasing your skills from time to time by presenting your work and ideas to your team, or offering tutorials for other statisticians and junior trainees in the group, is always helpful in integrating yourself with the science and making yourself visible. Selectively choosing a few important and unsolved research problems that originate from your collaboration and persistently pursuing them to develop new methodology can be very helpful. These tasks are not assigned to you specifically, but rather you have to actively identify these research problems, and form your own tasks and volunteer yourself to solve them. It is important to let others know your original contribution (e.g., you developed a new strategy for multiple imputation, or in obtaining causal estimates, or in modeling a complex dose response curve) that leads to better analysis, inference, and interpretation of the data. Some degree of self-promotion, advocating what statistics can bring to the table in light of the big picture science are critical skills for an emerging leader. Being flexible, and meeting deadlines when necessary, makes you a valued and respected member of the team.

It is important to identify multiple mentors, particularly senior female mentors, and to seek feedback and suggestions from them at each step of your career, particularly when taking a leap. Female mentors who have pursued a similar trajectory are likely to understand your situation better; ask about their experiences and lessons learned. When you receive honest input from your collaborators and mentors about your work, be mature in accepting constructive criticism.

Showing passion, energy, and enthusiasm for your statistical and scientific work is important to promote cutting-edge statistics in a team and to earn respect of others. It is important to seek input from the scientific colleagues that you are on the right track toward answering the scientific question, otherwise the method you develop may be complex and intricate, but without much use for the actual research question of substantive interest. Writing novel methodology grants based on the motivating data earns respect from the collaborator. It helps to demand senior authorship in papers where statistics played a key role. Publishing papers in subject matter journals as first or senior author commands respect of the entire team and is reflective of your leadership potential. Identifying meaningful questions in the subject matter area shows independence and reflects knowledge in the science.

Focusing on the true spirit and meaning of the word "collaboration" rather than "competition" can help with transitioning to better leadership. This is one of the key management principles, according to the famous statistician, engineer, and management consultant W.E. Deming (Deming, 1993). Make the scientific team your ally so that they support and believe in the adoption of the most accurate and cutting-edge statistical methods. Reinforcing the need for integrity as well as advancement can help set the work culture of the group you will be leading. Nurturing the career of research staff and junior trainees in the group reflects that you care about the team, not just yourself.

In trying to build up a leadership portfolio, it is often key how one dealt with complex situations and conflicts in the past. If you always work for the team to reach a better solution, keeping politics and individual agendas aside, people will start noticing your skills and you will earn their trust. There is nothing more compelling than to take on a leadership position when the entire team wants you to take on the leadership challenge and wants you to succeed. Mutual respect is the first step toward good leadership. For example, here is a short story. Dr. S helped with designing a study so that it reduced the number of bio-samples the lab had to analyze by 30%, compared to what they had proposed in the original grant. This saved her scientific collaborator a substantial chunk of the budget, and in spite of initial resistance to the complex design, he finally admitted that the "statistical trickery" led to a distinct improvement in study design and subsequent analysis.

Best Practices in Leadership Roles

There is an invisible barrier between mid-career roles to highest-ranking leadership roles at the top of the collaboration pyramid. New tips and skill sets need to be acquired as one progresses to climb the ladder (the Korn/Ferry institute career playbook on women in leadership is an excellent reference). As one prepares to take on leadership responsibility, recruiting judiciously becomes incredibly important, as you now have to be responsible for the work of others. One has to learn to hire a good team of staff and analysts, involve colleagues with the right expertise, and learn to let go of everyday details and be able to trust other people's work. There should be checks and balances within the team, and a rigorous structure to ensure reproducibility of the analysis, that are followed by default, and the documentation process of quality control and analysis should be automated, rather than supervised by you at each step. Establishing a solid hierarchy and a team that works well is the first step as you contemplate taking on a leadership role in a large study.

Being involved in a study/project since its inception helps to integrate with the team, with explaining and understanding what is feasible from the data from a quantitative perspective. Designing a study is a pivotal part that a lead statistician has to be involved with for meaningful downstream analysis.

Share not just your vision but your passion about statistics with the team. Women are often subdued in their expressions. However, the fact that you love what you do and really like to see the analysis completed and papers published has a contagious effect. Good statistics is a necessary component of good science. Thus, being efficient and providing superb analysis and consultation in a timely manner is essential. Women often lack assertion and confidence when confronted. Listening to others and being respectful are desirable qualities, but a leader often has to intervene and make an executive decision. One should have that strength and courage of conviction to do so. For example, Dr. K is in a meeting where a staff statistician, Mr. Z, always interrupts her before a sentence is finished, and he expresses his disgruntlement through facial expressions. Dr. K had amicably tried to resolve the situation by allowing him to speak and listening to his concerns, but at some point she had to put her foot down as she realized he was simply bullying her. Dealing with male colleagues who are working under you can often lead to subtle altercations as a result of latent but persistent gender bias (Ibarra et al., 2013). Establishing your authority is possibly the best action in such situations.

Communication is a critical component of working in a team. For example, being dismissive of your colleagues' opinion regarding a fancy but difficult to understand plot that you created is not always the best solution. It may very likely be true that the plot does not really reflect the question she/he is trying to highlight in the paper, or is not suitable for the target journal. A compromise can be reached where you both listen to each other and use your best judgment given the context. At the same time, you cannot always over simplify or dilute an analysis that was not straightforward. Embracing complexity in analysis when necessary is a principle that needs to be taught to the team you are working with. Good communications involves more about good listening than just effectively speaking. Listen to different voices, and bring consensus to the team. Tell the team your intentions, make your expectations clear, encourage your team to share their own ideas, and explain your decisions and vision to them with clarity.

Leadership in statistics/biostatistics cores often involves providing services. This implies one often needs to put one's own agenda (advancement or ego) aside and focus on the team and the project. It means that you, the leader, will need to do what is the best for the project that may or may not be aligned with your own agenda. Furthermore, you, the leader, need to focus on the needs of your the team members, give support and time, train, understand, appreciate, mentor, and inspire them to learn more, do more, and become more. In the process, you may find yourself losing some technical expertise, and one has to come to terms with the reality of balancing methodology versus collaboration.

Another important lesson as a leader, which is often hard for women to realize, is that it is acceptable (and often necessary) to talk about money and negotiate the financial needs of the group. The project investigator may be under the impression that you can do it all, from data cleaning, to quality control, to analysis, but it is important to be very realistic about what a 10% effort level means for you, the data management and analysis needs of the project and how much time it really takes to do all the work that needs to be done for a careful analysis. Women often have problems in saying no and tend to commit to multiple projects with 5% to 10% effort level, ending up with no time to deeply think about anything nontrivial.

Finally, as a statistical leader, it is very important to have strong ethics, integrity, and emphasize the need for reproducibility in clinical research (Baggerly et al., 2008) to your team. There are many clinical investigators who know statistics to some extent and may think they know more statistics than you do. Frequently, they insist on only using simple and inadequate statistical methods that they are comfortable with. These methods may be inferior to the methods you or your team proposed or, to make it worse, these methods can be the wrong choices. There are examples of clinical investigators who may want your team members to manipulate, torture, and distort the data to "prove" their hypothesis. In fact, there are many researchers who will pressure you directly or indirectly to find ways to produce noteworthy P-values so that they can write papers and grants. You need to stand up for your team and for good science and be willing to walk away from an investigator who insists on doing things without sound principles. As a leader taking responsibility for the work of others, it is extremely crucial to work in an environment that can mirror your values and allows you to create and maintain high ethical standards.

Challenges, Particularly for Women Leaders

The challenges that one faces in large projects are diverse, depending upon the field of collaboration. In a public health setting, there is often an automated sense of authority and respect because of your statistical expertise. What becomes crucial is to impart the statistical knowledge with respect, gauging the right level of your audience. It almost introduces the notion of personalized education. Creating a space for cross-fertilization of ideas in a team of junior and senior investigators, statisticians, and nonstatisticians in a respectful way is the primary challenge in collaborations. There is a fine line between this respect and being assertive when needed. One has to choose one's battles carefully and not be constantly vigilant. Just like in a large classroom with hundreds of students reluctantly signed up for introductory statistics or probability, it often helps to describe subtly how grades are going to be assigned and who is in charge of that final outcome in the very first lecture, it may be helpful to clearly delineate expectations and evaluation metrics in a large group when one assumes a leadership role, so that the authority is attributed in a latent way.

This balancing act is also needed as you try to devise new methods for a study and want the team to adopt them. There may be resistance at various levels. The study section receiving the grant may not be immediately receptive to complex and unproven analyses; it is our job to articulate the standard analysis and emphasize the additional advantages of new and refined methods. Presenting both mundane and interesting analyses side by side only helps one to appreciate the contribution of the methodology more.

These challenges are shared by every statistician leader. However, there are challenges that are more common among woman leaders than among their counterparts. Women are more likely to lack the confidence than men. As such, it can be more challenging for women to impose their decision and intervene, which is an essential part of leadership. It is good to be transparent with leadership style, but ultimately some executive and hard decisions need to be made in a time-sensitive way. There are many books and talks available on how to train to be confident. For example, one can change how others see us, which will change how we see ourselves, by learning to change our body positions (Cuddy, 2012). One can act confidently, speak in an authoritative voice, and real confidence will follow. As women, we may also be more uncomfortable in a large group expressing our ideas or confronting others. However, in collaborative settings, it often requires statisticians to promote our study designs, interpret the statistical results, express our opinions, challenge long-held beliefs, and defend our proposed better-but-less familiar methods. A strategy that works well for many women is to discuss ideas or point of views with a few members (preferably senior) in order to gain their support first before large group meetings. These one-to-one conversations will help them to see things differently and better understand your perspective. They can then become your advocates in the large conference room full of multiple players in action.

We as women are also less likely to negotiate. It takes time to realize that it is perfectly fine to talk about money and ask for more resources! In collaborative settings, negotiations are often constant and necessary because of a variety of personalities and jobs involved. While some clinical or public health collaborators value what statisticians bring to the team, there are research groups that have limited appreciation for the intellectual contribution of statisticians. Subsequently, they may not be fair with authorship policies and may not reward you by allocating adequate effort in grant applications. Hence, we often have to negotiate authorship positions, the right amount of effort, and support staff/research assistants. These negotiations are important not only for our own but also for our team members in terms of furthering their professional growth. For example, it is important for statistical leaders to negotiate authorships for themselves and their team members. It is also important to negotiate support staff to help with simpler statistical analyses and editorial assistance and for you to delegate tasks to; otherwise, your time and effort will be spent on things that keep you from fully using your strengths to further the projects, accomplish bigger goals, or tackle more important tasks. It is almost mandatory for women leaders to learn to negotiate (Babcock and Laschever, 2009).

Women are also more likely than men to devalue their skills and talents, feel inadequate, and second-guess themselves, even though evidence shows otherwise, namely, the well-documented phenomenon of "imposter effect" (Sandberg 2012). The more we recognize that it is common to have it and the psychosocial underpinning of why we have it, the better we can deal with it. We can overcome the syndrome by recognizing our skills, understanding our accomplishments, and internalizing our strengths. There are several services available, such as "coaching for professionals" that can give you one-on-one advice on how to deal with this issue (and other leadership issues) in a more personalized fashion.

Additionally, we as women will benefit more if we learn to be more proactive in promoting ourselves. We should try to recognize our own strengths and become comfortable talking about them. We can volunteer to give presentations that showcase our skills, strength, and contributions and not consider these as blatant acts of "bragging."

It is also generally more important for women to strive for work–life balance, because they are responsible for child bearing and possibly more for child caring than partners at home. It is essential for us to figure out the temporal sequence of our goals we want in both our personal and professional lives and then try to organize/synchronize our lives and careers accordingly. Fortunately, many existing services and books are available to help. Sheryl Sandberg in *Lean In* told women to abandon the myth of "having it all" and "do not leave before leaving." Often, it may not be possible to achieve a perfectly balanced life, but it is important to evaluate what's important in life, what is temporary, and what are acceptable trade-offs. Fortunately, for statisticians, even as leaders in collaborative settings, many positions do not necessarily require us to travel frequently, and one can have a somewhat flexible schedule. However, some leadership positions are more demanding than others, some require more frequent travel than others, and some roles are less flexible than others in terms of telecommuting or virtual presence at a meeting. We are capable of coordinating and timing the needs of different leadership positions with the different stages of our lives with advanced planning and a wholesome vision for our lives. We can be present for our work and family in meaningful ways. Ultimately, it is an individual decision and optimization, and there is no general prescription for who we want to be in our lives as a whole as each of us determine our own personal equation.

Paving the Way for the Next Generation of Women

The more women we have in leadership roles, the more role models we create for the next generation of women. But first, each one of us has to survive and reach that point where it is possible to help others. Once we become a leader, we can emphasize and recognize the strengths that women bring to leadership positions. Collectively working to increase the value that we assign to those strengths, we are indirectly supporting women. We do not need to be male-like. In fact, compared with male leaders, we female leaders tend to be more collaborative rather than being competitive. We seek to meld the thinking and ideas of others, and we build consensus with our teams. We are more likely to be sensitive to others' feelings, and better at team building and creating a sense of solidarity and cohesion. These qualities will serve well as a female statistician leader in a collaborative setting where the team science prevails and will lead to a happier and more secure environment. Once in a position of power, we also have more power to recruit junior women colleagues and facilitate their growth. We can achieve this in a number of ways: primarily by being willing to act as mentors to junior women statisticians and be their advocates. We should consider practicing what we preach when we build our own teams and recruit in gender-balanced ways. We should not merely compete with men, and most importantly, we should not compete overzealously with other women. We should actually be willing to help our fellow women statisticians and listen to their needs. A strong desire to inspire and lead by example is sign of a great leader. We should make every attempt to foster gender diversity in a workplace and look for opportunities to promote qualified female individuals. We reiterate the fact that when leadership roles are shared by both men and women in a gender-balanced way, women and men can complement each other and work well together. In fact, Broschak says, "when women are not able to move ahead in firms, and when we don't see women put into new positions, they become stagnant," and "A little bit of risk is a good thing" (Cohen and Broschak, 2013).

On the other extreme is the opposite issue of senior women being particularly difficult to junior women. The trailblazing women who rose to the top often had to work extra hard and had to change themselves to survive in a male-dominated world. However, being relatively few, they probably could not change the culture around them. Now that they are there and have transformed themselves to be more "male-like", it may be difficult for them to see why other women cannot (or choose not to) modify themselves to conform to that culture. The time has probably come when a generation of women has already done their job by proving women can rise to leadership positions, and it seems wiser to engage the next generation of women in helping to change the culture.

Ultimately, we are in charge of our own careers, and women should embrace leadership roles in collaboration (or otherwise) only if it matches with their career aspirations in the long term. High-quality work, confidence, respect for others, keen interest in the science of the team, and an overall courage of conviction can help them to reach milestones in leadership. A path less traveled is always hard to embark on, but often more rewarding in terms of the ultimate journey. We wish all women leaders in statistics the very best as they continue/start/plan their personal journies.

Acknowledgments

We would like to thank our colleagues Drs. Mousumi Banerjee, Brenda Gillespie, Amy Herring, and Brisa Sanchez for sharing their experiences and thoughts as we prepared this article.

Bibliography

Babcock, L. and Laschever, S. (2009). *Ask For It: How Women Can Use the Power of Negotiation to Get What They Really Want.* Bantam Books: New York.

Baggerly, K.A., Coombes, K.R., and Neeley, E.S. (2008). Run batch effects potentially compromise the usefulness of genomic signatures for ovarian cancer. *Journal of Clinical Oncology.* 26:1186–7.

Bosco, J.L., Silliman, R.A., Thwin, S.S., Geiger, A.M., Buist, D.S., Prout, M.N., Yood, M.U., Haque, R., Wei, F., and Lash, T.L. (2010). A most stubborn bias: No adjustment method fully resolves confounding by indication in observational studies. *Journal of Clinical Epidemiology.* 63:64–74.

Cohen, L.E. and Broschak, J.P. (2014). Whose jobs are these? The impact of the proportion of female managers on the number of new management jobs filled by women versus men. *Administrative Science Quarterly.* 58(4):509–541.

Cuddy, A. Your body language shapes who you are. [Video file]. Retrieved from http://www.ted.com/talks/amy_cuddy_your_body_language_shapes_who_you_are?language=en. March 21, 2015

Deming, W. E. (1993). *The New Economics for Industry, Government, and Education.* Boston, MA: MIT Press. p. 132.

Ibarra, H., Ely, R., and Kolb, D. (2013). Women rising: The unseen barriers. *Harvard Business Review*, 91(9):60–66.

Kahnweiler, J.K. (2013). *The Introverted Leader.* Berrett-Koehler Publishers: San Francisco.

Kellerman, B. and Rhode, D.L. (2007). *Women and Leadership: The State of Play and Strategies for Change.* Wiley and Sons: San Francisco.

McElwee, R.O. and Yurak, T.J. (2012). The phenomenology of the impostor phenomenon. *Individual Differences Research.* 8(3):184–197.

O'Konnor, S. (2007). Forword. *Women and Leadership: The State of Play and Strategies for Change.* Wiley and Sons: San Francisco.

Orr, J.E. (2013). *A Career Playbook for Women in Leadership.* Korn/Ferry Institute: Los Angeles.

Singer, T., Seymour, B., O'Doherty, J.P., Stephan, K.E., Dolan, R.J., and Frith, C.D. (2006). Empathic neural responses are modulated by the perceived fairness to others. *Nature.* 439:466–469.

Sandberg, S. and Scovell, N. (2013). *Lean In: Women, Work, and the Will to Lead.* Alfred A. Knopf: New York.

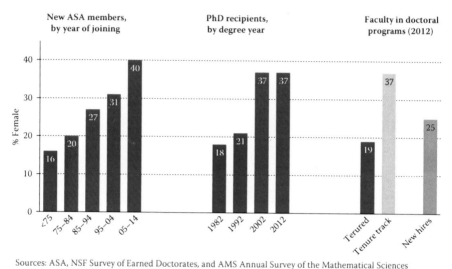

Sources: ASA, NSF Survey of Earned Doctorates, and AMS Annual Survey of the Mathematical Sciences

FIGURE 27.1
Some statistics on women in statistics.

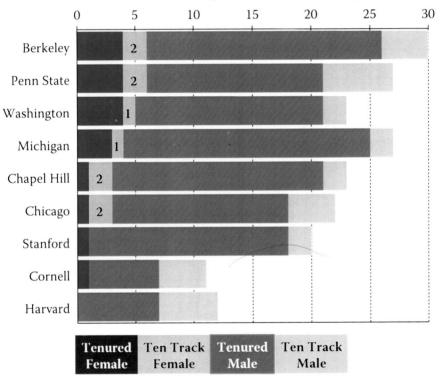

Number of ladder faculty in the "top"*
9 US statistics departments, by rank and gender

*As ranked by the National Research Council, 2010

FIGURE 27.2
Number of ladder faculty in the top 9 US statistics departments, as ranked
by the National Research Council (2010), by rank and gender.

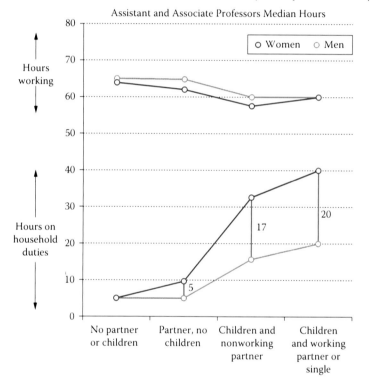

Work/Life Balance
Gender gaps in time spent on household duties for
Harvard faculty members with children

Analysis restricted to faculty who provided data on both hours worked and hours
spent on household duties. Source: Harvard Faculty Development & Diversity Office

Assistant and Associate Professors Median Hours

FIGURE 27.3
Gender differentials in work–life balance.

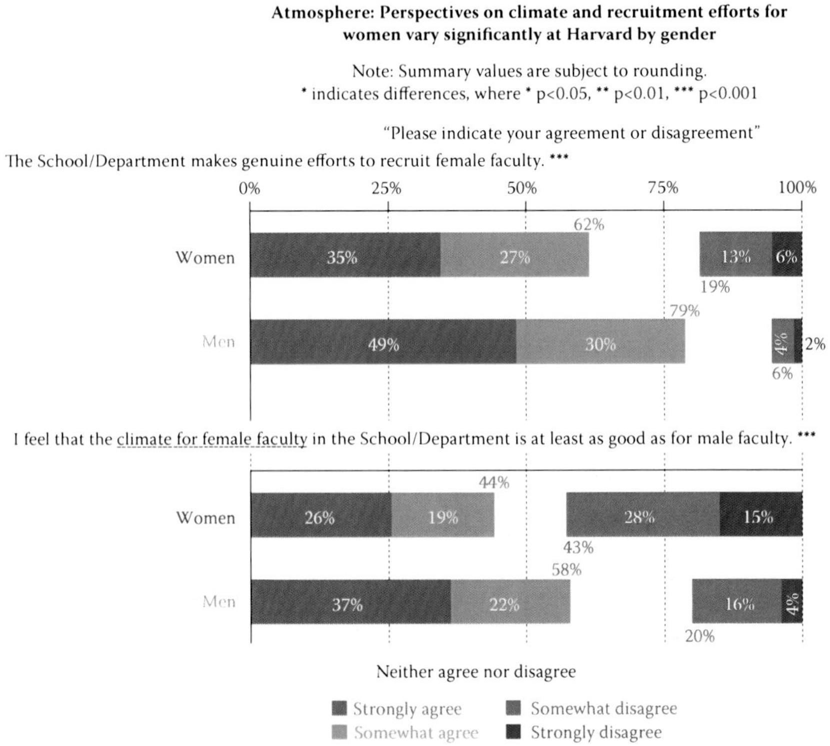
Atmosphere: Perspectives on climate and recruitment efforts for women vary significantly at Harvard by gender
Note: Summary values are subject to rounding.
* indicates differences, where * p<0.05, ** p<0.01, *** p<0.001
"Please indicate your agreement or disagreement"
The School/Department makes genuine efforts to recruit female faculty. ***
0%
25%
50%
75%
100%
Women
35%
27%
62%
13%
6%
19%
Men
49%
30%
79%
4%
2%
6%
I feel that the climate for female faculty in the School/Department is at least as good as for male faculty. ***
Women
26%
19%
44%
28%
15%
43%
Men
37%
22%
58%
16%
4%
20%
Neither agree nor disagree
Strongly agree
Somewhat agree
Somewhat disagree
Strongly disagree

FIGURE 27.4
Climate survey data.

10

Leadership in Statistical Consulting

Duane L. Steffey

Exponent, Inc.

Public consciousness of the statistics profession has perhaps never been greater than it is today. Big data, predictive analytics, data science — all these terms have been coined and have risen in popularity within the past decade or so. This rising visibility presents both opportunities and challenges for consulting statisticians. (Consistent with the American Statistical Association's Ethical Guidelines for Statistical Practice, the term "statistician" shall be read to include all practitioners of statistics, regardless of job title or field of degree.) Prospective clients in academia, government, and industry are now more likely to recognize the potential value of bringing statistical thinking to bear in the solution of their problems, but they are also more likely to seek transfer of the consultant's technology as part of the engagement. Responding to this growing consciousness and demand for services are consultants from diverse backgrounds and with varying levels of training and professional competence in statistical theory and methods.

How does the consulting statistician distinguish herself or himself in this dynamic marketplace? How does she or he demonstrate leadership by creating exceptional value for clients, by developing the next generation of consulting statisticians, and by building the foundation for a successful and sustainable consulting business?

Overview of Statistical Consulting

Statistical consulting takes place both within and outside academia. Consultants vary in background and experience, clients vary in sophistication and appreciation of statistical science, and organizations vary in how closely their culture and operating procedures are aligned with a consulting mission.

Consulting in Academia

For university statisticians, especially those with interests in applications or working outside traditional departments of statistics, consulting is a familiar but not always formally recognized role. Statisticians are routinely sought out by academics in other fields for assistance with the design and analysis of research studies. In exchange, the value offered may be scholarly in nature (e.g., a joint publication in the field of application), financial (e.g., salary support from a funded grant), or both. Statisticians working in graduate schools of business, medicine, and public health may be expected to provide such services, regardless of the degree of methodological novelty, either personally or through their students. In addition to relying on tenured faculty, some universities have hired statisticians into staff positions, often with core funding support, to improve the statistical quality of research done on campus.

Consulting vs. Advice vs. Collaboration

Academic statistical consulting varies not only in its location within the institution, but also in the breadth of services provided. Although this chapter adopts a very general definition for the term "consulting," some universities mark a distinction between consulting and collaboration — the latter intended to imply a larger role for the statistician in identifying and addressing the key questions motivating the research. In contrast, when the clients are graduate students in other departments seeking assistance in their dissertation research, the services provided may be limited to technical advice, with the student bearing personal responsibility for implementation.

Consulting Laboratories and Courses

In the United States, the creation of statistical consulting laboratories in academia dates back to 1933, when George Snedecor established the first such laboratory at Iowa State University in association with the inauguration of the nation's first academic statistics department. Today a plaque on the campus marks the historical site:

> Established in 1933, the Statistical Laboratory at Iowa State was the first research and consulting institute of its kind in the country. Early emphasis on agriculture data analysis has continued, however the laboratory's range of activities now includes statistical activities in social, physical, engineering and biological sciences for clients worldwide.

Jerzy Neyman also created a similar laboratory shortly after arriving in 1938 at the University of California in Berkeley. At the time, no academic departments of statistics existed in the United States, and laboratories were seen as a way to bring the discipline visibly into the university.

University consulting laboratories today vary considerably in their missions. Some laboratories serve the campus community exclusively; others accept engagements with off-campus clients. Laboratories receiving core support from an academic unit (e.g., school or college) may provide services at no charge. Self-supporting laboratories will typically operate on a fee-for-service basis.

An effectively operating consulting laboratory on a university campus provides value to all parties involved. Clients seeking statistical support have an identifiable source of competent technical advice and services. Statistics graduate students employed at the laboratory, instead of as research or teaching assistants, gain experience in a potential career path. Faculty and staff encounter new applied problems that may inspire the development of new methodology or new areas of application for existing methods. Revenues generated by laboratory activities constitute an additional source of funds to support the larger mission of the sponsoring academic unit.

Formal courses in consulting were added more recently but now are part of the curricula of many of the nation's most eminent graduate programs in statistics. The content and structure of such courses varies, but they can complement the activities of an on-campus laboratory, provide exposure to the process of consulting and communicating with nonstatisticians, and offer students an opportunity to apply their classroom knowledge on tangible research problems.

Consulting in Industry and Government

Most statisticians working in industry or government positions are not employed by organizations offering consulting services to external clients. However, many are functioning effectively as internal consultants to their colleagues who possess subject-matter expertise consistent with the mission of the organization. The similarities and differences between intramural and extramural consulting are instructive.

Intramural Consulting

For decades, many of the most successful American companies, from multinational conglomerates to industry leaders in particular sectors, have maintained in-house groups staffed by statisticians, most of whom hold advanced degrees in statistics or have acquired significant on-the-job expertise in one or more subject-matter areas of application. Frequently, such groups are housed within the company's research and development division, but they may also be found in such areas as manufacturing, product safety, or regulatory affairs. Under such an organizational structure, projects are typically staffed in a matrix format in which statisticians and other professionals are deployed as needed to execute individual projects. Alternatively, statisticians may be embedded within specific product lines instead of an identified central consulting unit. One potential disadvantage of such an arrangement is the effective limitation it may place on the statistician's access to peer advice.

Intramural consultants typically do not face external competition from statisticians at other firms, and there may be strong corporate incentives to leverage the resident statistical expertise. Nevertheless, the fundamental dynamics of the client–consultant relationship still apply, in that the statistician's ability to bring demonstrable value to projects will facilitate rapport in the present engagement and increase the volume of repeat business and referrals.

Extramural Consulting

Statisticians also hold positions in diverse companies organized to provide professional consulting services to external clients. In the United States, such firms include many of the largest companies engaged in business and management consulting, as well as smaller engineering and scientific consulting firms with more specialized technical expertise. In these organizations, statisticians typically constitute a small fraction of the consulting staff. Although they may engage directly with external clients seeking statistical support, these consultants may also build strong collaborative relationships with other consultants who perceive value in teaming with statisticians to execute projects and to pursue new business opportunities.

At the smaller endpoint of the size range are companies composed exclusively, or with a majority, of statisticians. Often, such companies have developed a recognized niche expertise and may be led by a consulting statistician with a prominent reputation in a particular industry sector or an extensive clientele cultivated over many years of experience.

Sole Proprietorships

Because most statistical consulting services can be provided effectively using only a personal computer, there are relatively few barriers to statisticians entering the market as individual consultants. Being your own boss brings a degree of freedom, of course, but also a series of practical issues associated with running a business. Sole proprietors must decide whether to incorporate, whether to carry liability insurance, and how to replace such common employee benefits as health insurance and retirement accounts. Typically, they will not have access to the administrative support available at larger firms for contract review, billing, and accounting. Also, the scale of projects in which sole proprietors are engaged may be effectively limited by their available resources.

Leadership in Consulting Practice

How do we exercise effective leadership in statistical consulting? In view of how many statisticians essentially work as consultants, interacting with nonstatisticians to solve applied problems, the question ought to be an important one for the profession.

The development of a successful and sustainable consulting practice in statistics is an organic, evolutionary process that can take years, if not a decade or more. The process shares many similarities with those in traditional professions, such as medicine and law. The consultant enters her or his first engagement with a client and, if all goes well, may reasonably expect additional work from that client and referrals of new clients. Such a process takes time.

Key presentations and publications, highlighting the consultant's expertise in addressing a frequent or significant problem, can accelerate the business development process. Whether advertising in lieu of professional activity is similarly effective can be debated, but some firms do invest in prominent advertising campaigns.

While leadership theorists may argue over whether great leaders are born or made, the more experience a statistician has acquired with the consulting process, the stronger is her or his foundation for assuming a leadership role in professional practice.

The Consultant/Client Relationship

Leadership has been described by Chemers (1997) as "a process of social influence in which one person can enlist the aid and support of others in the accomplishment of a common task." But who is leading the consulting project? The client, who has identified the problem and determined the statistical consultant to have the requisite skills to produce the solution? Or the consultant, on whom the client is relying for technical leadership? By this definition of leadership, the relationship between the client and consultant is more complex and defies simple characterization in terms of leader and follower.

One interpretation (Hahn, 1992) is that the statistician has reached a position of proactive leadership in statistical consulting when she or he is viewed by key clients as a counselor and source of conceptual as well as technical advice — someone who helps to develop a more precise definition of the problem and how to address it. Others (Derr, 2009) have argued that for consulting statisticians to exert leadership and have a significant impact in multidisciplinary settings, it is essential they possess the vision of a broad role for themselves in the enterprise, as expressed in the vision statement of an internal statistical consulting group in a manufacturing company: "To create a culture where statistical methods are routinely used in the decision-making process and statisticians are partners within the company organization."

Qualities of Effective Consultants

It is one thing to express a broad vision for the role of the consulting statistician. But how does one actually realize that vision? Effective consultants possess excellent technical and communication skills, a genuine interest in solving applied problems, and a sense of entrepreneurship — i.e., the ability to sell the value of one's profession and oneself. Many articles and books that have been written about statistical consulting expand on that summary. A recent list of traits possessed by successful statisticians (Hahn and Doganaskoy, 2012) clearly applies to those engaged in consulting:

- Strong analytical and technical skills

- Communications and related skills

- Ability to size up problems and see the "big picture"

- Flexibility

- A proactive mind-set

- Persistence

- A realistic attitude

- Enthusiasm and appropriate self-confidence

- Ability to prioritize, manage time, and cope with stress

- Team skills

- Leadership skills

- Ability to properly apply and adapt knowledge

- Passion for lifelong learning

To this list, one could add professionalism, responsiveness, judgment, situational awareness, good listening, an understanding of the roles and responsibilities of involved parties, a willingness to involve other consultants in service of the project, and a reputation for both technical excellence and integrity (Morganstein, 2012).

Leadership of Consulting Statisticians

Consulting statisticians in senior positions within academia, government, and industry — with the possible exception of sole proprietors — bear a professional responsibility for the development of more junior statisticians in their universities, agencies, and companies. Academic statisticians have the opportunity to introduce students to consulting, not only as a means of broadening their skills beyond the traditional classroom, but also as a viable career path after graduation. By helping their colleagues to reach their full potential, consulting statisticians in government and industry can make a positive impact on the performance of their organizations extending well beyond their individual contributions in direct service of clients.

Education in University Programs

The traditional undergraduate and graduate curricula in statistics are designed to impart the knowledge of statistical concepts and methods required for students to function effectively as professional statisticians. The best opportunities for students to learn complementary, nonstatistical skills come through involvement in multidisciplinary research teams, internships with organizations outside the university, and direct experience in statistical consulting.

Today, more than ever before, university programs in statistics include a consulting course, typically at the graduate level. Several well-received textbooks (Derr, 2000; Cabrera and McDougall, 2002) are available for use. Course elements may include video examples, review of case studies, classroom visits by experienced consultants and by clients, and structured consulting experiences. Such courses can offer students valuable opportunities to practice speaking, presenting, and writing about statistics to a nonstatistical audience (Derr, 2009).

Mentoring of Government and Industry Consultants

A key responsibility of senior consulting statisticians in government and industry positions is the recruitment and mentoring of new colleagues. Some organizations, such as industrial consulting firms, operate formal mentoring programs for new colleagues with limited prior experience, typically after an initial period of establishment (e.g., 12 to 18 months). In other organizations, training happens more informally as on-the-job learning during the execution of projects.

Although the details of implementation will vary, successful mentoring of new consultants can be characterized by several key elements:

- *Peer review* — the new consultant's technical work is reviewed by a more experienced colleague. Some organizations maintain a formal requirement for peer review of draft work products as part of their quality management system. Alternatively, the new consultant may perform peer review as a means of gaining experience with new problem areas and methods of solution.

- *Increasing direct contact with clients* — the new consultant may attend client meetings led by more experienced colleagues, in preparation for roles involving more significant responsibility for maintaining the client relationship.

- *Expanding autonomy over work content* — as the new consultant becomes familiar with technical aspects of typical problems encountered in the organization, she or he assumes greater independence in developing and implementing study designs and data analysis plans.

- *Growing personal reputation* — clients within and, if applicable, outside the organization develop an appreciation for the value and any distinctive expertise the new consultant brings to projects. A new consultant's external reputation can be enhanced substantially by conference presentations and peer-reviewed publications — an especially important consideration for firms engaged in extramural consulting.

A recent conference proceedings paper (Walsh and Hooker, 2009) succinctly describes the mentoring process at one government agency:

> Newer team members generally assume an observatory role and are guided by more senior members. A team member's level of involvement in project-related decisions naturally increases as they gain experience and confidence.

Professional Activities and Communications

Statistical consulting has long been recognized as an important activity of the profession. W. Edwards Deming, "probably the best-known statistician, in the eyes of the general public, who ever lived," achieved prominence largely because of his effectiveness and influence as a consultant (Hahn, 2002). A recent search of the Current Index to Statistics produced a listing of more than 250 articles related to consulting published over a period of nearly 60 years. This published literature, reflecting the collective wisdom of hundreds of experienced statisticians, constitutes a valuable resource in the leadership and education of aspiring consultants.

Attention to consulting by statisticians has increased considerably over the past three decades. A newsletter titled "The Statistical Consultant" appeared in 1984. As explained in the first issue ("The Statistical Consultant," 1984), "The newsletter was an outgrowth of the November 1982 Wisconsin workshop on statistical consulting internship programs and a follow up meeting at the 1983 ASA (American Statistical Association) national meeting in Toronto. The participants at these meetings recognized the importance of university based statistical consulting laboratories and internship programs in education."

After 30 years, the newsletter continues today as a publication of the ASA's Section on Statistical Consulting, which was officially recognized in 1991 and now counts more than 1,500 members.

In February 2012, the ASA launched an annual Conference on Statistical Practice and, later in that same year, included a featured section on consulting in its newsletter's annual career issue. Such activities and communications complement other efforts by statisticians, perhaps encouraged by greater public awareness of the profession, to seek a greater voice on significant public issues (American Statistical Association, Annual Operating Plan, 2009). The opportunities for statistical leadership appear to be multiplying.

Applicable Theories of Leadership

The exertion of leadership in statistical consulting requires the strong personal, organizational, and visionary skills that characterize successful leaders (Hahn and Doganaksoy, 2012). But how does a statistician develop and apply these skills in the consulting environment? What insights can be drawn from the general study of a subject that has occupied researchers for centuries?

Individual Traits

Early studies of leadership proceeded from the assumption that leadership is rooted in the characteristics possessed by certain individuals — known as the "trait theory of leadership." Although trait-based perspectives have shortcomings and limitations (Zaccaro, 2007), they have reemerged as a prominent component of contemporary research. One popular work (Gardner, 1990) identified 14 attributes as being associated with leadership:

1. Physical vitality and stamina

2. Intelligence and action-oriented judgment

3. Eagerness to accept responsibility

4. Task competence

5. Understanding of followers and their needs

6. Skill in dealing with people

7. Need for achievement

8. Capacity to motivate people

9. Courage and resolution

10. Trustworthiness

11. Decisiveness

12. Self-confidence

13. Assertiveness

14. Adaptability/flexibility

Prefacing this list is the observation, "the attributes that follow are not present in every leader. The importance of the attribute to effective leadership varies with the situation." The proposition that people who are leaders in one situation may not necessarily be leaders in other situations has motivated research not only on traits of leaders but also on effective leader behaviors.

Behaviors and Styles

The situational theory of leadership (Hemphill, 1949), which appeared as a reaction to the trait theory, holds that different situations call for different characteristics, implying no single optimal psychographic profile of a leader exists. Synthesizing the trait and situational approaches, some researchers in more recent years have developed contingency theories in which the style of leadership is defined as contingent to the situation.

One prominent contingency leadership theory, path-goal theory (House, 1996), argues that leaders must engage in different types of leadership behavior, depending on the nature and the demands of a particular situation. "It is the leader's job to assist followers in attaining goals and to provide the direction and support needed to ensure that their goals are compatible with the organization's goals." The original path-goal theory (House, 1971) identifies four different types of leader behaviors:

- *Directive* — the leader lets followers know what is expected of them and tells them how to perform their tasks.

- *Achievement-oriented* — the leader sets challenging goals for followers, expects them to perform at their highest level, and shows confidence in their ability to meet this expectation.

- *Participative* — the leader consults with followers and asks for their suggestions before making a decision.

- *Supportive* — the leader shows concern for followers' psychological well-being, and her or his behavior is directed toward the satisfaction of subordinates' needs and preferences.

High-Performance Teams

For statistical leaders employed in consulting organizations, a key element of their mission is to build high-performing teams of colleagues to accomplish the organization's goals. The work environment of high-performance teams has been described (Burgin, 2001) as comprising six essential components:

1. *Clarity* — employees understand unit/team goals, policies, and job requirements, and they feel things are well organized and run smoothly.

2. *Commitment* — people are committed to goal achievement and continually evaluate their own performance against goals.

3. *Excellence* — management sets high standards of performance and applies pressure to continue improvement.

4. *Responsibility* — employees feel they are personally responsible for their work and also that individual initiative is encouraged.

5. *Recognition* — employees feel they are recognized and rewarded for doing good work.

6. *Teamwork* — employees feel they are part of a team and take pride in belonging to the unit.

One contemporary approach to leadership training, therefore, involves determining which styles of leadership contribute to building a strong team climate and then working to adopt those styles consistently in the organization. In this formulation, the *directive style* of leadership (an apparent rephrasing of the achievement-oriented leader behavior in the path-goal theory) is the principal driver of clarity and excellence. The *participative style* is the principal driver of commitment and teamwork, while the *coaching (supportive) style* is the principal driver of responsibility and recognition.

Centering Principles

Another contemporary theory of principle-centered leadership has been developed in a series of popular books (Covey, 1989 and 1990) on personal development. As an alternative to the "personality ethic," this theory advances the "character ethic," based on the fundamental idea that there exist principles governing human effectiveness. These principles are described as deep, fundamental truths having universal application — guidelines for human conduct proven to have enduring, permanent value. Although not intended as an exhaustive list, numerous principles are named in the books: fairness, integrity, honesty, human dignity, service, quality, excellence, potential, growth, patience, nurturance, and encouragement (Covey, 1989).

This theory holds the more closely a leader's values and paradigms are aligned with these principles or natural laws, the more effective will be her or his leadership. The model for principle-centered leadership is depicted graphically as a set of four expanding circles, from internal to external, representing effectiveness at four levels: personal, interpersonal, management, and organization. Key concepts are trust, empowerment, and alignment — the latter referring to an overlap between individual and organizational goals. The theory postulates an organization structured around trust and empowerment will produce effective individuals who share a common mission of working to support their own personal growth and the growth of the organization (Shriberg, Shriberg, and Lloyd, 2001).

Ethical Considerations in Statistical Consulting

If a consulting statistician aspires to principle-centered leadership, how should she or he proceed? How do general principles, regarded as deep and fundamental truths, find specific expression in the practice of statistics?

Ethical Guidelines of Professional Societies

In the United States, the effort to develop a code of ethical statistical practice began 65 years ago, when an ASA committee in 1949 first recommended the development of such a code. With contributions from many statisticians over more than three decades, the ASA's ethical guidelines for statistical practice were approved in 1981 and revised most recently in 1999. Other professional associations — including the International Statistical Institute, the Royal Statistical Society (United Kingdom), the Statistical Society of Australia, and the Statistical Society of Canada — have similarly established codes of ethics. The stated purposes of the ASA guidelines are (1) to help statisticians make and communicate ethical decisions and (2) to inform employers of statisticians and those relying on statistical results of expected standards. Others (Jowell, 1981) have noted development of a professional code may create a stronger professional identity among statisticians and correct the misperception of some who view statistics as a "mischievous and meddlesome discipline."

The preamble to the ASA guidelines expresses the view that all statisticians have a social obligation to perform work in a professional, competent, and ethical matter. Also included is an acknowledgment that application of ethical guidelines generally requires good judgment and common sense. The subsequent guidelines address professionalism and seven distinct areas of responsibility:

- to funders, clients, and employers;

- in publications and testimony;

- to research subjects;

- to research team colleagues;

- to other statisticians or statistical practitioners;

- regarding allegations of misconduct; and

- of employers.

All these areas of responsibility are potentially applicable to consulting statisticians. Summary versions of the guidelines have been presented elsewhere (Hogan and Steffey, 2014) and are repeated here for the first key area of responsibility. With respect to funders, clients, and employers, the ethical statistician shall:

1. Present choices among valid approaches varying in scope, cost, or precision.

2. Clearly state relevant qualifications/experience.

3. Clarify respective roles of study contributors.

4. Explain consequences of deviation from study plan.

5. Apply data collection and analysis procedures without predetermining the outcome.

6. Make new statistical knowledge widely available.

7. Guard privileged information.

8. Fulfill all commitments.

9. Accept full responsibility for performance.

How well are the general principles identified in the discussion of principle-centered leadership embodied in these ethical guidelines?

Case Studies from Statistical Practice

The first ad hoc committee created by the ASA to formulate a code of ethical practice recommended (Clausen et al., 1954) use of a method based on the "critical incident" technique developed and used successfully by the American Psychological Association. The procedure would involve soliciting from members brief, written descriptions of specific incidents in their experience that raised ethical issues. From these descriptions, preliminary principles and standards would be derived, published, and discussed at professional meetings. The committee reasoned that standards, to be effective, would require substantial agreement among members of the association and judged this procedure as the one best designed to assure consensus.

Today, the ASA Committee on Professional Ethics continues to maintain a set of exemplar case studies (Gardenier, 1987), contributed by the association's membership:

> As a professional statistician, you are called by a colleague to examine and "bless" a biomedical experimental report. You are urged to do it quickly because the report has already been submitted and accepted for publication in a prestigious journal in the author's field. One of the reviewers, however, had suggested that a quick review by a statistician might be in order. To your horror, the report appears to be utter statistical nonsense. The data were not sampled according to any plan, but rather were drawn from various similar experiments done for different purposes. There is no reason to assume the observations were random or independent within or among data sets. There was no definition of how many data points had been originally available or how those used had been selected. The scatter plots within the paper were plainly skewed, but the computer statistical tests which had been run would have presumed a normal distribution. You explain gently that the statistical work is not an asset to the paper and could prove embarrassing to the author and the institution if published. You suggest that he eliminate the

statistical portions and describe his work based on the qualitative reasoning which he obviously used. Initially very angry, he calms down and says, "I'll leave the contents alone, but I will add you as a coauthor. How's that?"

How we respond as consultants to challenging situations reflects on us as individuals and also, to some degree, on the culture of the statistics profession. By adhering to a code of ethical practice based on enduring ideals and principles, we may be placing ourselves in the best position to exert effective leadership in statistical consulting.

A Look Ahead: Challenges and Opportunities

This chapter began with the observation of the general public's growing awareness and appreciation of the influence of statistical concepts and methods on diverse aspects of contemporary life — ranging, for example, from targeted online advertising to new discoveries in medicine. Such recent successes have brought more recognition to statisticians but also drawn competing approaches to problem-solving from other disciplines, perhaps most notably from computer science.

A recent editorial (Matloff, 2014) by an academic statistician characterized many of his colleagues, as well as the ASA leadership, as worried that the field of statistics is being eclipsed by these other disciplines and "is headed for a future of reduced national influence and importance." The editorial proceeds to advance systemic reasons why attempts by computer scientists to apply statistics tend to produce poor results, including limited knowledge of the statistics literature, a cavalier attitude toward underlying models and assumptions, and a tendency toward oversimplification. The larger concern is that resources for research are being allocated disproportionately to computer science, even though statistics is arguably better equipped to make use of them. Among suggestions to address the portrayed state of affairs, the editorial recommends that statistics researchers be much more aggressive in working on complex, large-scale, "messy" problems.

A similar message seems appropriate for consulting statisticians, who are among those on the front lines applying the tools of the discipline to solve important, tangible problems for clients. Facing competition from consultants in other fields, statisticians should be aggressive in pursuing consulting engagements when they can offer superior value to prospective clients. Consultants should be skilled technicians, articulate communicators, and ethical professionals in their dealings with colleagues and clients. In addition to delivering specific solutions for their current clients, consulting statisticians should strive to present and publish general versions of their solutions in places where they can reach a larger audience. Such qualities and actions essentially describe effective leadership in statistical consulting — to the mutual benefit of the client, the consultant, their respective organizations, and society in general.

Bibliography

American Statistical Association, Annual Operating Plan 2010—Visibility and Impact, Alexandria, VA, June 19, 2009.

Burgin, A.L. (2001). Leading Units and Teams for High Performance. http://www.hrmginc.com/.

Cabrera, J. and McDougall, A. (2002). *Statistical Consulting*. Springer-Verlag, New York.

Chemers, M. (1997). *An Integrative Theory of Leadership*. Lawrence Erlbaum Associates, cited at http://en.wikipedia.org/wiki/Leadership.

Clausen, J.A. et al. (1954). Report of the Ad Hoc Committee on Statistical Standards. *The American Statistician*, 8, 19–23.

Covey, S.R. (1989). *The Seven Habits of Highly Effective People: Restoring the Character Ethic*, Simon & Schuster, New York.

Covey, S.R. (1990). *Principle-Centered Leadership: Strategies for Personal and Professional Effectiveness*. Simon & Schuster, New York.

Derr, J. (2000). *Statistical Consulting: A Guide to Effective Communication*. Duxbury, North Scituate, MA.

Derr, J. (2009). Professional Development for Statisticians: Useful Skills in a Multi-Disciplinary Setting. *Statistical Society of Ottawa, Spring Symposium*, April 8.

Gardner, J.W. (1990). *On Leadership*. Free Press, New York.

Gardenier, J. (1997). Toward a Statistical Ethics Casebook. *ASA Proceedings of the Joint Statistical Meetings*.

Hahn, G.J. (2002). Deming and the proactive statistician. *The American Statistician*, 56, 290–298.

Hahn, G.J. and Doganaskoy, N. (2012). Traits of a successful statistician. *Amstat News*, June, 25–28.

Hemphill, J.K. (1949). *Situational Factors in Leadership*, Columbus: Ohio State University Bureau of Educational Research, cited in http://en.wikipedia.org/wiki/Leadership.

Hogan, H. and Steffey, D. (2014). Professional ethics for statisticians: an organizational history, *ASA Proceedings of the Joint Statistical Meetings*.

House, R.J. (1971). A path-goal theory of leader effectiveness. *Administrative Science Quarterly*, 16, 321–339.

House, R.J. (1996). Path-goal theory of leadership: lessons, legacy, and a reformulated theory. *Leadership Quarterly*, 7, 323-352, cited in http://en.wikipedia.org/wiki/Path%E2%80%93goal_theory.

Jowell, R. (1981). A Professional Code for Statisticians? Some Ethical and Technical Conflicts. International Statistical Institute, 43rd Biennial Session, Buenos Aires, December 1.

Matloff, N. (2014). Statistics losing ground to computer science. *Amstat News*, November, 25–26.

Morganstein, D. (2012). Consulting best practices. *Amstat News*, September, 26–27.

Shriberg, A., Shriberg, D., and Lloyd, C. (2001). *Practicing Leadership: Principles and Applications*, 2nd edition, Wiley, New York.

The Statistical Consultant, Vol. 1, June 1984, available for download at http://community.amstat.org/cnsl/documents/newsletters/archive/1984-1999.

Walsh, R. and Hooker, R. (2009). Statistical consulting within the Internal Revenue Service. *ASA Proceedings of the Joint Statistical Meetings*, 2246–2252.

Zaccaro, S.J. (2007). Trait-based perspectives of leadership. *American Psychologist*, 62, 6–16.

11

Women Leaders in Federal Statistics

Marilyn M. Seastrom

National Center for Education Statistics

Introduction

Women leaders in the field of statistics have made important contributions over the past 50 years by encouraging new professional connections, by challenging the status quo, by promoting new technologies, by applying modern concepts in project management and organizational structure, and by addressing policy and legislation around critical topics such as confidentiality and data sharing. This chapter on the role of women statisticians as leaders in the federal government will look at how statisticians, especially women statisticians, can best position themselves to help solve our nation's challenges. The goals and organization of the federal statistical system, the role of the US chief statistician, issues in leadership of federal statistical agencies and that of women especially, and the journeys of five women leaders will be recounted in this chapter. The chapter concludes with my own career path and the ways these women role models contributed to my success and the success of the initiatives that I have promoted through my work as chief statistician at the National Center for Education Statistics. It is my hope that women who are contemplating a career in statistics, those who may be wrestling with the question of when to step up and when and where to gain support, those women who are currently in federal service, and women generally interested in the challenges of organizational leadership will gain something from this reading.

Considering what the principal federal statistical agencies should and should not do as the providers and keepers of objective, impartial, and high-utility information collected for statistical purposes, the National Academy of Science Committee on National Statistics (CNSTAT) first released "Principles and Practices for a Statistical Agency" in 1992, with the most recent edition published in 2013. The principles specify that federal statistical agencies must function in a manner that maintains a strong sense of credibility and trust with data users and providers. Related to this, federal statistical agencies must provide objective, accurate, and timely data that are relevant to issues of public policy, while at the same time remaining independent from political and other external influences. These principles are relevant to the question of how statisticians can best position themselves to help solve our nation's challenges. Statisticians in federal statistical agencies must always balance the need for solid, policy-relevant information with the requirement that they stay above the fray of policy recommendations and decisions. Women who find themselves in a leadership role in such agencies bring their own value and challenges to these activities.

Background

To set the stage for considering the role of women statisticians as leaders in the federal government, I would like to take a step back to set the context of how the federal statistical system is organized. First, in many other countries, our colleagues work in centralized statistical agencies, with one head of the country's statistical system, one statistical/survey methodology unit, one dissemination unit, and additional units covering major subject matter topics. In contrast, the US federal statistical system is comprised of some 13 separate principal statistical agencies and the Chief Statistician's Statistical and Science Policy Branch in the Office of Management and Budget (OMB). As a result, the US has the chief statistician at OMB, 13 statistical agency heads, 13 statistical/survey methodology units, 13 dissemination units, and myriad senior staff in leadership positions (163).[1] In addition, there are another 116 programs in the Executive Branch that conduct statistical activities within a program agency.[2]

At the center of federal statistics, the US chief statistician is responsible for coordinating the activities of the Federal statistical system, to promote efficiency and effectiveness within the system; and to ensure the integrity, objectivity, impartiality, utility, and confidentiality of information collected for statistical purposes. The Statistical and Science Policy Branch evaluates the statistical agency budgets and reviews and approves federal agency information collections that involve statistical methods. The US chief statistician chairs the Interagency Council on Statistical Policy (ICSP). The council's membership includes the heads of the 13 principal statistical agencies. The ICSP identifies priorities for improving programs, and promotes integration, coordination and collaboration across the federal statistical system. The Statistical and Science Policy Branch, in collaboration with ICSP, establishes government wide statistical policies and standards. The Statistical and Science Policy Branch and ICSP also support additional statistical activities and collaborations through the work of the Federal Committee on Statistical Methodology (FCSM), the Statistical Community of Practice and Engage-

---

[1] This count of 163 is based on a review of the "About Us" sections of websites of the 13 federal statistical agencies.

[2] Statistical Programs of the United States Government: Fiscal Year 2014. OMB, Washington, DC.

ment (SCOPE), and a number of subject-specific interagency committees and working groups.

Taken together, the 129 statistical agencies and programs are responsible for all aspects of data collection, processing, tabulation, analysis, research, evaluation, program management, and information dissemination. Despite this broad mission, in nondecennial census years, the budget resources spent of federal dollars is only about 0.04 percent of the gross domestic product (GDP). This relatively modest investment provides the data that are used to drive the expenditures of billions of federal and private sector dollars through measures of prices and the cost of living programs, and through population counts of the various populations to be served by specific programs. This investment in the federal statistical system also provides information to inform policy makers and the public about the social and economic health of the nation and its components.

Role of Women in Leadership Positions in the Federal Statistical System

For the last 12 years, the position of chief statistician of the United States has been held by Katherine Wallman. As noted above, she leads the federal statistical system. In a 2010 interview in *Science News*, Wallman talked about the role of the chief statistician, emphasizing the fact that her office in the Executive Office of the President at OMB does not produce data (*Science News*, October 22, 2010). Instead, she noted that her office "provides oversight, coordination, priority settings and standard settings for [national statistics]." Wallman described her role as focusing on "improving our statistical programs and how the budget is affected by that."

In a presentation on the "United States Federal Statistical System: Coordination, Cooperation, and Collaboration" Wallman noted that she and her staff identify priorities for improving programs, establish statistical policies and standards, and evaluate statistical programs for compliance with OMB guidance.[3] Wallman stressed the role of coordination, cooperation, and collaboration in leading the decentralized federal statistical system. In this vein, she reported that priorities for the federal statistical system are set in collaboration with the statistical agencies that comprise the ICSP. Wallman reported that she also supports and fosters an environment of cooperation and collaboration when it comes to the establishment of standards, noting that she and her staff carefully coordinate the participation of key players drawn from across the statistical agencies, whether the goal is developing core standards for statistical surveys, guidelines on protecting confidential information, classi-

---

[3]Speech at the November 2007 international Symposium on the Development of Official Statistics held at the United Nations University in Tokyo, retrieved June 13, 2014: from http://www.stat.go.jp/english/info/meetings/sympo/pdf/sympo6.pdf.

fication standards, or data release standards. Although the chief statistician's office does not directly collect data that can be used to help solve our nation's challenges, the responsibilities and actions of the chief statistician provide the underpinnings that are necessary for the development of data that are objective, accurate, timely, and relevant to issues of public policy. Rather than using the ample authority of OMB to direct the federal statistical system, Wallman uses the tools of coordination, cooperation, and collaboration to lead the federal statistical system. The chief statistician and the staff of the Statistical and Science Policy Branch interact with staff throughout the federal statistical system primarily through the auspices of the ICSP. As noted above, the membership of the ICSP includes the agency leader of each of the 13 principal federal statistical agencies. In the spring of 2014, the agency leaders in three of the thirteen (23%) principal federal statistical agencies were women.[4] A few months earlier, in the winter of 2013–14, 5 of the 13, or 38%, of the agency leaders were women. Over the history of the federal statistical agencies, all 13 of the principal statistical agencies have had a woman leading the agency at one or more times.

Women Leading Federal Statistical Agencies

Each of the five women who led one of the principal statistical agencies in the winter of 2013–14 was interviewed as part of a series of articles in *Amstat News* that introduced the head of each principal statistical agency to the ASA membership.[5] They were asked a series of questions, such as:

- What about this position appealed to you?

- What do you see as the biggest challenges for your agency?

- What are the top 2–3 priorities you have for your agency?

- How can the statistical community help you?

- Prior to your tenure, what do you see as the biggest recent accomplishment of the agency?

The responses to these common questions provide insights into the leadership priorities of each of these women. Although the questions posed are broad enough to allow a considerable amount of latitude in their responses, there were strong similarities in the elements identified as important to their agencies. Each of these five leaders identified factors related to the relevance,

---

[4]One left to retirement, and one moved to a different position in the same department.

[5]American Statistical Association Membership Newsletter, Amstat News, Erica Goshen, April 1 2013; Mary Bohman, May 1, 2012; Cynthia Clark, December 1, 2011; Patricia Hu, July 1, 2011; Susan Boehmer, May 1, 2011.

objectivity, and quality of their agency's data, with phrases such as "maintain the confidence of our customers and stakeholders in the objectivity of the ERS's [Economic Research Service] work" (Mary Bohman), "the work product of the agency will be recognized as transparent and statistically sound" (Cynthia Clark, National Agricultural Statistics Service (NASS)), and "At BLS [Bureau of Labor Statistics], our data should always meet the following criteria: accurate, objective, relevant, timely, and accessible" (Erica Groshen).

All five women also pointed to the need for their agencies' data to inform critical public policy issues; for example, Patricia Hu, head of the Bureau of Transportation Statistics (BTS), pointed to the need to use transportation data "to better inform policy, investment, planning, and operations decisions"; Susan Boehmer noted that through the use of Statistics of Income Division (SOI) in the Internal Revenue Service data to shape economic and tax policy "SOI plays an important role in achieving good government"; and Erica Groshen stressed the relevancy of BLS data in "informing economic policy and decision making." Mary Bohman, the head of ERS, and Cynthia Clark, the head of NASS, each identified agriculture-related topics that are informed by the data collected and analyzed by their agencies — the challenge of feeding a growing world population, monitoring global and domestic food insecurity, ensuring a safe food supply, promoting health and nutrition, providing water for societal needs, responding to climate change, and maintaining an adequate supply of energy.

The CNSTAT principles that call for providing objective, accurate, and timely data that are relevant to issues of public policy are evident in each of these leaders' visions for their agencies. Thus, one answer to the question of "how statisticians, especially women statisticians, can best position themselves to help solve our nation's challenges" emerges. To accomplish this task, statisticians must have a keen sense of ongoing and emerging policy issues, and strive to provide high-quality, objective, timely, relevant data. Additional priorities identified by these leaders show an awareness of the fact that producing solid policy-relevant data is by itself not sufficient. There is a need to leverage new and emerging technologies in the collection, analysis, and dissemination of data. They also cite a need to ensure that policymakers and stakeholders know that the data are available. These leaders suggest that this can be done by reaching increasingly broader audiences, by increasing the accessibility of their agencies' data through increased clarity in their releases, and by maximizing collaborations with stakeholders inside and outside of government.

Three of the five leaders discussed both the importance of human capital and operating under budget constraints. Cynthia Clark, NASS, pointed to increasing staff career opportunities through restructuring and meeting programmatic needs through the implementation of a project management certificate program for staff. In discussing the need "to provide more data and provide it faster" in an environment of shrinking budgets, Susan Boehmer, IRS/SOI pointed to the creativity of SOI staff in introducing changes through the use of technology improvements that resulted in a 17 percent reduction in

SOI's corporate tax program cycle time. In discussing the biggest challenges facing BLS, Erica Goshen pointed to "maintaining quality of data and operations in a challenging budgetary environment," to "recruiting and retaining the best talent," and to the importance of having talented, dedicated staff who are working on ways to maintain and improve essential functions.

In the face of a challenging budgetary environment, these leaders recognize the need to do more with less and the importance of leveraging staff resources, whether it is through increasing opportunities, fostering creativity, recruiting and retaining the best talent, or some combination of each of these approaches. To be in a position to help solve our nation's challenges, each agency must find a workable approach to managing both fiscal and human resources.

This analysis of the interviews provides insights into the priorities and challenges facing these women as they lead the federal statistical system and the individual statistical agencies, but it is important to note that the sphere of leadership in statistical agencies extends beyond the single person at the top of each agency. Women statisticians play key roles at every level of the federal statistical system. For example, most agencies' organizational charts include the name and title of staff in leadership positions between the head of the agency and the agency's branch chiefs.[6] In total, 163 senior leadership positions were identified across the 13 principal statistical agencies, with women holding 53, or one-third, of these positions.[7]

Women Providing Federal Statistical Leadership

The interviews with women leaders of statistical agencies only provide glimpses into their leadership styles, but the interviews do illustrate the fact that there are different facets to statistical leadership from an awareness of new and emerging policy issues, to setting statistical policy and standards for high-quality, objective, timely, relevant data, to leveraging interagency cooperation to meet common goals, to addressing key measurement issues, to maintaining and supporting ongoing and new data collections, to reaching increasingly broader audiences, to making difficult decisions in the area of fiscal and human resources, to effecting change in government wide policies. Each facet of leadership provides a different example of how statisticians, especially women statisticians, can best position themselves to help solve our nation's challenges. They are all important and serve to demonstrate the fact that leadership is multifaceted. Examples drawn from these facets of leader-

---

[6]This information was retrieved from organizational charts and lists of agency leadership on the websites of the 13 principal agencies in the spring of 2014. The analysis was limited to positions above the level of branch chief, because most agencies do not include names of branch chiefs, making it impossible to identify which positions were held by women.

[7]The number of senior leaders listed and the percentages who are women varies by agency. The modal number of senior leaders by agency is 9, with numbers ranging from 5 to 25. The modal percentage of women leaders was 27, with percentages ranging from 14 to 67.

ship can provide better insights into different aspects of leadership. There are many examples that offer a more in-depth look at the role women have played in contributing to federal statistical leadership and thus helped to solve our nation's problems. However, I selected examples that meet two criteria. First, they cover the different facets of leadership identified in the interviews with women leaders of federal statistical agencies. Second, each of the women selected has made an important contribution to Federal statistics that has been significant to my own professional and leadership development. The leaders included are Margaret Martin, Dorothy Gilford, Katherine Wallman, Nancy Potok, and Rochelle (Shelly) Martinez. First, I will explore some of the contributions of each leader, and then I will conclude with a discussion of the roles these women and their contributions have played in my career. While I have had the pleasure of meeting each of these women, and in some cases working with them on interagency efforts, I expect that some of them are unaware of the influence that they have had on my career.

Margaret Martin

Martin's contributions combine several facets of statistical leadership. Her work on the measurement of employment and unemployment from household interviews compared to payroll statistics and unemployment insurance claims were key to the decision to implement the Current Population Survey. Upon her retirement from Office of Management and Budget (OMB), Martin was recognized by the heads of a number of agencies for her long-term significant contributions to the collection of labor and income statistics. After retirement, Martin returned to work, serving from 1973 to 1978 as the first executive director of the CNSTAT. Over the years, that committee has convened panels of experts that have helped the federal statistical system tackle numerous measurement and policy concerns. Whether it was her own direct contributions to the measurement of labor and income statistics, or her leadership in the establishment of CNSTAT, Martin's accomplishments helped solve our nation's challenges.

Margaret Martin started her career in 1938 at the New York State Division of Placement and Unemployment Insurance, which was established as part of the New Deal (Muko, 2011). While there, she worked as a junior economist on the classification of employers covered in the unemployment insurance program. The War Manpower Commission (WMC) was established during World War II to take over the research and statistics offices in the state unemployment insurance agencies and to run the rationing of the labor force for the country. With this consolidation, Martin took a position as a senior economist at the WMC. In 1942, Martin started her federal career in the US Bureau of Budgets (now OMB) in the Division of Statistical Standards (DSS) (now the Statistical and Science Policy Branch), where she was responsible for the improvement and coordination of statistics and the review and approval of forms.

In her early work at OMB, Martin was involved in the early development of the Current Population survey. The Works Progress Administration (WPA) had a team that developed a population-based survey of employment and unemployment, with the first estimates available for publication in the early 1940s. The estimates from this population-based survey were at odds with the administrative statistics from establishments. Martin's unit brought in Gladys Palmer, a research professor at the University of Pennsylvania, to lead meetings of representatives from the various federal agencies that had a stake in the measurement of employment and unemployment statistics to make a recommendation concerning the use of the WPA population-based survey. After months of deliberations, Palmer's group recommended publishing the population-based survey data as the Monthly Report on the Labor Force.

When the WPA was disbanded, Palmer was again asked to lead a committee to decide whether the Monthly Report on the Labor Force should be continued; and, if so, which agency would assume responsibility for the data collection. Ultimately, it was decided that the Census Bureau would operate the survey, with advice and direction from a Policy Committee made up of the secretaries of the Departments of Agriculture, Commerce, and Labor, and from a technical committee chaired by Palmer. Martin worked with a subcommittee of the technical committee that was tasked with understanding and explaining the differences between estimates of employment from the population-based survey versus payroll samples, and between differences in the household survey unemployment estimates compared to unemployment estimates based on the number of workers claiming unemployment insurance. Martin took the lead on preparing the committee's statement on differences between household and establishment surveys. When differences were identified, clarifying questions were added to the population-based labor force survey. This expanded survey became known as the Current Population Survey. Today, this survey is a major source of data on labor force characteristics and other social and demographic statistics in the United States.

Throughout her career at OMB, Martin developed a strong track record for coordinating efforts across agencies to ensure that the data collected were the best available, minimizing respondent burden to the extent possible, and controlling data collection costs. Martin retired from OMB after 30 years of service. Following her retirement, the heads of a number of agencies came together to honor Martin for her long-term significant contributions to the collection of labor and income statistics.

Having retired from OMB, in 1973 Martin returned to work for five more years, serving as the first executive director of the CNSTAT of the National Academy of Sciences' National Research Council.

Dorothy Gilford

Gilford made a number of contributions to the federal statistical system during her career. The issues that Gilford worked on and championed in the 1970s are important to today's national conversation on the state of education in the United States. The relationships that Gilford fostered between National Center for Education Statistics (NCES) and state and local education authorities are important to the quality of the data that are used to measure high school graduation rates and other characteristics of the education enterprise. The high school longitudinal studies that Gilford launched with the National Longitudinal Study of the High School Class of 1972 have fueled the national debate over the quality of high school education in the United States over the last 40 years. Gilford's work through the Board on International Comparative Studies in Education (BICSE) to improve the quality of the international education studies strengthened the quality of the international assessment data that provide the basis for the conversation that continues today concerning our students' ability to compete in a world economy.

Dorothy Gilford started her career in the federal government in the 1950s at the Civil Aeronautics Administration, where she served as the chief of the Biometrics Branch in the Medical Division (Wegman and Martinez, 2007). She served as the deputy director of the Financial Statistics Division within the Federal Trade Commission for two to three years, before moving to the Office of Naval Research (ONR) in 1955. At ONR, Gilford served in a number of leadership positions, rising from the chief of the Statistics and Logistics Branch to director of the Mathematical Sciences Division. Gilford's time at ONR predates the establishment of the National Science Foundation; as a result, during her tenure at ONR, her unit was the primary funding source for external researchers doing basic mathematical and statistical research. Gilford's unit provided funding to scholars such as Jerzy Neyman, Harold Hotelling, and William Kruskal.

In 1968, Gilford moved to the Office of Education to lead the NCES. During her six-year tenure at the helm of NCES, Gilford was responsible for launching the series of longitudinal studies that start with high school students and follow them into postsecondary education and the workforce. The National Longitudinal Study of the High School Class of 1972 (NLS-72) has been called the "grandmother" of the NCES longitudinal studies. Study participants were seniors in high school when the study started in 1972. These students were reinterviewed on five separate occasions before the study ended in 1986, when the study participants were in their early 30s.

This study has been replicated in each subsequent decade; the most recent in the series, the High School Longitudinal Study, was launched in 2009. This series of studies provides a baseline for monitoring the progress and achievement of America's high school students — making it possible to trace the experiences of each study's cohort as they progress into young adulthood, and to compare the experiences of these five cohorts to examine differences over the 40-year period. The success of the series of high school longitudinal studies resulted in subsequent decisions at NCES to move toward a multi-cohort longitudinal sequencing model for the entire range of P-20 studies: early childhood, middle grades, high school, baccalaureate, and post-baccalaureate longitudinal studies.

A number of NCES studies, then and now, rely on the good will of people working in state offices of education. Recognizing this, Gilford reported that she spent a lot of her time working on improving federal and state relations; she focused her efforts on building strong relationships with the Chief State School Officers in an effort to encourage participation in NCES surveys.[8] Gilford noted that this effort was complicated by the fact that many of the chiefs' positions were political appointments, and as a result, every year she had about 12 new chiefs to convince of the importance of surveys. Working with state and now local education agency staff to support and improve the collection of high-quality education data continues today. The importance of the relationships with state and local education data providers has been codified in the NCES authorizing law since 1988 when language was introduced calling for the establishment of "one or more national cooperative statistics systems for the purpose of producing and maintaining, with the cooperation of the states, comparable and uniform information data ... that are useful for policymaking at the Federal, State, and local levels" (Education Sciences Reform Act of 2002, Section 157). NCES established the National Forum on Education Statistics that provides a foundation for ongoing working relationships between representatives of the state and local education agencies and staff and managers at the federal level.

After retiring from NCES in 1974, Gilford worked at the National Academy of Sciences. While there, she read a report from a state governors' meeting in which the governors cited studies done by the International Association of Education (IEA) as a source of concern about mathematics and science education in the United States. Gilford was upset to see governors making national policy based on what she knew to be a set of flawed studies. In a 2007 interview, Gilford said, "The IEA had not done anything about nonresponse. They had not done anything to ascertain whether the questions on the test fairly represented the curricula in the countries, and so forth."[9] Gilford talked with managers at the Department of Education and the National Science Foundation about her concerns and came up with the idea of setting up a Board on International Comparative Studies in Education (BICSE). Funding was provided, and Gilford served as the director of that board for 12 years. Gilford's efforts and those of BICSE paid off, the IEA studies developed high standards to ensure the quality of their data, the quality of the data improved, and in 2007 the academy determined that BICSE had successfully accomplished its mission and BICSE was dissolved. The United States continues to participate in the IEA studies — Trends in Mathematics and Science Study (TIMSS) and Progress in International Reading Literacy Study. These studies provide

---

[8] Ibid.
[9] Ibid.

the basis for understanding how our 4th, 8th, and 12th graders' performance compares to that of their peers in other countries.

Katherine Wallman

Reliable data and confidence in those data are the basis for informed decision making by our nation's policymakers. Through her 12 years directing the Council of Professional Associations on Federal Statistics (COPAFS), Wallman strengthened the flow of information about federal statistical data by establishing a successful forum for communication and collaboration between federal statistical agencies and their constituents and policymakers. Wallman also recognized the need for strong data confidentiality laws for statistical data with limited data sharing within the federal statistical system. The goal was a law to provide the American public strong confidentiality protections for the data they share with federal statistical agencies and through codifying limited data sharing the law would reduce the data collection burden on the American public. From the inception of COPAFS, through her first 10 years as chief statistician at OMB, Wallman worked unflaggingly for the passage of confidentiality and statistical data sharing legislation for the federal statistical system. That goal was attained in 2002 with the passage of legislation that provides strong confidentiality protections to all federal statistical data and authorizes limited sharing between three of the thirteen federal statistical agencies. Although not described here, as chief statistician of the United States, Wallman is also responsible for the development and promulgation of statistical standards and guidelines and a number of government wide policies all intended to strengthen the information base available to help understand and solve our nation's challenges.

Early in her federal career Katherine Wallman worked in the National Center for Education Statistics and in the Office of Federal Statistical Policy and Standards. In 1981, Wallman founded the Council of Professional Associations on Federal Statistics (COPAFS) with the goal of building a bridge between individual researchers, educators, public health professionals, civic groups, and businesses that rely on high-quality and accessible statistics and the federal statistical system (*http://www.copafs.org/about/default.aspx?*). Wallman served as the executive director of COPAFS from 1981 through 1992. Established on a foundation of communication and collaboration, included among COPAFS functions are providing government policy decision makers with information that demonstrates the value of federal statistics and a strong federal statistical system, organizing forums in which federal statisticians interact with the users of their statistics, and holding workshops on challenges and opportunities facing groups that rely on federal statistics and those facing the federal agencies that produce them. COPAFS continues to thrive, with a current membership that represents more than 300,000 individual researchers, educators, public health professionals, civic groups, and businesses.

During Wallman's tenure at the helm of COPAFS, she championed the need for a consistent set of strong confidentiality protections for federal statistical data. Then when Wallman moved to OMB in 1992 to become United States chief statistician, in addition to the duties described earlier in this chapter, she maintained her support for strong statistical confidentiality protections.

In 1983, the administration sought to have the Congress enact statistical enclave legislation, which would provide a single set of confidentiality policies for all federal agencies and their components that collect data for statistical purposes. If passed, this legislation was to provide "(a) a statutory basis for the traditional promise of confidentiality long given respondents to statistical collections, and (b) restricted sharing of individually identifiable records ('protected statistical files') for exclusively statistical agencies ('protected statistical centers') whose confidential records are provided statutory protection under this legislation" (Parker, 2006).

The Statistical Policy staff at OMB and Wallman, as the executive director of the newly formed COPAFS, worked to garner support for the statistical enclave legislation from the federal statistical community and their data users. While these efforts served to increase awareness of the need for uniform confidentiality provisions coupled with controlled data sharing, the Treasury Department raised objections to increasing access to individually identifiable data for statistical purposes. Ultimately, these concerns prevailed and the legislation was not passed.[10]

Concerns over protecting confidential statistical data, while at the same time allowing access to those data for statistical purposes, did not diminish with the failure of the 1983 legislation. In 1989, CNSTAT and the Social Science Research Council formed a Panel on Confidentiality and Data Access charged "with developing recommendations that could aid Federal statistical agencies in their stewardship of data for policy decisions and research" (National Research Council, 1993). The panel met between November 1989 and January 1992; its deliberations focused on "protecting the interests of data subjects through procedures that ensure privacy and confidentiality, enhancing public confidence in the integrity of statistical and research data, and facilitating the responsible dissemination of data to users."[11]

Among the many recommendations that emerged from this effort are two that are directly relevant to the continuing concerns over the protection of confidential statistical data:

- **Recommendation 1** Legislation that authorizes and requires protection of the confidentiality of data for people and organizations should be sought for all federal statistical agencies that do not now have it and for any new federal statistical agencies that may be created.

---

[10]Ibid.
[11]Ibid.

- **Recommendation 2** There should be legal sanctions for all users, both external users and agency employees, who violate requirements to maintain the confidentiality of data.

Wallman became the chief statistician at OMB in 1992, and renewed the push for a comprehensive set of confidentiality provisions for federal statistics. Bolstered by the CNSTAT panel's ongoing focus on the protection of confidential statistical data, Wallman kept the focus of her OMB Statistical and Policy Branch on the importance of the proper protection, handling, and uses of statistical data. Garnering support for this change in law required coordination and collaboration with the federal statistical community, especially with the heads of the principal statistical agencies. The ongoing monthly meetings of the ICSP were an important mechanism for obtaining and maintaining the continuing support of these key leaders in the federal statistical agencies. Presentations and discussions at professional societies were used to get the message out to the broader federal statistical community and to the network of researchers and other users of federal statistical data. During this period, Wallman's successor at COPAFS, Edward Spar, maintained that group's support for a set of comprehensive confidentiality provisions for federal statistics.

In 1995, the Clinton administration sent a new proposal for statistical confidentiality legislation to Congress. The proposed legislation was introduced in the House of Representatives as the Statistical Confidentiality Act in the summer of 1996.[12] The proposed legislation required that data or information acquired by an agency for purely statistical purposes could only be used for statistical purposes and could not be shared in identifiable form for any other purpose without the informed consent of the respondent. The legislation also designated eight federal agencies as statistical data centers: the Bureau of Economic Analysis, the Census Bureau, the Bureau of Labor Statistics, the National Agricultural Statistics Service, the National Center for Education Statistics, the National Center for Health Statistics, the Energy Consumption Division of the Energy Information Administration, and the Science Resources Statistics Division of the National Science Foundation.

Coinciding with this legislative initiative, the OMB Statistical Policy Office developed and promulgated an "Order Providing for the Confidentiality of Statistical Information".[13] This Order was released for public comment in 1996 and released as a final Order in 1997. Section 2 of the final Order states the following:

a Information that a statistical agency or unit acquires for exclusively statistical purposes may be used only for statistical purposes, and shall not be disclosed, or used, in identifiable form for any other purpose unless otherwise compelled by law.

---

[12]62 Federal Register, 35044.
[13]62 Federal Register, 35044.

b When a statistical agency or unit is collecting information for exclusively statistical purposes, it shall, at the time of collection, inform the respondents from whom the information is collected that such information may be used only for statistical purposes and may not be disclosed, or used, in identifiable form for any other purpose, unless otherwise compelled by law.. . .

This OMB Order promulgated by Wallman kept the focus on the protection of confidential statistical data. Following five separate attempts to get a bill through Congress to protect data collected under a pledge of statistical confidentiality, a statistical confidentiality bill was finally passed in the 107th Congress. The Confidential Information Protection and Statistical Efficiency Act of 2002 "CIPSEA," included as Title V in the E-Government Act of 2002,[14] was passed unanimously by both chambers of Congress on November 15, 2002, and subsequently signed into law by President George Bush on December 17, 2002. CIPSEA provides a uniform set of confidentiality protections and extends these protections to all federal agencies that collect individually identifiable data for statistical purposes under a pledge of statistical confidentiality. CIPSEA also permits the sharing of business data by 3 of the 13 principal federal statistical agencies: the Bureau of Economic Analysis, the Bureau of Labor Statistics, and the Bureau of the Census.

While it took a number of years to get the current law enacted, this law provides strong confidentiality protections to all data collected under a pledge of statistical confidentiality, it authorizes the use of agents for statistical uses of the covered data, it adds strong Class E Felony penalties to any staff or agents who disclose identifiable information that was collected under a pledge of statistical confidentiality, and it allows for the sharing of business information among three designated agencies (Census, Bureau of Economic Analysis, and Bureau of Labor Statistics). This accomplishment involved not only Congress, but the cooperation of many within the federal statistical community and among the supporters and users of federal statistics. Wallman's strong leadership, perseverance and adept skills at building networks were effective in providing this much needed legislation for the federal statistical community.

Nancy Potok

The decennial census provides valuable information about the American population that is used to insure equal representation in the conduct of the country's government. The decennial census is also the largest and most expensive undertaking within the federal statistical system. Thus, it is important that the Census Bureau "gets it right" and maintains the confidence of the American public. During her work on the 2000 Census, Potok responded to criticism over missed schedules and cost overruns in the buildup to the 2000 Census by working to establish a training program that led to a master's certificate in project management. This emphasis on strong project management contributed to the success of the 2000 Census and the 2010 Census. Planning

---

[14] 107th Congress, H.R. 2458.

for the 2020 Census started soon after the 2010 Census finished data collection. Faced with rising costs and tighter budgets, the Census Bureau leadership was directed by Congress to reduce the cost of the 2020 Census. In addition to plans to leverage the use of technology, the Census Bureau leadership concluded that there is a need to restructure both the field operations and the headquarters organizational structure. As deputy director of the Census Bureau, Potok has a major role in the ongoing restructuring of the agency. Potok's work leading the restructuring of the largest of the federal statistical agencies to meet the fiscal constraints imposed by America's economic condition is a different example of solving a national challenge.

Nancy Potok, deputy director and chief operating officer of the US Census Bureau, has more than 30 years of public, private, and nonprofit senior management experience. Her public service includes working in the Judicial and Legislative Branches, as well as the US Department of Transportation, the Office of Management and Budget, and the US Department of Commerce. At Commerce, Potok previously served as the deputy under secretary for Economic Affairs, the Census Bureau's associate director for Demographic Programs, and the principal associate director and chief financial officer in charge of Field Operations, Information Technology, and Administration during the 2000 Census.[15]

Potok is well aware of the role that data produced by Census plays in supporting public policy. In a 2012 interview in a Spotlight on Commerce series, Potok noted that "Much of the data we [Census] produce is used by State and local Economic Development Authorities to bring businesses to their area. Businesses use the information to make relocation decisions and to target their marketing appropriately. We also report, at various geographic levels such as States, counties, cities, and small towns, on educational attainment, income, poverty, how people make various use of government assistance programs, and other critical information needed to inform our communities on how we as a nation are doing and where we need to invest our resources to strengthen our future. Without the data collected by the Census Bureau, we could not have the information we need to grow our economy, create jobs, improve our schools, build roads, and other activities critical to our civil society."

In the build up to the 2000 Census, the bureau received a report from the National Performance Review (NPR) that called for improved performance due to problems that included budgetary overruns, scheduling lapses, and missed deadlines. In response, the Census Bureau partnered with ESI International, an industry leader in project management training, to develop training that was tailored to the needs of the bureau (ESI International Inc., 2014). The resulting training program included a series of courses on that would lead to a master's certificate in project management for participants. These courses were offered on site for Census Bureau

---

[15]Census Newsroom, *http : //www.census.gov/newsroom/releases/pdf/bio_potok_nancy.pdf*

employees who were nominated by their supervisors to join training groups that moved through the program together. Upon completion of the program, each employee received a master's certificate in project management from ESI and The George Washington University.

The fact that the 2000 Census was an operational success that was completed on time and close to $2 billion under budget is attributable, at least in part, to strong project management. The impact of improved project management was evident in other Census Bureau programs as well, with a 40 percent reduction in time to complete a new project in one division. In summarizing the impact of the training program, Potok said, "A common vocabulary and work approach really helps people be more productive, we find we're getting projects done on time, with fewer mistakes because we are doing more planning up front. Therefore, we get a better product delivered on time and under budget."[16]

These successes were repeated in the 2010 Census that was completed on time and under budget by nearly $2 billion.[17] Shortly after the completion of the 2010 Census, in the face of direction from Congress for a leaner Census Bureau budget in 2020, Census Bureau leadership announced plans for a reorganization of the Census Bureau field offices, noting that despite technological changes in the way work is accomplished, the regional office structure was substantially unchanged for 50 years. The field reorganization was completed in early 2013, with a reduction in the number of field offices from 12 to 6 and expected annual savings to the government of $15 million to $18 million annually starting in FY 2014. The agency is also testing technological upgrades for the 2020 Census.

The field reorganization, coupled with restructuring in headquarters and an increased use of technology, is also geared to improve service delivery to the other federal statistical agencies for which Census Bureau conducts surveys. In an era of flat or shrinking budgets, survey sponsors require lower costs, improved efficiency, and increased responsiveness. Although such reimbursable work accounts for approximately 20 percent of the bureau's annual budget, the bureau has experienced a decline in their share of federally sponsored survey work. The Census Bureau plans to leverage its ongoing restructuring and increased use of technology to maintain, or increase, its current share of federally sponsored survey work.

With the departure of Director Robert Groves in 2011 and the addition of Director John Thompson, Potok, in her current role as deputy director and chief operating officer, continues to provide continuity to the ongoing restructuring efforts. Part of this effort involves staff self-assessments of their job skills. Noting that current staff may have worked on censuses for as many as three decades, Potok reports that "we have to look forward and do something non-traditional in a sense. So we want to make sure we have the right skill

---

[16] Ibid.

[17] From a July 2012 interview with the then-Census Director Robert Groves on "The Federal Drive" on Federal News Radio by Tom Temin and Emily Kopp.

sets here."[18] The goal is to have managers and employees work together to develop training plans that will allow employees to get the skill sets they need. Another part of this strategy is described by Potok as "deploying people to the places where they can be best-utilized within the whole agency."[19]

In a December 2012 presentation at the FCSM Statistical Policy Conference, Potok talked about the Census Bureau's efforts to respond to a challenging future. She described the previous organizational structure at the Census Bureau as having expensive infrastructure in the field operations and stove piped functions throughout the organization that resulted in various bureaucratic barriers to innovation and a series of redundant data collection, processing, and dissemination systems. Potok pointed to the field reorganization, the restructuring in headquarters, the approach to 2020 Census research and testing, the use of a Census Center for Applied Technology, and the application of adaptive survey design as focal points for improvements. Potok also highlighted a number of factors that are essential to the successful implementation of the anticipated changes. Leading the list is the need for a major culture change; other factors include improved internal communications; practical solutions to improve efficiency developed from working teams; and the development and implementation of aggressive schedules.

The restructuring efforts within Census Bureau headquarters are based on a matrix management approach in which staff members are organized by function but managed programmatically to provide support for individual or related surveys. This requires an organizational culture that supports collaboration across organizational boundaries for the success of the matrix management approach. The underlying principle here is to create a data driven, adaptive, agile organization. This principle is represented in current efforts to use real time administrative data (i.e., paradata) from ongoing data collections to make adjustments to improve data collection efficiency and improved data quality (i.e., adaptive design). Another priority of the restructuring is to establish structures that will promote continuity within programs, produce an increase in knowledge, and enhance ongoing and new operations. The goal here is to enhance, reinforce, and support staff knowledge and skills, be they subject matter, methodological, or technical. Finally for this effort to be fully successful, Census Bureau leadership recognizes the need to establish a structure that maximizes the ability to standardize and simplify processes, methodology, and systems with the end goal of creating efficiencies and reducing program costs.[20]

---

[18] From a September 2012 Federal News Radio with Nancy Potok, deputy director.

[19] Ibid.

[20] The description of the restructuring underway at headquarters is based on information conveyed to federal survey sponsors in a January 2014 meeting, and on discussions in quarterly meetings concerning the survey work sponsored by my agency, NCES.

Census leadership's plans include using the Center for Applied Technology (CAT) as a hub for Census innovation. The CAT is tasked with connecting innovative ideas with subject matter experts, facilitating collaboration, providing a forum for thought leadership, supporting the adoption of new technology, and fostering outreach and information sharing.

Another part of the Census Bureau's leadership's plan is to incorporate technical and technological improvements in the ongoing data collections, with the goal of employing some of the changes with demonstrated success in the 2020 Census. Some of the changes envisioned as part of the full restructuring plan are already being set in place — for example, the staff skills assessments, the use of a matrix management approach for cost-reimbursable sponsored survey work, and the use of an adaptive design approach to make adjustments to survey operations in real time. Other changes that are tied to the structural realignment of organizational units are progressing through the departmental, OMB, and Congressional approval process.

The multitiered leadership approach promulgated and supported by Potok includes a combination of leveraging new technology and leveraging staff resources to do more with less. Positive steps are underway, the final measure of success will be in whether the Census Bureau maintains or increases its share of the federally sponsored survey portfolio, and whether the 2020 Census meets the Congressional charge of costing less per household than the 2010 Census.

Rochelle (Shelly) Wilkie Martinez

Rochelle (Shelly) Martinez's work in the Statistical and Science Policy Branch at OMB promoting the use of administrative data for statistical purposes builds on the strong 2002 confidentiality and data sharing laws. Like Wallman, Martinez also champions the use of data sharing with strong confidentiality protections. In an era of tighter budgets and declining participation rates in federal data collections, Martinez led an interagency effort to increase the statistical use of administrative data in federal statistics. Throughout this seven-year effort, Martinez's early experiences as a staff member in the US House of Representatives and then as a program analyst and manager at the Census Bureau provided useful insights. Martinez successfully guided an interagency effort through the work involved in moving from the initial step of establishing an interagency working group on the statistical use of administrative data through a strategically planned set of activities in support of the endeavor to the release of a governments wide OMB memorandum that provides formal guidance to all departments and agencies for providing and using administrative data for statistical purposes. The implementation of the guidance will contribute significantly to our national challenge for increased transparency and the efficient use of data for informed decision making.

Rochelle (Shelly) Martinez began her career in 1989 in the US House of Representatives, working first as a staff member and then as a senior professional staff member on the Subcommittee on Census, Statistics, and Postal

Personnel. In the spring of 1995, Martinez joined the staff at the Office of the Inspector General in the US Department of Commerce as a program analyst. While there, Martinez identified a need to revitalize a dormant effort to build a record linkage capacity at the Census Bureau, and led an independent assessment of the design phase for the 2000 Census. That assessment and others resulted in a number of design changes to the 2000 Census.

Martinez moved to the Census Bureau deputy director's staff in the fall of 1997. During her nine-and-a-half-year tenure at the Census Bureau, Martinez moved from her first position as a program analyst in the Office of the Deputy Director to the position of team leader in the Policy Office to the position of branch chief and acting assistant division Director in the Data Integration Division. Across these positions, Martinez worked to support data quality and data integrity. She developed and promulgated privacy and quality assurance programs; Martinez's contributions to the Census Bureau were marked by her interest in problem solving and her strong leadership.

In the spring of 2007, Martinez moved to her current position as a senior statistician in the Statistical and Science Policy in the Office of Information and Regulatory Affairs within OMB. Once there, in addition to her ongoing work serving as a statistical consultant for several agencies under the Paper Work Reduction Act and monitoring Executive Branch and legislative initiatives that might impact those agencies or the broader federal statistical community, Martinez took on the topic of encouraging the use of administrative data for statistical purposes.

This topic turned into a seven-year effort that culminated in the release of a government wide OMB Directive on Guidance for Providing and Using Administrative Data for Statistical Purposes (OMB; M-14-06). The path over the seven years provides an interesting example of leadership through a carefully thought-out process that included identifying an issue in need of resolution, increasing the general level of awareness of the issue in the federal statistical community, and developing a constituency that cares about the issue, to garnering consensus and support from a leadership base resulting in a government wide change in policy.

The issue of using administrative records for statistical purposes is not a new concern. In fact, there are a number of instances across the federal statistical system where aggregations of administrative records have been in use for a number of years. For example, the National Center for Health Statistics (NCHS) receives reports of births and deaths that start with local registrars who report data to state Offices of Vital Statistics, which in turn report the state-level data to the Vital Statistics Program at NCHS. A similar process is in place through the Common Core of Data (CCD) at the National Center for Education Statistics (NCES), where data on P-12 students, teachers, and schools are reported from schools to school districts to states to NCES. In a like manner, the Census Bureau's Census of Governments includes data that are reported from the lowest levels of government through a hierarchy to the national level at the Census Bureau. Every federal statistical agency relies upon data that are aggregated from administrative data for at least a portion of their work.

Using administrative data aggregated to different reporting levels is just one of the possible ways that federal statistical agencies can use administrative data for statistical uses. Recall that during Martinez's tenure in the Commerce Department's Inspector General's Office, she identified a need to build record linkage capacity at the Census Bureau. While record linkage projects are not limited exclusively to administrative data projects, statistical agencies can leverage the use of administrative data through record linkage either between administrative data and records from a sample survey, or through record linkage across two or more administrative data systems. Again, there are examples of projects using these techniques (e.g., the National Postsecondary Student Aid Study uses information from the National Student Loan Data System), but there is not a shared knowledge of these individual projects or of the difficulties encountered along the way in each of these efforts.

In the summer of 2007, Martinez and John Eltinge, from BLS, proposed the formation of an FCSM Interagency Working Group on Statistical Uses of Administrative Records to the ICSP. Their proposal was based on the premise that declining survey response rates and rising survey costs both contribute to the need for broader usage of administrative data. Acknowledging that opportunities for the expanded use of administrative data and issues of quality, confidentiality, and legal barriers all need further consideration, the proposal was based on the premise that collaborations on each of these topics across the federal statistical system would be beneficial. The ICSP members endorsed the formation of the proposed working group, and Martinez and Eltinge became co-chairs of the interagency working group.

All ICSP agencies were asked to nominate staff members to join this new working group. As a result of information sharing across the participating agencies, specific needs were identified and three subgroups were formed — one to develop a model agreement for data sharing, one to identify and address informed consent barriers to sharing and linking data, and one to examine ways to assess the quality of data used in data sharing projects. The regular meetings of the working group, an initial working group report[21], and meetings of the three subgroups were the first steps in increasing awareness of the use of administrative data for statistical purposes. The participants on the working group and its three subworking groups resulted in a constituency that cares about the issues surrounding the use of administrative data for statistical purposes, and the fact that the membership included staff from agencies across the federal statistical system helped to ensure a broad base of interest.

Each of the three subworking groups produced a report. The report, "The Unique Method for Obtaining Data: Model Agreement to Share Administrative Records," provides a model agreement that is annotated with examples

---

[21] "Profiles in Success" documented barriers and success factors for a set of projects that successfully accessed administrative data.

to facilitate its use. The report "Informed Consent: Requirements and Practices for Statistical Uses of Administrative Data" includes a review of federal legislation governing informed consent for statistical uses of administrative records, provides examples of how federal agencies achieve informed consent, and concludes with a review of existing research regarding informed consent and public opinions toward data sharing and linking for statistical purposes.[22] The report "Data Quality Assessment Tool for Administrative Data" provides a means for a new user to evaluate the "fitness for use" of administrative data for the new user's data needs. Martinez was a co-author on the first two of these three reports, and a strong proponent of the work required to develop the third report.

These reports increased awareness of the use of administrative data for statistical uses in the federal statistical community. By organizing sessions and presenting papers on their work and related topics at the American Statistical Association annual meetings and the FCSM Statistical Policy and Research conferences, the working group kept this topic front and center in the federal statistics arena. These efforts increased awareness of statistical uses of administrative data and developed a constituency interested in these issues within the federal statistical community.

Martinez and cochair Eltinge provided periodic oral and written reports to the ICSP members to keep them informed of the working group's activities and to seek their support and endorsement of ongoing and planned activities. This frequent contact with the leadership of the federal statistical system was essential to garnering consensus and support from the directors of the federal statistical agencies.

While the FCSM Working Group activities continued, the use of administrative data and data sharing was under discussion within OMB. For example, OMB memorandum M-11-02 "Sharing Data While Protecting Privacy" (November 2010) states, "The purpose of this Memorandum is to direct agencies to find solutions that allow data sharing to move forward in a manner that complies with applicable privacy laws, regulations, and polices. These collaborative efforts should include seeking ways to facilitate responsible data sharing for the purpose of conducting rigorous studies that promote informed public policy decisions." The memorandum includes a statistics initiative that states, "Our ability to contain costs and reduce burdens on respondents, while increasing the quality and quantity of statistical information, depends on the untapped potential of data sets held by program, administrative, and regulatory offices and agencies."

Simultaneous with the work of the FCSM Working Group, the Government Accountability Office (GAO) conducted a performance audit of the federal statistical system at the request of Senator Thomas Carper from December 2010 to February 2012. Among other things, the resulting report recommended

---

[22]This subgroup was cosponsored by the FCSM Privacy Working Group, chaired by M. Seastrom.

that OMB develop "comprehensive guidance for both statistical agencies and agencies that hold administrative data to use when evaluating and negotiating data sharing".[23]

In response, Martinez led the drafting of a new OMB memorandum and moved it through the internal OMB review processes. On February 14, 2014, OMB issued M-14-06 "Guidance for Providing and Using Administrative Data for Statistical Purposes." The memorandum encourages greater use of administrative data for statistical purposes; it provides "guidance for addressing the legal, policy, and operational issues" that arise when administrative data are used for statistical purposes. The memorandum includes guidance that delineates how departments and agencies are to proceed to fully implement the requirements in the memorandum. The detailed guidance is intended, in part, to bridge some of the barriers identified through the work of the FCSM Working Group. In fact, Appendix A to the guidance, "Further Guidance on Privacy Act Requirements Related to the Provision of Administrative Data for Statistical Purposes," reflects some of the efforts of the subworking group on informed consent; and Appendix B, "Model Agreement for the Provision of Administrative Records for Statistical Purposes," reflects the efforts of the subworking group that developed a model agreement for data sharing.

While much work remains in supporting departments and agencies in the implementation of the OMB guidance, Martinez's multiyear effort produced a government wide change in policy.

Summary

I have been fortunate to know, and in some cases to have worked and collaborated, with these remarkable statisticians and leaders. My experiences have been influenced by the five women I chose to highlight as examples of women providing federal statistical leadership. While in graduate school at the University of Illinois at Urbana-Champaign, I worked on a social indicators project that included identifying data sources, assembling the annual data for the post-World War II period, and conducting time series analyses that were used to examine correlates and project estimates of key social indicators. Much of the data used in this effort were drawn from the Current Population Survey (CPS). My dissertation was related to the social indicator project, but used a social accounting framework that was based on the methodology used in economic accounts. CPS data on voter participation across two national elections provided the data required to demonstrate the utility of the social accounting framework. None of this work would have been possible without the CPS data and the early efforts of Margaret Martin in establishing the CPS.

My first full-time job was in the federal government. I worked as a statistician analyzing issues related to mortality and morbidity for the first nine

---

[23]Federal Statistical System: Agencies Can Make Greater Use of Existing Data, but Continued Progress is Needed on Access and Quality Issues. Page 36, GAO-12-54.

years of my career, before moving to the NCES in 1988. Ten years into my tenure at NCES, I became the chief statistician and director of the Statistical Standards and Data Confidentiality programs, with responsibilities for the NCES statistical standards, technical review, statistical consulting, and data confidentiality. I continue in that role today; although I also served as the acting deputy commissioner of NCES from March of 2011 through November of 2013.

My move to NCES was indirectly influenced by Margaret Martin. As the first executive director of the CNSTAT, Martin was instrumental in setting the course for the role that CNSTAT would fulfill for the federal statistical system. In the mid-1980s, CNSTAT conducted an independent review of NCES at the request of the assistant secretary of education's Office of Educational Research and Improvement due to a perceived lack of confidence in NCES and concerns over the quality and timeliness of its products. After their review, the CNSTAT panel recommended one of two courses of action. The first was for the Department of Education to make a strong commitment to undertake "wide-ranging actions to change both the image and reality of the center," including establishing an Office of Statistical Standards and Methods led by a chief statistician, developing and codifying statistical standards to guide the conduct of all phases of NCES work, improving the timeliness of NCES products, increasing the focus on the quality of NCES products, and initiating an active recruiting program to obtain staff with the needed skills. The second recommended course of action was to give serious consideration to "abolishing the center and finding other means to obtain and disseminate education data" (CNSTAT, 1986). NCES and ED leadership chose the first option and recruited established statisticians from other agencies. I moved to NCES as part of that recruitment on the advice of the CNSTAT panel study director.

Since becoming chief statistician, my work has been strongly influenced by the advice from the CNSTAT panel. Although my predecessors paved the road by promulgating statistical standards, those early standards were developed using a top-down approach. In my first few years in the job, I realized that staff viewed the standards as being forced on them. In an effort to develop a sense of ownership of the standards among NCES staff, I established teams of NCES staff who worked on revising existing standards or writing new ones. In all, we had 16 working groups that involved more than half of the NCES staff. We followed the advice of the CNSTAT panel and developed standards around all phases of NCES work, from the initial planning of a data collection to the production of data products and reports. In the process, we took care to address specific issues identified by the panel (e.g., monitoring and evaluating all sources of survey error, requiring significance testing to support assertions in reports, defining a range of report types and specifying the methodological components to be included in each type of report). We convened an expert technical panel to review the proposed standards, and posted the standards for public comment prior to issuing the 2002 NCES Statistical Standards.[24] The 2012 revisions were not as extensive; we followed the practices of providing opportunities for staff participation and posted the proposed revisions for public comment prior to a 2012 release.[25] Both the 2002 and 2012 standards reflected ongoing efforts to standardize the calculation of response rates across the portfolio of NCES data collections and to implement nonresponse bias analyses when needed. Related to the NCES standards effort, I coauthored the department's Information Quality Guidelines, helping to bring increased rigor to data collected and used for different purposes throughout the Department of Education.[26] In keeping with the two other concerns from the CNSTAT panel, I established and maintain the NCES Handbook of Survey Methods that describes each NCES data collection along with basics of the methodology of each one[27], and continue to work with staff and monitor progress toward the NCES Government Performance and Results Act (GPRA) measure on the timeliness of initial release reports. The work of CNSTAT continues to have a direct impact on the day-to-day operating procedures at NCES.

When I moved to NCES in 1988, Dorothy Gilford had not been at the center for some 14 years, but the legacy she left NCES persists today. During her time leading NCES, Gilford worked to foster good working relationships with state education officials. The importance of that effort was recognized when it was codified in the 1988 authorizing law for NCES through the establishment of one or more cooperative systems to produce and maintain "comparable and uniform information and data on early childhood education, elementary and secondary education, postsecondary education, adult education, and libraries." These cooperative systems require collaboration between federal and state data stewards to produce data that are useful for policymaking at the federal, state, and local levels. The work at the elementary and secondary levels is supported through the National Forum on Education Statistics, which strives to provide states, districts, and schools with helpful advice on the collection, maintenance, and use of elementary and secondary education data. To achieve these goals, forum members work collaboratively to address problems, develop resources, identify best practices, and consider new approaches to improving data collection and utility, all while remaining sensitive to privacy concerns and administrative burden. Early in my career at NCES, I staffed a committee tasked with charting the course for forum activities. Now, the administration and NCES leadership of the forum is in my office. In addition to contributing to improved data quality, the forum provides national leadership opportunities to forum members from state and local education offices.

---

[24] 2002 NCES Statistical Standards and Guidelines, http://nces.ed.gov/statprog/2002/stdtoc.asp.

[25] 2012 NCES Statistical Standards and Guidelines, http://nces.ed.gov/statprog/2012/.

[26] US Department of Education Information Quality Guidelines, http : //www2.ed.gov/policy/gen/guid/iq/iqg.html.

[27] NCES Handbook of Survey Methods, http://nces.ed.gov/pubs2011/2011609.pdf.

As a result of Gilford's early efforts introducing the first high school longitudinal study, NCES currently conducts longitudinal studies of students in the elementary grades, the middle school grades (under development), the high school grades, and the college level and beyond. The third high school longitudinal study, the National Education Longitudinal Study of the 1988 eighth grade cohort, was underway when I joined NCES. At the present time, the fifth such study, the High School Longitudinal Study, is still in collection. As the senior statistician for the Elementary/Secondary and Postsecondary Division and then as the NCES chief statistician, I have had the opportunity to be involved in different aspects of each of the NCES longitudinal studies.

In addition to my statistical and methodological responsibilities as the NCES chief statistician, I direct the data confidentiality and data licensing programs for NCES and the Institute of Education Sciences. This work requires a full understanding of the confidentiality laws that apply to the data that NCES collects, as well as additional laws that apply to data that NCES uses from other sources. My work with the Disclosure Review Board, defining the types of disclosure avoidance techniques that are used with NCES data, and my work with the restricted-use data licensing procedures the NCES has in place, have resulted in collaborations with Katherine Wallman, Rochelle (Shelly) Martinez, and others on the SSP staff.

As my career has progressed, I have had a number of opportunities to interact with Katherine Wallman. I am a member of the FCSM, and as the NCES chief statistician, I often represent NCES at the monthly meetings of the ICSP. Thus, I have had the opportunity to work with Wallman and to observe and learn from her approach to leadership first hand. As a result of Wallman's leadership and through opportunities provided by the ICSP and the FCSM, I have been involved in a number of OMB-sponsored standard and policy setting activities.

On the statistical side, I was an active participant in the working group that established the framework that statistical agencies agreed to use in developing their legislatively required Information Quality Guidelines and of the working group that developed the Draft Standards and Guidelines for Statistical Surveys that were adopted and promulgated by OMB for use across the government. I was also on the steering committee that established SCOPE that encourages collaboration and sharing of statistical protocols across federal statistical agencies with the goal of improving data quality, information security, and operating efficiency through improvements in data interoperability and reductions in duplication of efforts among the principal statistical agencies.

On the privacy and confidentiality side, I participated in the working group that developed a draft of the guidance that OMB issued on the implementation of the 2002 CIPSEA; I was also a member of the working group that prepared the federal statistical system's response to the governmentwide policies on the use of the label Confidential Unclassified Information; and I chair the FCSM/ICSP Privacy Working Group.

Through the Privacy Working Group, I collaborated with Martinez in establishing a joint working group comprised of members from the FCSM Statistical Uses of Administrative Records Working Group and the FCSM Privacy Working Group to examine privacy protections offered to individuals providing data to federal agencies. Specifically, the joint working group looked at whether the statement of anticipated routine uses that is published in the Systems of Record Notice (SORN) and in informed consent statements provided to prospective respondents are barriers to using administrative data for statistical purposes. The joint working group identified the array of laws and regulations that impact privacy, confidentiality, data sharing, and data linking; identified common practices for informing the public and respondents about statistical uses of data; and reviewed the extant literature on public attitudes towards different components of informed consent notifications.

In January of 2011, Jack Buckley was confirmed by the Senate as the Commissioner of NCES. Having served a few years earlier as the NCES deputy commissioner, Buckley was aware of a number of inefficiencies in the way NCES was organized and in the way work was conducted and had definite ideas about reorganizing NCES. A few months into his term, Buckley asked if I would serve as the NCES deputy commissioner. I agreed and became the acting deputy commissioner. In that role, I worked with Buckley on the NCES reorganization. I was involved in planning the NCES reorganization and in drafting the supporting documentation. One of the elements of the reorganization involved moving a unit that collects program data for K–12 programs throughout ED from another Office in ED to NCES. The other main elements of the reorganization involved organizing programs around the types of data collected, rather than education level, and flattening the organization to yield an organizational structure with fewer supervisors.

When NCES started planning its reorganization, the Census Bureau leadership was already working on its own restructuring. Because of the volume of NCES projects conducted by the Census Bureau, NCES and Census Bureau senior managers meet quarterly to discuss the status and progress of NCES projects at the Census Bureau. These quarterly meetings have provided NCES senior managers with periodic updates and discussions on the Census Bureau restructuring in general, but especially as it relates to Census Bureau work on NCES projects. This was first time the commissioner or I had led a major reorganization, and it was very helpful to learn from the experiences of a larger statistical agency. Nancy Potok was very helpful in this regard. One of my earliest recollections of the process was a friendly conversation with her in which she explained the approval process that laid ahead, including the likely requirement for Congressional approval. She was entirely correct; our estimate of six to nine months start to finish was unrealistic. The actual time was much longer, as she predicted. In fact, bureaucratic approvals were finalized in the fall of 2013, and the physical moves were completed in March of 2014.

With the spring 2014 release of a government-wide set of guidelines for Using Administrative Data for Statistical Purposes (OMB Directive M-14-06), potential new avenues for statistical uses of administrative data are likely to open up. To facilitate this, the Statistical Uses of Administrative Data Working Group laid out a plan of action. One element of the plan is to identify one or two instances where data sharing could be used to enhance the available data in a statistical database. Given the current interest in the returns to the investment in postsecondary education, NCES is interested in sharing data with the Social Security Administration (SSA) to learn more about current and cumulative earnings as they relate to the educational experiences of students in the various high school and postsecondary longitudinal studies that NCES has conducted.[28] If data sharing is approved for this use, NCES will be able to look at whether the returns to investment in postsecondary education have changed over the 40 year period encompassed in the NCES portfolio of studies. Martinez attained the resources needed to support one or two initial case studies that will undertake a data sharing project with SSA, with the understanding that these early case studies will carefully document the process to pave the way for future data sharing projects. NCES has volunteered to participate as one of the initial case studies. I am working closely with Martinez and the NCES staff on this project and eagerly looking forward to the results of both the data sharing activity and the analysis.

One of the requirements in the OMB Directive M-14-06 is for the identification of administrative data sets with potential statistical uses. Several years prior to the release of M-14-06, or a companion Directive M-13-13 that requires an inventory of each department's data holdings, the Department of Education leadership called for an inventory of all Department of Education data holdings — administrative program data and statistical data. At the request of the ED Data Strategy Team, I cochaired a departmental working group to identify the metadata elements that should be included in the inventory, and then worked with a team of government and contractor staff to develop the database of metadata and a searchable webbased tool to make the information about the Department of Education's data holdings accessible to the Department of Education and the general public.[29] This database and tool will support the department in identifying administrative program data that may be useful for sharing.

Conclusion

This chapter describes women leaders who participated in improving the federal statistical system and their roles, actions, and views as they worked within the system. The contributions of five women were presented and several themes were noted: awareness of new and emerging policy issues; setting

---

[28] The high school cohorts are each followed beyond high school into the workforce, postsecondary education, or both.

[29] ED Data Inventory, http://datainventory.ed.gov/

statistical policy and standards for high quality, objective, timely, relevant data; leveraging interagency cooperation to meet common goals; addressing key measurement issues; maintaining and supporting ongoing and new data collections; reaching increasingly broader audiences; making difficult decisions in the area of fiscal and human resources; and effecting change in governmentwide policies. Although there are other examples that could be drawn from every statistical agency, the women selected represent a cross-section of experiences across time in post-World War II America and provide examples of various aspects of leadership, from establishing new data collections that continue to play an important role in measuring the state of the American public, to establishing committees and boards that have played vital roles in improving the quality of federal statistical data and in monitoring the performance of statistical agencies, to work that spans a number of years to get legislation enacted and guidance issued that impacts how the federal statistical system does its work, to the restructuring of the largest federal statistical agency to meet the constraints imposed by level or decreasing budgets. One common theme is that in each case, the women involved recognized and acted on a specific need that influenced the future direction of the federal statistical system.

Bibliography

American Statistical Association Membership Newsletter, (2011). Meet the Commissioner of the Statistics of Income: Susan Boehmer, *Amstat News*, May 1.

American Statistical Association Membership Newsletter, (2011). Meet the Commissioner of the Bureau of Transportation Statistics: Patricia Hu, *Amstat News*, July 1.

American Statistical Association Membership Newsletter, (2011). Meet the Commissioner of the National Agricultural Satistics Service: Cynthia Clark, *Amstat News*, December 1.

American Statistical Association Membership Newsletter, (2013). Meet the Commissioner of the Bureau of Labor Statistics: Erica Groshen, *Amstat News*, April 1.

American Statistical Association Membership Newsletter, (2012). Meet the Commissioner of the Economic Research Service: Mary Bohman, *Amstat News*, May 1.

Burns, S. (Ed.) (2011–continuously updated online). *NCES Handbook of Survey Methods* (NCES 2011-609). US Department of Education, National Center for Education Statistics, Washington, DC: US Government Printing Office. http://nces.ed.gov/pubs2011/2011609.pdf.

Committee on National Statistics, *Creating a Center for Education Statistics: a Time for Action,* 1986 National Academy Press, Washington, D.C.

Federal Statistical System: *Agencies Can Make Greater Use of Existing Data, but Continued Progress is needed on Access and Quality Issues.* GAO-12-54.

Muko, M. (2011). Margaret Martin: A Leader in the Federal Statistical System. *Amstat News,* September 1, Statisticians in History feature, http://magazine.amstat.org/blog/2011/09/01/margaretmartin/.

National Research Council. (1993). *Private Lives and Public Policies: Confidentiality and Accessibility of Government Statistics.* Washington, DC: The National Academies Press.

Parker, R.P. (2006). Data-Sharing History and Legislation: Background Notes, in National Research Council. *Improving Business Statistics Through Interagency Data Sharing: Summary of a Workshop.* Washington, DC: The National Academies Press.

Science News, October 22, 2010, Society for Science and the Public, Washington, DC.

Seastrom, M.M. (2002). *NCES Statistical Standards and Guidelines,* US Department of Education, National Center for Education Statistics, Washington, DC. http://nces.ed.gov/statprog/2002/stdtoc.asp.

Seastrom, M.M. (2012). *NCES Statistical Standards and Guidelines,* US Department of Education, National Center for Education Statistics, Washington, DC. http://nces.ed.gov/statprog/2012/.

Statistical Programs of the United States Government: Fiscal Year. (2014). OMB, Washington: DC.

US Department of Education Information Quality Guidelines, http://www2.ed.gov/policy/gen/guid/iq/iqg.html.

Wegman, E.J. and Martinez, W.L. (2007). A Conversation with Dorothy Gilford, *Statistical Science,* Vol. 22, No. 2, 291–300.

Part IV

Leadership Competencies

12

Competencies Needed for Statistics Leadership from an International Perspective

Motomi (Tomi) Mori and Rongwei Fu

Oregon Health & Science University

In this chapter, we will discuss challenges and opportunities in leading a diverse group of statisticians and biostatisticians in an academic medical center and share our personal stories and lessons learned. We emphasize the importance of cultural awareness and sensitivity in statistical leadership.

Statistics Is a Culturally Diverse Profession

Close your eyes and think back to your own graduate school days. Can you see yourself when you were a teaching or research assistant? Who were your office mates and fellow students in your class? Even 25 years ago, many statistical graduate students were international students, mostly from Asian countries, such as Korea, China, and India. The proportion of international students remains high over the years. The National Foundation for American Policy (2013) released a brief report called "The Importance of International Students to America" in July 2013, which included Table 12.1.

According to Table 12.1, almost 45% of the graduate students in mathematics and statistics are international students, and this number may be even higher for PhD students in statistics. So it is not surprising that statistics is a diverse profession, consisting of people from all over the world. This is also confirmed by the most recent Student and Exchange Visitor Information System (SEVIS) report (SEVIS by the Numbers July 2014, the US Immigration and Naturalization Services). The most recent report (April–July 2014) (US Immigration Services, 2014) shows a total of 981,440 international students in the US; of those, 344,299 (36%) students major in science, technology, engineering and mathematics (STEM) fields, with 85% from Asian countries (see Figure 12.1). As we know from our own graduate school days, international students, from Asia in particular, are overrepresented in

Table 12.1

Full-time Graduate Students and the Percent of International Students by Field (2010).

Field	Percent of International Students	Number of Full-time Graduate International Students	Number of Full-time Graduate US Students
Electrical Engineering	70.3%	21,073	8,904
Computer Science	63.2%	20,710	12,072
Industrial Engineering	60.4%	5,057	3,314
Economics	55.4%	7,587	6,117
Chemical Engineering	53.4%	4,012	3,504
Materials Engineering	52.1%	2,660	2,891
Mechanical Engineering	50.2%	8,352	8,273
Mathematics & Statistics	44.5%	7,840	9,766
Physics	43.7%	5,716	7,369
Civil Engineering	43.7%	6,202	7,989
Other Engineering	42.1%	7,279	9,992
Chemistry	40.3%	8,059	11,952

Source: National Science Foundation, Survey of Graduate Students and Post-doctorate, webcaspar.nsf.gov. US students include lawful permanent residents.

statistics, while American students are underrepresented. Increasingly, this is also reflected in the composition of the faculty in statistics and biostatistics departments.

These observations point to the importance of cultural diversity awareness as a key component of leadership in statistics and biostatistics. In particular, women face even more difficult challenges because in most cultures women are traditionally expected to take a subservient role and focus more on family and housework. While these attitudes are continuously changing and evolving over time, the impact of this traditional expectation is nonetheless profound and long-lasting, on both women and men. As a relatively extreme example, an Asian female graduate student was once asked by her US advisor whether it was true in her country that women were not supposed to make their own decisions. While she and her advisor were both surprised by such a statement, it came from her fellow, male student from her country. The statement was, of course, untrue, and Asian countries have been trying (albeit with divergent levels of commitment and success) to promote gender equality for some years. To be a woman leader in statistics, we have additional hurdles to overcome. The impact on highly educated women can be complex, and the effort to break free from those traditional expectations can be a battle in itself. In academic settings, department leaders are called upon to mentor graduate students and junior faculty from a wide range of cultures that they may not be familiar with. Unfortunately, the majority of graduate programs do not teach us how

U.S Immigration
and Customrs
Enforcement

Stem Students, Schools and Fields Of Study
How many STEM F & M students are in the
United States and where and what do they study?

Percent of Stem F & M Students in the
United States Per Continent

Europe (4%)
Australia &   Africa (5%)
Pacific Islands (0.26%)   South America (2%)
North
America (4%)

Asia (85%)

Total # of Stem and All F & M Students

Students (F-1 & M-1)			
	Stem	**All F & M**	**Percentage In Stem**
F-1 & M-1 Students	344,299	981,440	36%
Top Three Stem Schools for F-1 and M-1 Students			
School	**# Of Stem Students**		
Purdue University	6,314		
University of Southern California	5,845		
University of Illinois	5,348		
Top Three Stem Majors for F-1 and M-1 Students			
Major	**# Of Stem Students**		
Engineering	148,164		
Computer and Information Sciences and support Services	75,309		
Biological and Biomedical Sciences	35,580		

* Continent information was compiled using the United Nation's composition of macro geographical (continental)
regions, geographical sub-regions, and selected economic and other groupings.

FIGURE 12.1
STEM students, schools, and fields of study, US Immigration Services, 2014.

to be good managers, mentors, or leaders, and nor do they teach awareness of, and sensitivity to, cultural differences. We usually learn these skills on the job, on our own through trial and error, mishaps, and occasional triumphs. So we would like to share our stories and lessons learned, and hopefully these events echo some of the experiences of our readers. For confidentiality reasons, we had to alter the names and specific situations, so consider it as a docudrama rather than a reality show.

Performance Evaluation: Is Modesty a Virtue?

Many Asian cultures, in particular in the East Asian countries, value group accomplishments versus individual accomplishments. Modesty is highly valued, and those who publicly announce their accomplishments are often considered to be bragging and showing off. There is a Japanese saying that a noble eagle hides its claws, meaning that those who are intelligent do not show it off in public. Modesty is a virtue, and speaking of one's success and accomplishments in public is considered arrogance. However, this is not necessarily true in US culture, where individual accomplishments are valued, and

assertiveness and self-promotion are in many ways viewed positively or sometimes a "must."

As a manager and supervisor, I (MM) have conducted many annual performance evaluations. A typical review process involves employee self-evaluation,
independent supervisor evaluation, a joint conference, and the postconference
evaluation. The final evaluation form then gets sent and filed with human
resources. Over the years, I found people at two extreme ends, those who
rate themselves "exceptional" on every single item ("egomania"), and those
who rate themselves "needs improvement" on every single item ("extreme
modesty"). While most people are somewhere in-between, there is definitely
a cultural tendency for lower or higher self-rating. Given an equivalent performance, I would say that most of my non-American employees tend to rate
themselves lower, while most of my American employees tend to rate themselves higher. This is one example, and such differences are actually reflected
in many levels in everyday communication or whenever there is a chance to
talk about the outcomes of any given project. Of course, there is always an
individual variation. I think of the observed rating as a sum of $\mu + \alpha_i + \beta_j$,
where μ is the universal norm, α_i is the i-th cultural norm, and β_j is j-th
individual variation. As it is in statistical modeling, there is an identifiability
problem for α_i and β_j, and for those who lack a specific cultural knowledge, it
will be almost impossible to tease out how much is due to cultural background
and how much is due to that particular individual.

Generally speaking, however, in the US, we are expected to be our own
advocate, and self-promotion and self-marketing are considered normal and
sometimes essential, especially in certain situations, such as seeking a promotion. On the contrary, in many Asian cultures we take a more passive approach,
believing that, if we work hard and do an excellent job, others will notice and
reward us, and there is no need to publicly announce our accomplishments. I
wonder how many of us made this mistake, waiting to be noticed, appreciated,
and rewarded, and realizing that we were ignored or even allowing someone
else to take credit for our work. The bottom line, here in the US, is that we
need to speak up and be our own advocates! One of the challenges is to know
when, who, where, and how to speak up. I found some non-US colleagues
who tried to overcompensate for this difference by going to the other extreme,
and pursuing the "egomaniac" strategy, thinking that this is what one is supposed to do. Too much assertiveness can be viewed as aggressive, noncollegial
and noncollaborative, even in the US. Finding the right balance is tricky and
depends on the organization. For example, an academic university setting
tends to be quite traditional and hierarchical; one needs to learn what level
of assertiveness is optimal, when to demand and negotiate, and when to pull
back, accept the situation, and be a good citizen of the department and the
university. What I learned over time is that we actually possess a skill set to
do this correctly. We are trained to be data-driven, objective, scientific, and
to go beyond stereotypes, biases, and predisposing expectations. We need to
apply our statistical skill sets and principles to our own self-assessment and

evaluate each situation objectively. I still believe that modesty is a virtue, but one can simultaneously be modest and be an advocate.

When in Doubt: Just Ask (Calmly)

I was in the late stages of graduate school working on my dissertation. The university offered summer scholarships for PhD candidates specifically to enable students to focus on their dissertations. However, the university could not award these scholarships to everyone who applied. Therefore, if a department had more students than scholarships, preference would be given to those students who were most advanced in their dissertation. That summer four of us applied, and I (RF) was the only one who did not receive an award. I was surprised and disappointed, as I had made more progress on my dissertation than the other three (eventually completing it sooner than the others). So I decided to ask the department chair the reasons for the decision — indeed, I only meant to ask why, and I was not angry or protesting against the decision. Although I do not recall his immediate answer, he later admitted to me that he had made a mistake. While he was unable to change the decision, he was able to secure equivalent funding for me from another departmental source. Incidentally, while it may have been purely coincidence, I was the only female student among the four.

As a graduate student, the summer scholarship was a major issue for me, and therefore I felt compelled to ask for an explanation, while at the same time remaining calm. I was surprised by the outcome, as I was unaware that there might be an alternative source of funding. However, I learned that there is nothing to lose by stepping forward and asking questions in this kind of situation, and that there are reasonable ways to work things out, even between a departmental chair and a graduate student!

Prioritizing Your Work: Learning to Say NO

We are all familiar with the term "terrible two's" to describe a 2-year-old child learning to say NO and MINE to everything. It is during the socialization process that we learn to say YES and to accept our limits and boundaries. Somehow that socialization process made non-US women (especially from Asia) even more prone to say YES and to develop a tendency to want to please everyone. This may be in part due to the fact that in some Asian cultures, young children are expected not to talk back in front of parents when criticized. Every day we get bombarded with requests for statistical consultation, guest lectures, manuscript reviews, letters of recommendation for faculty going up for promotion, committee service, etc. We (MM and RF) are terrible at saying NO, and if we do, we feel obligated to find an alternative solution for each request. When we analyze our own actions, they are derived from our own sense of obligation and responsibility combined with a strong desire to please everyone, to serve, and to be a good citizen. What we fail to acknowledge is our own limitation in what we can take on in order to attain the level of performance we expect from ourselves. Yes, we can probably take 50 consulting requests and do a fairly cursory job, or take on 20 consulting requests and do an outstanding job. We would rather take on less and do an outstanding job. What we learned is that saying NO does not mean leaving people hanging out to dry, but rather it emphasizes the need to delegate work effectively to others in our group, so that the researcher still receives the help needed, just not from us but from someone else. So saying NO and delegation often go hand in hand, and that means that we must be able to delegate.

One of the important lessons we learned over the years is to develop a trusting relationship with our master's level biostatisticians and delegate much of data management and programming tasks to them. It was hard at first, since many of us learned to be hands-on, and enjoy working with the data and programming. But as we move further into a career, we cannot spend all our time on a couple of projects, but rather we need to delegate time-consuming tasks, such as data management and initial analyses, to our staff. An effective leader knows how to delegate, how to work as a team, and trust and empower each member of the team to do her/his best.

Another important element of saying NO is prioritization. At the beginning of each week, we list all the tasks we need to accomplish during the week and prioritize them according to their importance and deadline. We then map out which day we need to work on which tasks, thus making up a weekly schedule. If we need to block out work time in the calendar, we will do so to insure that no one can book a meeting during that period. A brain needs a block of thinking time, and it is impossible to think deeply during 10-minute breaks between meetings. During the prioritization process, we may find other projects that we should have said NO to, and we will know which projects should be NO when a new request comes to our inbox. When there are competing deadlines, and when the collaborators are overly anxious, we also find it very helpful to explain to them about our workload and the "realistic" timeline that we could get results and our part to them. We think that it makes us, as well as the collaborators, less anxious.

Building Collaboration: Price of Misunderstanding

Once I [MM] was brought in to a situation where a junior Asian biostatistics faculty (Karen) and young American researcher (Tom) were upset with each other. I interviewed Karen and Tom separately, trying to figure out what happened. It turned out that Karen was upset with Tom because he asked her to perform a specific statistical analysis suggested by his mentor. His mentor wanted him to do similar analyses as her previous paper. Karen felt offended by Tom's request, because she felt that it was her job to decide on the appropriate statistical analysis, and she felt that Tom did not trust her ability or competence. Without explaining why she was offended, Karen told Tom that he was not respectful and did not know how to collaborate with a biostatistician. This made Tom confused, upset, and not wanting to collaborate with Karen any longer. When I found out what happened, I was visualizing two parallel lines that never cross, complete misunderstanding by Karen of Tom's intent, and both seeing the world in an entirely different way. Once I explained to Karen and Tom my understanding of the situation, Karen realized that she had misinterpreted the request and apologized to Tom. However, the relationship was already damaged, and Tom decided to work with another biostatistician. So the opportunity for collaboration was lost for Karen. If she wants to collaborate with Tom in the future, it would take a lot more convincing than a simple apology.

Much of a biostatistician's work depends on successful collaboration, and building a good relationship with research collaborators is a critical part of the job. We try hard to avoid miscommunication and misunderstanding, and make sure that we correctly understand the situation before we act on it. Open and honest communication, and frequent checking of our assumptions and perceptions are important components of successful relationships, whether it is research collaboration, mentor–mentee, or student advising. We have to invest in developing a good, trusting relationship, and sometime it takes going above and beyond what is essential. Otherwise, it will be another case of an important opportunity lost.

Work Family Balance: Time for Everything

I (MM) feel very lucky that I was able to have a family and career and did not have to choose one or the other. However, it was not easy, and I had to make different choices along the way. Over the years, I learned that, like a season, there is a time for everything, and I do not need to balance everything at the same time. For example, when my children were very small, I did not attend professional conferences or participate in grant reviews. However, once the children became teenagers, I started going to the conferences and served on National Institutes of Health (NIH) study sections and other review panels. My point is that we do not need to have a family–work balance at any one time, and it is OK to be unbalanced at a particular time point, as long as we can achieve a balance over the course of 15–20 years. Your short-term career goals can be adjusted according to the needs of your family. So it is important to take a long-term perspective in regard to one's career, and always remember that children grow up fast and you will not be able to make up the time with them. A few bumps and slowdowns in career progress can almost always be compensated for later.

One of the biggest challenges for non-US women is to meet the cultural expectations of being a wife and mother. In many cultures, women are expected to be the primary caretakers of the family, responsible for most of the housework including cleaning, cooking, washing, shopping, and caring for children and the elderly. It can be overwhelming and virtually impossible to meet these expectations while simultaneously keeping up with the demands at work. We need help from our spouses and must be prepared to make full use of the wide range of available community resources, such as babysitters, daycare, cleaners, online grocers, and take-out restaurants to make our lives a little easier. We should openly discuss challenges with our significant others and engage them in helping and sharing with the housework. We must acknowledge the unreasonable pressure and stress that accompany the expectation of being a "supermom," and be prepared to seek help and support from our community and those closest to us.

One often does not intentionally choose a family or career. Sadly, it just happens. Recently, I met a Nigerian woman with a PhD in nursing, and she told me the story of Alice (not her real name), her student in her 40s who recently completed her PhD in Nigeria. Alice was deeply depressed and almost regretting her PhD degree, because she felt that it made her undesirable as a wife and mother. She wanted to start a family but could not find a husband. She felt that Nigerian men prefer marrying a younger woman with less education than they have, and that therefore no Nigerian man would marry her now since she has a PhD and is older. She felt that men would be intimidated by her intelligence and education. I have heard similar stories from colleagues from other Asian countries as well as in the US. My American colleague, who received her PhD from Harvard, said that she had a hard time dating when she was an assistant professor because men were intimidated by her Harvard PhD. Luckily, she is now married, has two children, and is successfully balancing family and work, although she notes that she started her family late because she was not able to find a husband until later in her 30s. Many people assume that single women with a successful career choose a career over family, but in many cases, it is not necessarily the result of conscious choice. Although the world is slowly changing, with more women going to the university and graduate schools, old prejudices and cultural expectations stubbornly persist, and unfortunately, in some countries and cultures, highly educated women encounter more difficulties in finding suitable mates and starting families.

Archetype of Biostatistician: Is There a Biostatistician Personality Type?

I attended the International Biometric Conference (IBC) in Dublin, 2000, and there were close to 800 biostatistics and statistics professionals from all over the world. As a part of the IBC, we had a tour to the Merry Ploughboy, a traditional Irish music pub. That evening there were 40–50 IBC biostatisticians, and two other tour groups: a group of younger, casually dressed Spanish tourists, and another group consisting of older, well-dressed Italian tourists. In the pub, there were at least five rows of long, connected tables, looking like very long wooden picnic tables, each table seating somewhere between 30–40 people. There were Irish folk singers with guitars and pints of Guinness, singing old Irish folk songs on the stage. Toward the end of the concert, the musicians instructed us to sing certain tunes and to do certain movements with the cue, i.e., holding hands with our neighbors and waving left to right, singing certain lyrics, etc. The IBC biostatisticians did exactly what we were asked to do, clapping hands, holding hands with our neighbors, and waving left and right, singing certain tunes and lyrics when instructed. On the contrary, the Italian tourists quietly watched the musicians and sat there doing nothing, totally ignoring the instructions as if they did not understand a word of English. The Spanish tourists were completely the opposite; they eagerly participated in the music and dancing, but ignored the instructions and did their own version of songs and dances, obviously having a great time but quite disruptive to the musicians and the rest of the tourists. I sat there and observed the IBC group with great curiosity. I realized that, regardless of where they were from, there was perhaps an archetype biostatistician personality, that is, a personality type and traits that we all share. We are generally respectful, fairly conformist, highly collaborative, and try to avoid conflict. We are like well-behaved children who do what we are told to do. Just as a certain body type is associated with a certain sport (e.g., basketball players are tall), a certain personality trait may be associated with a certain occupation or profession. It was amazing to realize that there are common traits among biostatisticians, regardless of where we come from.

Just as Asian culture predisposes us to be more collaborative and team-oriented, so also being a biostatistician does the same. Further, being a woman makes us even more passive, so there is a triple hurdle we must overcome when we want to speak up and be our own advocates. No wonder this is so difficult for us and takes so much courage, as well as emotional and intellectual energy.

Cultural Competencies and Cultural Intelligence

In the book *Culture, Leadership, and Organizations: The GLOBE Study of 62 Societies* (House et al., 2004), the GLOBE Study researchers surveyed almost 20,000 middle managers in 62 countries. They found many leadership characteristics that are universally valued and also many that are valued differently among different countries. Universally valued leadership characteristics include: trustworthiness, optimism, motivation, decisiveness, intelligence, kindness, and altruism. Those that varied by country include: independence, ambition, cunning, intuition, logic, and risk-taking. Some of the stories in this chapter are rooted in the lower value that many Asian cultures place on independence, ambition, and risk-taking.

In the book titled *Leading with Cultural Intelligence: The New Secret to Success* (Livermore, 2010) [New York, NY: AMACOM], the author David Livermore describes the fourdimensions of cultural intelligence, which is measured by CQ (cultural quotient analogue of IQ): CQ drive (showing interest and willingness to adapt cross-culturally), CQ knowledge (knowledge and understanding of cross-cultural issues), CQ strategy (being able to develop cross-cultural strategies based on the knowledge), and CQ action (cross-culturally appropriate behaviors). He states that today's top business executives need CQ in order to understand diverse customers; manage diverse teams; recruit and develop cross-cultural talent; effectively adapt leadership style; and demonstrate respect. Sound familiar? These could easily apply to statistics or biostatistics units in academic, government, or industry settings. To be effective leaders in a culturally diverse environment, we need to acknowledge and appreciate cultural diversity, be willing to learn and understand, develop a departmental culture of inclusion, and create business plans that leverage and take advantage of diversity, with each member of the department behaving in a cross-culturally appropriate manner. For example, the faculty and students would appreciate a simple courtesy like not scheduling a departmental retreat or social event on a major Jewish or Islamic holiday.

As part of writing this book chapter, I (MM) interviewed Leslie Garcia, director of the Center for Diversity and Inclusion (CDI) at the Oregon Health & Science University (OHSU). She reminded me of the first goal of the OHSU Vision 2020 Strategic Plan: "Be a great organization, diverse in people and ideas." After giving me the overview of various resources the CDI offers to students, staff, and faculty in both research and patient care arenas, she said, "You know, no one is ever 100 percent knowledgeable or competent in every culture of the world. I do not actually expect people to be culturally competent or knowledgeable in all cultures. But I would like people to develop cultural awareness and sensitivity." Then she gave me a small card, the size of a business card, with the B.E.L.I.E.F. Model (Dobbie et al., 2003) printed on it:

B.E.L.I.E.F. Model

Beliefs about health (What caused your illness/problem?)
Explanation (Why did it happen at this time?)
Learn (Help me to understand your belief/opinion.)
Impact (How is this illness/problem impacting your life?)
Empathy (This must be very difficult for you.)
Feelings (How are you feeling about it?)

Although the B.E.L.I.E.F model is used for a clinician treating culturally diverse patients, the same principles can apply to a department leader managing culturally diverse students, staff, and faculty. A leader's willingness to listen and understand a situation from a cross-cultural perspective goes a long way toward building and sustaining a trusting relationship with the individual, and ultimately proves to be a successful business strategy for recruiting and retaining global talents.

Summary

So what are the competencies needed for statistics leadership from an international perspective? We discussed the importance of cultural awareness and sensitivity, being your own advocate, prioritization and delegation, and communication. We acknowledge particular challenges faced with being Asian women biostatisticians and the cultural barriers to taking a leadership role.

To be an effective leader, one needs not only to be a leader in statistics, but also a leader in all aspects of managing people and managing the department and business. Statistics and biostatistics graduate programs only provide technical training in statistics and biostatistics and do not teach human and business management. Yet, these latter skills are the underpinnings of effective leadership, with technical statistical knowledge representing only a small piece of the puzzle. As we know, the fields of statistics and biostatistics are very diverse, and there is no reason to expect this to change. This means that those who will lead the field must be culturally competent and appreciative of cultural diversity, or, as David Livermore terms it, have a high CQ.

A high CQ is requisite for international student recruitment and retention, faculty recruitment and retention, faculty mentoring and promotion, and promotion of the fields of statistics and biostatistics within the university community and in the wider community. We need leadership not only at the level of the department and institution, but also nationally and internationally. One of the reasons why the statistical profession feels left behind by large federal government initiatives, such as the "bioinformatics" and "BD2K" (Big Data to Knowledge) initiatives (*http://bd2k.nih.gov*) is that we do not engage in sufficient self-promotion and feel uncomfortable in doing so.

So, to conclude, our recommendations are as follows:

1. Take control of your own destiny and be your own advocate. Remember that you alone are in charge of your own life and career.

2. Do not compare yourself with your own neighbors or colleagues from your own country. Listen to your own voice and set your own internal goals and standards that are realistic and achievable. These goals and internal standards may change as you go through different stages of your life and career. It is OK and perfectly normal.

3. Do not be afraid to ask. Remember, the worst thing that can happen is being told no. If you do not ask, you do not gain.

4. Be a statistician, i.e., data-driven and objective in assessing your own performance and environment.

5. Be flexible and willing to learn new skills for leadership and adapt to a new situation.

6. Embrace changes and challenges, and convert them into opportunities for personal growth. Uncomfortable and difficult situations are frequently only

the result of misperception. You can make the effort to view them differently. Remember, a glass can be half full or half empty.

7. Be aware of your own cultural background, biases, and expectations. Be sensitive to others who come from different cultures. Be watchful of not stereotyping others, and make certain that you acknowledge unique individual variations and characteristics.

Sheryl Sandberg's book *Lean In: Women, Work, and the Will to Lead* (Sandberg, 2013) begins with a chapter titled "The Leadership Ambition Gap: What Would You Do If You Weren't Afraid?" We should all ask ourselves this question whenever we feel afraid or feel like shrinking back into our inner cocoon. Then we should take a leap!

Bibliography

Dobbie, A.E., Medrano, M., Tysinger, J., and Olney, C. (2003). The BELIEF Instrument: A preclinical teaching tool to elicit patients' health beliefs. *Family Medicine.* 35(5): 316–9.

House, R.J., Hanges, P.J., Javidan, M., Dorfman, P.W., and Vipin, G. (editors) (2004). *Culture, Leadership, and Organizations: The GLOBE Study of 62 Societies.* Thousand Oaks, CA: Sage.

Livermore, D. (2010). *Leading with Cultural Intelligence: The New Secret to Success.* New York, NY: AMACOM.

National Foundation for American Policy (2013). The Importance of International Students to America. URL: http://www.nfap.com/pdf/ New%20NFAP%20Policy%20Brief%20The%20Importance%20of% 20International%20Students%20to%20America,%20July%202013.pdf.

National Institutes of Health (NIH) Big Data to Knowledge (BD2K). URL:http://bd2k.nih.gov/ #sthash.UCpuKYmH.dpbs.

Sandberg, S. (2013). *Lean In.* New York, NY: Alfred A. Knopf.

US Immigration and Naturalization Services (2014). Student and Exchange Visitor Information System (SEVIS) report: SEVIS by the Numbers July 2014. URL: https://www.ice.gov/doclib/ sevis/pdf/by-the-numbers.pdf.

13

Organizational and Business Acumen: Observed and Latent Attributes

Sally C. Morton

University of Pittsburgh

Introduction

The field of statistics is faced with more opportunities, and more challenges, than ever before. The recent 2013 International Year of Statistics capstone event in London produced a report that stated the discipline "is as healthy as it ever has been, with robust growth in student enrollment, abundant new sources of data, and challenging problems to solve over the next century" (London Workshop, 2014, p. 5). To meet these challenges, the profession must bring all talent to bear, embrace, and engage diversity, and lead. Women as leaders are a dimension of that diversity, and are essential to the future success of the discipline. The goal of this book is to explore the role of statisticians, and particularly of women statisticians, as leaders.

The purpose of this chapter[1] is to provide some practical advice on the critical skill of acumen, which is essential for effective leadership. This skill can be cultivated if an individual identifies it as important, preferably early in a leadership position, and seeks individual and professional development opportunities. I will first define acumen, and distinguish between the related concepts of organizational and business acumen. I will then focus on three aspects: understanding an organization internally; placing that organization in the "big picture"; and tools for developing and optimizing acumen. These issues will be discussed both with respect to observed dimensions, as well as latent ones, drawing a parallel with observed and latent variables in a statistical model, i.e., those variables that can be observed directly and those that cannot. While I admit this is a bit of a disciplinary ploy, after 25 years of

---

[1]This chapter is based on a presentation given at the American Statistical Association (ASA) Workshop "Preparing Statisticians for Leadership: How to See the Big Picture and Have More Influence," held at the Joint Statistical Meetings in Boston in August 2014. This workshop is part of the "Developing Training in Statistical Leadership" presidential initiative of the 2014 ASA president Nat Schenker (2014).

experience, I am convinced that both perspectives are inherently important to leadership success.

Organizational and business acumen are critical for the success of any leader. From personal experience, I can attest they are particularly important for an individual new to leadership, or to someone who is unique to the role, such as a woman leader in an organization that has not had a woman leader before. Abilities that may be more common among women, such as collaboration, team engagement, listening skills, and openness to learning, are particularly useful in developing acumen.

My own background in statistics is rather eclectic in that I have moved between the industry and academic sectors, while focusing on public policy, primarily in health, an emphasis which has connected me to the government sector. After attaining my PhD in statistics, I worked for fifteen years at the RAND Corporation, which is a private, nonprofit, policy "think tank," and led the RAND Statistics Group. I then joined RTI International, and led a large department of statisticians and epidemiologists. I am currently professor and chair of Biostatistics in the Graduate School of Public Health at the University of Pittsburgh. Throughout my career, I have led projects and supervised staff, both scientific and administrative. I recount my history for context rather than interest — to emphasize that I have led groups of various sizes consisting of different types of members with a variety of backgrounds, purposes, and goals. In this chapter, I will draw some distinction, if appropriate, between the different sectors in the statistical discipline, but in general, "organization" will refer to a group in any setting, i.e., an academic department, an industry group, or a government group.

Definitions

Leaders "inspire people to take a specific direction or action when they truly have the freedom or choice to do otherwise" (LaFleur, Seastrom, and Sullivan, 2014). As elucidated at the American Statistical Association (ASA) 2014 Workshop on "Preparing Statisticians for Leadership: How to See the Big Picture and Have More Influence," statistical leaders purposefully put themselves at the center of issues that are critical to their organizations, and motivate others to initiate change, develop solutions, and create value by acting on data and statistical reasoning.

Leadership opportunities are becoming more available for statisticians, whether via a position at a new employer, or in a new role at one's current institution. The former situation can be particularly challenging, and considering prior to starting a position how to be as effective as possible is well worth the investment (Ciampa and Watkins, 1999, pp. 121–139). The latter situation has the added complexity that one's position changes relative to former peers. Leadership opportunities may be for a specific group of individuals, e.g., an academic department; for a specific task, e.g., a scientific project; or more nebulous, e.g., being the senior statistician at a company. All of these situations require acumen.

In this section, I will define acumen broadly and then distinguish between organizational and business acumen. At the ASA workshop, *acumen* was defined as

> A keenness and quickness to understanding and dealing with a situation. Synonyms are awareness, discernment, grasp, insight, perception, understanding, and wisdom. Acumen includes seeing the big picture (LaFleur, Seastrom, and Sullivan, 2014).

In the remainder of this chapter, I shall focus on several pieces of this definition: first, understanding; second, seeing the big picture; and third, tools for dealing with a situation.

University Advancement at the University of Washington defines organizational acumen as

> The ability to size up a situation, balance reason and the interest of others, and act in a decisive, timely and appropriate manner that is congruent with the organization's values, goals and mission to achieve success (University of Washington, 2014).

If I expand the definition of organizational acumen to include insight about "governance, environment, culture, processes, procedures and how decisions are made" (California State University, Fullerton, 2014) for an organization, then business acumen may be defined as a focus on external understanding of market trends, the business environment, and how the organization functions with respect to competitors. Statisticians must be both inwardly facing in understanding one's organization, but also outwardly looking in order to position their organization for success.

I note that business acumen applies to organizations in any statistical sector: academe, government, or industry. "Business" does not necessarily mean producing a product, or existing in a for-profit setting. Academic entities are in the business of education, research, and service; government agencies produce products for their stakeholders. To be an effective leader, organizational and business acumen are both essential. For the remainder of this chapter, no distinction is made necessarily between the two.

Acumen — A Means to Leadership

In order to achieve acumen, both organizational and business, a leader must (University of Washington, 2014):

- Have a deep and thorough understanding of the organization, including its strengths, weaknesses, and competitors;

- Stay aware of current trends, policies, and technologies that can impact the organization; and

- Understand leadership priorities, anticipate reactions, and successfully communicate with leadership.

In particular, communication plays an important role in attaining acumen and effectively using that knowledge in leadership. Recently, the profession has identified communication as a key skill for statisticians, as evidenced by the ASA focus on the development of this expertise (Rodriguez, 2012). My own department at the University of Pittsburgh has integrated the teaching of communication into our graduate student curriculum (Buchanich, 2012), as have other academic programs. The discipline's historical emphasis on good interpersonal skills (Morganstein, 2012) in consulting and collaboration is useful here, as those skills can parlay into success at attaining acumen. Statisticians should realize they have these skills already, but may not have utilized them in this context or with this audience before. Professional development throughout a leader's career should include development and improvement of communication skills, both written and verbal, and for a variety of audiences. Myers (2014, p. 129) emphasizes the need for clarity in communication by a leader; in fact, this author uses the word "vivid" to convey the passion and focus necessary in this clarity. Throughout the positions I have occupied in my career, the ability to communicate, particularly with nonstatistical audiences, be they CEOs, policymakers, or students outside of statistics, has been highly prized.

Statisticians understandably pride themselves on technical skills and expertise. However, as 2012 ASA President Robert Rodriguez puts it, we need to "move to the middle" (Rodriguez, 2014) to be part of the decision-making at institutions. Statisticians are too often at the technical end of the table, with business (or academic or government) people at the other end of the table. To be effective leaders, statisticians need to move to the middle and bring their colleagues along with them. Acumen is a means to make that move.

Women leaders may be particularly adept at moving to the middle, given they are often perceived as less threatening or adversarial than men. Women often are better listeners and consensus-builders. Myers (2011, p. 37) emphasizes the need for all leaders to be authentic in their approach in order for others to have trust and confidence. This author also counsels letting "others have their own voice," even when it is difficult to do so (Myers, 2011, p. 99). Statisticians should value their own consulting experience and expertise, and realize they are transferable to the leadership setting.

Understanding Organizational and Business Acumen

In terms of observed understanding, obviously one must be cognizant of the organizational chart of one's organization. Reporting lines, as well as procedural responsibilities such as who has signatory authority for different decisions, should be understood. As a leader, defining clearly who is responsible for what and to whom sets the stage for transparent communication about responsibility and authority, and can increase efficiency and effectiveness. In my experience, organizational charts are not as common in academic settings as in industry, and, in fact, individual faculty may be uncomfortable with the idea of direct reporting to supervisors, given such authority may seem contrary to the concept of academic freedom. However, it is possible, and from my perspective preferable, through the establishment of committees with specific authority and responsibilities to effectively determine and lay out an efficient mode of operation that furthermore involves all via a participatory approach and promotes decisions by consensus (Leaming, 2007, pp. 63–67).

A related organizational chart is the latent one. This latent organizational chart may be similar to the observed organizational chart, but the two charts are almost never exactly the same. The latent chart is often driven by history, which may be difficult for someone new to an organization to uncover. Personal connections, family ties, friendships, or animosities may also underlie and affect the latent chart. As a very simple example, often married couples do not share the same name, so a person new to the organization may not even be aware of such relationships initially. Good advice is to tread carefully at the outset. The latent organizational chart does not hang on a wall, and indeed, two people in an organization may disagree as to exactly what relationships exist, or what they mean, in this organizational chart. This underlying organization may also be called the "shadow organization" or the "informal organization" (Ciampa and Watkins, 1999, p. 16).

Women leaders, who may be excluded from some personal interactions due to gender, either purposively or by accident, may have to work particularly hard to gain a thorough understanding of the latent organizational chart. In my experience, albeit limited to three employers, the latent organizational chart is more challenging to uncover in an academic setting, because one may not even start with a detailed observed structure, and thus it is harder to have a clear conversation about the latent traits behind the observed.

The Donabedian health services research conceptual model for evaluating quality of care may be a good paradigm to consider (Donabedian, 1988). In this model, "structure" is the context in which health care is delivered, "process" is the manner in which it is delivered, and "outcome" is the health outcome the patient experiences. In the acumen setting, structure is the actual organizational chart hanging on the wall, process is the latent organizational chart whose relationships impact process, communication, and action at the organization, and outcomes are the results you wish to produce as a leader.

For the leader at an organization, understanding how you fit into both the observed as well as latent organizational charts is important. Obviously, the observed chart is clear in terms of reporting authority, but making sure that your responsibilities are agreed upon by both your supervisor and you is a necessity. The latent flow chart is more difficult. Though you may report to one individual, whom else do you need to keep informed? One must beware of a hidden expectation that is not made clear and which if one does not meet it, spells disaster. A good tactic is to "manage up," that is, to ask your supervisor, and perhaps other senior members of the organization, what constitutes success early in one's tenure as a leader. Eliciting feedback early rather than waiting for a reaction after a longer period of time, perhaps at a formal review point, is useful for, if nothing else, you have time to adjust before a performance evaluation. In my own experience, the performance review process is more developed in industry, so this approach of proactive engagement of one's supervisor may work better in that type of employment setting.

The elevator conversation is one often discussed as part of preparing for a job interview or conference attendance. Could you succinctly state in a three-minute elevator conversation why someone should hire you, or what your conference talk or thesis is all about? Now imagine that your supervisor enters the elevator with his/her supervisor. What might happen, and is your supervisor ready with the speech you want said? Part of organizational and business acumen is keeping your supervisors, both observed and latent, informed so that all elevator rides are successful from your perspective. Essentially, you would like the outcome to be as if you were there, along for the ride, and able to speak for yourself and your group. In my experience, advocating for your group's or department's members well is of primary importance as a leader, particularly if you are leading statisticians whose purpose, relevance, and contributions may not be commonly understood. And if you perform this advocacy well, your members will greatly appreciate your efforts, and the group will move forward as well.

Seeing the Big Picture

The observed big picture can be succinctly defined via a strategic plan. Useful building blocks of such a plan include:

- A *vision* for the future;

- A *mission* to define what the organization does;

- *Values* to guide the actions of the organization and its members;

- *Goals* to achieve the vision and accomplish the mission; and

- *Metrics* to measure progress toward the goals.

A strategic plan can be a useful guide that enables all to see where the organization is trying to go, and provides a common language to elicit and assess ideas. The ASA underwent a strategic planning activity in 2008, with subsequent revision in 2012, which resulted in a succinct and robust plan that is easily presented on two pages (American Statistical Association, 2012). The plan is organized around the two themes of inclusion ("The ASA as the Big Tent of Statistics") and visibility ("Increasing the Visibility of the

Profession"), with eight dimensions, for example, membership growth under inclusion, and public awareness under visibility. All ASA board initiatives are organized around the plan and progress is easily measured versus the plan. Even ideas that do not fit easily within the plan are welcomed, with the acceptance that the plan is a dynamic document that is to be updated and changed periodically.

Leaders are well-served to establish a strategic plan to transparently and effectively rally their members to common themes with clear and frequent communication (Morton, 2009). An important element to consider in the construction process is succession planning, particularly for tenured faculty in an academic setting. When success is achieved, it is easily identified in the context of this plan, and decisions can be explained and justified via the plan as well. Though I am cognizant that strategic planning is often dismissed as ineffectual, a strong and visible strategic plan can be enormously helpful to a leader.

The latent big picture that is sometimes forgotten when one becomes focused internally on an organization is the constellation of competitors, whether market-driven or not. Leaders may wish to conduct a SWOT analysis, which stands for strengths, weaknesses, opportunities, and threats. By honestly assessing the organization's placement in the wider world, the group can effectively determine how best to proceed. The elements of this exercise may be very different for an industry group versus an academic one, but the purpose and effectiveness of this exercise will be useful in either sector. In my experience, a SWOT is less common in academe, but this approach is increasingly being seen as worthwhile outside of industry.

Personally, a leader should conduct a SWOT analysis as well. Though perhaps such an assessment is uncomfortable, a leader should determine his/her value-added. As Myers states, "Leadership begins with ourselves" (Myers, 2011, p. 11). This assessment includes considering successors. One's own departure from an organization should not be detrimental to that organization, and effective planning allows one's impact to have a long-lasting, and in fact positive, trajectory into the future. I've found that some supervisors are comfortable with candid planning and others are not. Regardless of support or lack of it, the leader should pursue such planning for the good of the organization. In addition, if your group or department has members that are interested in leadership themselves, developing these colleagues will engage them and allow them to contribute their best to the organization.

Statisticians necessarily focus on technical issues and often do not possess the organizational awareness, that is, the acumen, to answer the following:

- How does the organization define success?

- How does the organization measure success?

- How do you contribute to the organization's success overall and with respect to specific goals?

- How are decisions made at the organization?

Statisticians can make a unique scientific contribution to metric choices and measurement decisions. By the nature of our training, statisticians have the ability to formulate questions, and provide suggestions for measurement definition, collection, and analysis. Thus, one can contribute both scientifically as a statistician and as a leader to achieving the organization's goals. I was fortunate to work at an employer, RAND, who embraced this concept and welcomed such input from its scientific staff, recognizing and allowing each discipline to contribute in its own unique way.

As an aside, resources are essential to achieving goals. Particularly, when considering a new leadership position, whether internally at one's current organization or at a new employer, assess the resources thoroughly. Understanding what is available and how it is controlled are essential. If a potential employer is unable, or, indeed, unwilling, to share this information, I would strongly consider not accepting the position. Once one has resources, transparency about the criteria that are used to distribute those resources and the decision-making process are paramount to good leadership and your group's trust in you.

Learning Useful Tools

A leader can and should learn specific tools and skills to increase organizational and business acumen. First, in terms of observed skills, statisticians should learn how finances work at their organization. This objective may seem obvious in a for-profit institution but is applicable regardless of one's employment sector. In academic and nonprofit settings, the intricacies of funding attained from federal, foundation, and private sources are becoming ever more complex in terms of overhead rates, fees, and flexibility. Understanding how money is attained at an organization, what it can be used for, who controls it, and how finances impact meeting the strategic vision overall is essential (Leaming, 2007, pp. 188–205). If a leader does not understand the money, he/she cannot make effective and timely decisions.

As a second observed skill, a leader should understand the human resources component of the organization. For example, the ability to provide a useful performance review is essential in terms of guiding and supporting individuals in their career progression. Constructing one's own performance review well is also paramount, as well as being able to review the organization one leads in the context of meeting the strategic plan. In addition, understanding the policies and procedures of an organization is critical. Policies, such as what qualifies for family leave, exist to ensure the fair and equitable treatment of all. Setting up policies, or perhaps guidelines, to distinguish them from the parent organization's legal policies, allows consensus to be built and agreed upon in the process. Guidelines on how departmental resources will be allocated are often the most contentious, and constructing such guidelines in an open and transparent manner allows not only all to participate and affect the use of these common resources, but also builds consensus and trust.

Professional development as a leader is ongoing in one's career. Formal courses are useful, as well as finding opportunities for coaching and mentoring. Determining a course of study with one's mentor or supervisor, with both short- and long-term goals, as well as an understanding of the support available to achieve the educational objectives, is an ideal way to begin a leadership position. Though all of us have limited time, elevating this professional development to the same level as ongoing statistical training is key.

In terms of latent tools, the most important is self-awareness. No one has the skills, or, for that matter, the time, to do everything. Identify what you are good at, and surround yourself with others who balance your strengths and weaknesses. Delegation is essential and delicate. If you cannot let go of control of an activity you have delegated, you will run into trouble. Not only must you cede the delegate the responsibility of the task but also the authority. Rest assured, the delegate will not do the task the exact way you would have, and perhaps will do it even better. Let it go. This is difficult, and it's not something I am very good at, I will freely admit, but I keep trying. And finally, be sure to give credit to the delegate. For example, allow this person to present material at meetings. At your organization's meetings, you should not be doing the majority of the talking. Meetings should be effective and efficient via an inclusionary process.

Conclusion

In this chapter, I have emphasized that acumen, both business and organizational, is a means to the end of leadership. Both observed and latent aspects of three key acumen aspects were discussed: understanding, seeing the big picture, and acquiring useful tools. Leadership is a lifelong ambition, and good leaders dedicate themselves to continual quality improvement. Knowing one's strengths and remembering constantly the potential to impact, both positively and negatively, is essential to success.

There is no doubt that statistics has potential leaders, and those leaders can have an important impact in an ever more data-driven world. Leadership is both an opportunity for, and a responsibility of, the discipline. Statistical consulting and analytic skills are transferable to the issues that leaders face, and thus statisticians should be confident they can tackle a leadership role, given their relevant background. While facing some biases in taking on leadership roles, women statisticians should also be confident that they can become effective leaders, as many bring a natural ability to collaboration and consensus-building, skills that are indispensable in leadership. Identifying and acquiring acumen is imperative in any statistician's journey to successful leadership, and should be a developmental objective of both the new and experienced statistician leader.

Bibliography

American Statistical Association (2012), "Strategic Plan." Available at http://www.amstat.org/about/strategicplan.cfm (accessed September 1, 2014).

Buchanich, J.M. (2012), "Scientific Course Strengthens Students' Communication Skills," *Amstat News*, February.

California State University, Fullerton (2014), "Organizational Acumen/Insight." Available at http://hr.fullerton.edu/ professionaldevelopment/ubi/UnivLeadAcademy/AboutULA/ CoreCompetencies/ OrgAcumen.asp (accessed September 1, 2014).

Ciampa, D., and Watkins, M. (1999), *Right From the Start: Taking Charge in a New Leadership Role*, Boston, MA: Harvard Business School Press.

Donabedian, A. (1988), "The Quality of Care: How Can It Be Assessed?" *Journal of the American Medical Association*, 121(11), 1145–50.

LaFleur, B., Seastrom, M.M., and Sullivan, G. (2014), "Preparing Statisticians for Leadership: How to See the Big Picture and Have More Influence," Workshop held at the Joint Statistical Meetings in Boston in August 2014 as part of the Schenker (2014) Presidential Initiative.

Leaming, D.R. (2007), *Academic Leadership: A Practical Guide to Chairing a Department, 2nd Edition,* San Francisco, CA: Anker Publishing Company, Inc.

London Workshop (2014), "Statistics and Science: A Report of the London Workshop on the Future of the Statistical Sciences." Available at http://www.worldofstatistics.org/wos/pdfs/Statistics&Science-TheLondonWorkshopReport.pdf (accessed September 1, 2014).

Morganstein, D. (2012), "Consulting Best Practices," *Amstat News*, September.

Morton, S.C. (2009), "Plan Your Work, and Work Your Plan," American Statistical Association President's Corner, *Amstat News*, February.

Myers, B. (with Mann, J.D.) (2011), *Take the Lead: Motivate, Inspire, and Bring Out the Best in Yourself and Everyone Around You*, New York, NY: Simon and Schuster.

Rodriguez, R. (2012), "Career Success Training for Statisticians: An Update," American Statistical Association President's Corner, *Amstat News*, October.

Rodriguez, R. (2014), "Preparing Statisticians for Leadership: How to See the Big Picture and Have More Influence," workshop held at the Joint Statistical Meetings in Boston in August 2014 as part of the Schenker (2014) Presidential Initiative.

Schenker, N. (2014), "Developing Training in Statistical Leadership" Presidential Initiative Workgroup. Members of the Workgroup are Janet P. Buckingham, John Eltinge, Amanda L. Golbeck, James L. Hess, Bonnie LaFleur, Colleen Mangeot, J. Lynn Palmer, Robert Rodriguez, Marilyn M. Seastrom, William Sollecito, and Gary R. Sullivan.

University of Washington (2014), "University Advancement Competencies." Available at http://depts.washington.edu/uwadv/administrative-resources/hr/university-advancement-competencies/#OrganizationalAcumen (accessed September 1, 2014).

14

Leadership: An Untold Story

Sallie Keller

Virginia Institute of Technology

Jude Heimel

Jude Heimel & Associates

Introduction

Thousands of stories about leaders and leadership lessons have been told. This paper focuses on what has not been said. The first part of the untold story is that a leadership journey is never a solo trip. The statistical leadership journey in this article is about Sallie. It has been influenced and shared by many colleagues, including Jude. Jude, as a professional coach, has been able to sharpen the telling of the story and convey the leadership principles around many of the concepts. So, let us begin.

> Who trains leaders for unforeseen catastrophes, unexpected life-changing events, and unanticipated violations to professional norms?

Insights and summaries from Sallie Keller and Jude Heimel are presented in third person. This leadership story discusses four of Sallie's leadership stories as case studies that provide her insights as a leader and show her evolution as a leader over time. These case studies are presented in first person.

No one trains leaders for the unexpected; it is up to us to train ourselves. Leaders, statisticians in particular, look for evidence, based on all kinds of data, to support personal and organizational decisions. We can't help ourselves! However, when faced with the unexpected, our emotions and feelings need to be drawn upon as well as grounding ourselves in data. We have learned this from our on-the-job experiences. In this chapter, we share some of these experiences and anticipate they may be valuable to others when confronted with the unexpected.

Coaching, Mentoring, and Counseling

Our view on the distinctions among coaching, mentoring, and counseling are important to help frame the rest of the discussion. Table 14.1 gives an overview of these concepts. Each of these methods results in different types of outcomes. As a leader, finding the right balance across these three when supporting different individuals is important. Equally important is recognizing which of these methods you are engaged in and being cognizant not to confuse them. Over the years, we have evolved definitions of these key concepts that may not be the typical textbook versions, but we have found them to be useful.

Table 14.1
Coaching, Mentoring, and Counseling

	Coach	**Mentor**	**Counselor**
Distinction	*Individual* is presumed healthy	No assumptions for *protégé*	*Patient* is presumed unhealthy
Physical Image	Sits next to *individual* and looks at the world together	Reaches down to help *protégé* up	Sits across or behind *patient*
Domain of Support	Helps *individual* achieve professional and personal goals	Helps *protégé* progress in a mutual profession	Helps *patient* address psychological, emotional barriers
Foundation of Knowledge	Experience; able to consider context and environment; unbiased perspective	Their own experience	Considerable training and certified or degreed
Conversation	Supports and challenges; problem solves; addresses the present and future	Recommends, introduces, gives ideas; addresses the future	Asks deep, probing questions, "why?"
Direction	Action	Action	Healing
Outcome	Coach's success is dependent on the success of the *individual*	Mentor's success is independent of the success of the *protégé*	Counselor's success is dependent on the quality of the treatment, which may or may not lead to *patient's* adoption of the treatment

The first step in defining the distinctions between coaching, mentoring, and counseling is to think about the individual being coached, mentored, or counseled. For coaching, the individual is assumed to be a healthy person, whereas with mentoring, there is no assumption on the state or well-being of the protégé. Counseling includes a patient that is assumed to be unhealthy.

There is a mental image that goes along with how these activities take place. A coach will sit next to the individual and the two will share their world views. A mentor will reach down to help the protégé up. A counselor positions herself or himself in a patient and physician setting where you are sitting either behind your desk or across from the patient. Keeping these images in mind helps a leader recognize what role she or he is playing. For example, if you are coaching, perhaps you need to come out from behind your desk and converse in a different way. When you, the leader, are sitting next to your staff member, you are saying that you are willing to look at the world as they see it.

The domains of support and the knowledge that are brought to bear can further help clarify distinctions. For the coach, you are helping the individual try to attain his or her professional and personal goals within the environment you both share. The coach brings personal experiences within that context and is able to provide unbiased perspectives. An analogy would be that if you do not know the game of baseball and you have no experience with it, you are not going to be a very effective coach. A mentor will also use personal experience as the knowledge base, but it may not necessarily be in the context of the protégé's environment. The mentor will share something about a mutual profession with the protégé, based on the mentor's experience within that profession. A counselor leverages specific training and works to help the patient address some issues that could be psychological or emotional.

The conversations that take place within these different methods are distinct. With the coach, the conversation is one that is both supportive and challenging. The goal is to develop an atmosphere of joint problem-solving, addressing both present and future issues. The leader-coach listens carefully, and asks searching, nonjudgmental questions. This may be difficult for the coach, yet powerful when executed appropriately. The individual is expected to formulate options and take action based on the conversation. The mentor, on the other hand, is there to introduce and expose the protégé to new ideas. The leader-mentor will choose experiences to share that could help the protégé think about how to address his or her professional future. The protégé would be wise to consider following the mentor's advice and taking some defined next steps. The counselor will ask deep, probing questions, largely addressing the past. The counselor tries to help provide some pathways for healing. Unless properly trained and certified, there is no acceptable leader-counselor role. One of our case studies presents the importance of bringing in counselors when a catastrophe happens at work, as a leader cannot always lead and counsel.

Perhaps the most important distinction between coaching, mentoring, and counseling is to look at the outcomes of the processes from the coach, mentor, and counselor points of view. From the coach's perspective, her or his own success will be measured by the success of her or his organization (team). This is completely dependent on the success of the individuals. For a mentor, success is completely independent of the success of the protégé. If the protégé succeeds, she or he will be happy, but the future success of the mentor is not dependent on that outcome. For the counselor, she or he has no control over the actions of the patient. Therefore, success is dependent on the quality of treatment that she or he can provide, which may or may not be adopted.

There is an important distinction between the leader-coach and a professional coach. The amount of training received in the principles and lessons learned that we enumerate below is one. This creates a broader set of knowledge and experience in the coaching domain. A leader-coach may well benefit from having a professional coach. The professional coach can step into a "trusted advisor" role, by coobserving the leader's environment and aiding in analysis of the environment and strategizing techniques with which to respond. This is a deeper coach relationship than is appropriate or sustainable between a leader and her or his staff member. Many leaders have found it useful to engage a professional coach when embarking on a new task or position, when involved in a particularly large or contentious project, or entering a new environment. The remainder of this chapter will focus on the leader-coach and leader-mentor roles.

Balances

A leader must both mentor and coach her or his staff. We contend that keeping these methods distinct is critical, especially when navigating unexpected and unforeseen events. The leader-mentor is a constant role model within a context broader than an immediate and specific situation. It is a strategic relationship with the staff, focused on long-term professional development. A leader must be fair to the entire organization in the delivery of mentoring to avoid the impression of preferential treatment of individual staff. Most importantly, the leader-mentor relationship is not about performance or performance assessment.

The leader-coach focuses on specific situations and helps lead individuals (staff) down a path of self-discovery and learning. This is a tactical activity and is derived from the wealth of experience of the leader. It can have strategic implications, but it is really about providing specific tactical direction. Coaching is directly related to performance, performance enhancement, and ultimately performance assessment. Hence, the leader-coach must be incredibly careful to take the time to coach the entire organization. The remainder of this chapter is about the leader as coach.

There are several balances to keep in mind when taking on the role of leader-coach. We will describe them here and then illustrate them with specific case studies in the next sections.

- First is the balance between what is good for a single individual versus what is good for the entire organization. This can be challenging, because frequently we end up in leadership positions within an organization we have been in for a while. We know these individuals quite well; we know their personal lives; we know their weaknesses and their strengths. This is when it is important to use the coaching technique of deep listening. It is critical to listen without judgment — listen for the purpose of understanding. Failing to listen carefully and relying on your past experiences with your colleagues is dangerous. This is particularly acute in catastrophic, unexpected situations where differences in meaning are heightened and taking time to listen is often forgone.

- Creating a balance between your personal value system and respecting that of others is another great challenge of effective leadership. Those systems will rarely be the same. They are not even the same within households, between siblings. These differences tend to get magnified in the workplace.

- One of the greatest challenges is balancing between being a sympathetic enabler versus trying to help provide and engender an environment of collaborative problem-solving. It is almost effortless to do the former and takes discipline to do the latter. However, doing the latter is how the leader empowers people to take action. It allows the leader to support an individual without a perception of favoritism among the others that are watching the situation unfold.

Understanding how these issues can emerge and play out is best demonstrated through a set of examples. These are given in the next sections.

Unforeseen Catastrophes

The first set of examples comes under the heading of unforeseen catastrophes. These are natural and manmade uncontrollable disasters, such as fires, floods, hurricanes, 9/11, campus shootings, or massive security breaches. While these catastrophes happen infrequently, throughout my (Sallie) leadership journey, more experience has been gained with each, and, in some instances, multiple occurrences of the same type of event. Therefore, one can argue these are unforeseen, but definitely not rare, events. Two events will be highlighted here, a fire and a collection of hurricanes.

Los Alamos Fire — Case Study

This event occurred in the spring of 2000 in Los Alamos, New Mexico. I was the group leader of Statistical Sciences at Los Alamos National Laboratory at that time. A wildfire grew out of a government-issued controlled burn in the Bandelier Wilderness. Within a few days, it was out of control. Los Alamos and the neighboring communities were required to evacuate. Although warnings had been given, few people were ready. None of us believed that the government would allow this to happen. Surely, with the national laboratory there, they would call in every fire department in the country, if necessary, and the National Guard as well. In hindsight, this was a ridiculous way of thinking, for many reasons, not the least of which was that the wind was too high and the fire was too strong to be controllable. When the calls with the order to evacuate came to our homes and the sirens went off, many of us realized that we were completely unprepared, and it was time to quickly pack and depart. The actual evacuation occurred in the middle of the night and was in Los Alamos style, very orderly. Everyone scattered to Santa Fe, to Albuquerque, and to friends and relatives in other locations. Eight days later, we were allowed to return to town, and four days after that, we returned to work. We all knew people who had lost their homes, pets, and vehicles. It was clear that one of the first tasks at hand, as the leader, would be to meet with each of the staff individually and make sure that they knew what work they needed to get back to and to be sure that their office environment was up and running. I thought naively that this would be simple. I allocated 10 minutes with each person — surely I would be done in a day. This was a perfect moment to make a major leadership mistake. However, in the first session, I quickly realized I literally needed to sit with each individual and patiently listen to them replay their entire evacuation event. In real time, I saw myself shift into a coaching role. I also found myself way outside my comfort zone.

Today, this still remains one of the most exhausting ten days of my career. I talked with individuals who had to move their grandparents and parents from nursing homes two and three times during the course of the evacuation. There were other individuals who had forgotten to take a precious heirloom with them when they evacuated. There were others who did not allow their children to take any toys, and others who only filled the car with toys. The stories were serious, and people were struggling to find the energy and courage to look forward. My role as a coach was to listen carefully and not to pass judgment on anyone's choices during that evaluation time or to judge the complexity of their current stress.

I needed to find a way to create a balance between my own personal value system and respecting that of others. My role was to identify the barriers keeping them from resuming a productive work posture. Breaking down the barriers included telling staff to take time off and spend the time with their parents and grandparents, or to go home and make a list of everything that they need to take in the event of another disaster, or to join the crews that were out planting trees and to take time to do service to the community.

In retrospect, Jude pointed out that what happened is known as "compassionate detachment." The detachment side of the equation is about keeping yourself reserved, not taking responsibility for solving the individual's problem, simply respecting that it is a serious problem for them. Leaders are problem solvers, so this can push them way out of their comfort zones because they need to accept the situation, not knowing the outcome, accept that they do not have the answer. The individual needs to find her or his own solution; the leader can only support them and help identify obstacles. This requires believing in the health and resilience of the person sitting next to you.

Katrina, Rita, and Ike — Case Study

Everyone remembers Hurricane Katrina. I was the William and Stephanie Sick Dean of Engineering at Rice University in Houston at that time. Following the event, those of us at the university opened our classrooms, offices, and residences to colleagues and students from New Orleans. Just as we were feeling safe and secure, along came Hurricane Rita, barreling down on Houston. Another interesting twist is that the Dalai Lama was scheduled to visit campus in 36 hours, and the forecasts said we were 48 hours away from Rita making landfall in Houston.

I was called to an emergency management meeting with the university president, the other deans, and the vice presidents. I was ready to be advised on emergency processes. However, oddly, when I arrived to the meeting, everyone was telling their favorite flood stories. We were a day and half away from the Dalai Lama visiting and 48 hours away from the hurricane arriving! I asked a simple question — "who's in charge?" Heads popped up and the looks were very puzzled. I asked a second question — "who makes the decision to close the university?" Our president looked at the main safety lead, and there was no answer. Frantically, I kept asking questions. Jude would point out that I was "yelling up" at that point. "What am I supposed to do?" "What is our process?" "The police, who were part time, had left the campus to patrol other parts of Houston." "Who's going to take care of the Dalai Lama?"

When we left that meeting, people were a little bit confused, and I was very frustrated. My leadership colleagues were exhibiting the same blindness to potential disaster that I had felt at Los Alamos. By the time I got back to my office, I realized there was much in my control. I gained comfort realizing it was within my authority and my responsibility to protect my faculty, my staff, and our facilities. I could not make the decisions on the Dalai Lama or closing campus. I had to just let those decisions go. But I could make a lot of decisions in my own environment. I made the decision to start powering down our labs. I made the decision to create a call list immediately. In the end, Rita did not hit Houston, everyone was safe, and the Dalai Lama event was canceled.

Two years later, Hurricane Ike hit Houston. The university leadership team came together much more quickly and jointly put the right plans in place. We had learned to work collaboratively during the intervening time. The university shut down, students sheltering in place, and afterward the university reopening went quite smoothly. There were some controversial decisions along the way, but they were respected and implemented by the leadership team.

Both the wildfire and the hurricanes are examples of balancing what is good for an individual (or subgroup) versus what is good for the organization. These experiences also demonstrate that organizations can learn. The moral of the leadership story is to identify what is in your control and to take that control. Jude points out that this is exercising a basic coaching technique, which is to recognize the power you have individually, set goals, and create a specific plan to achieve those goals. This helps one take charge in times of critical need. Frequently, taking the first small but incremental steps helps build momentum, and can break the frozen tableau we saw in the first university leadership meeting.

Unexpected, Life - Changing Events

Unexpected, life-changing events are events that change the life of the organization. These are events, such as someone leaving the organization due to death, a different position, or perhaps someone who has been terminated from employment. There are other more subtle situations involving physical changes, such as aging, menopause, and prostate cancer. Leaders will experience all of these across their organizations. Generally, there is no time to study up when you are confronted with such a situation.

Death — Case Study

Imagine a beloved colleague who seemed perfectly healthy one day, receiving a disastrous diagnosis and dying in less than five days. From the leadership point of view, this was not an ordinary colleague. This person built bridges (smoothed out relationships) between group members when necessary, was someone responsible for a major portion of the portfolio of work, and was a leading voice for the group to the broader institution. Many people in the group had spent their entire careers working with and being mentored by this individual. The end came so quickly that no one had an opportunity to say goodbye to this person.

I could see the organization beginning to crumble emotionally, and I did not know what to do. I reached out to our institution psychologist to ask for help. He offered to talk with the group. I was relatively new to the organization and decided it prudent to seek the counsel of some of the senior colleagues. They felt this was a bad idea, yet they had no alternatives. Once again, feeling outside my comfort zone and experience, I made the decision to have him come, and I made the meeting mandatory. This was one angry room, angry for their loss and angry about my decision to bring them together, but I had great confidence in our psychologist, a trained counselor, to be able to address the emotions in the room and start the healing process. It was an incredible session. He worked with each of us to figure out how we were going to come to closure for ourselves individually. He also worked with us together to figure out how we could gain closure as a group. We began to build confidence in each other, and, jointly, we began to move our organization forward.

In the days and weeks that followed, new spaces and opportunities opened, and we were able to fill them. This could not have been accomplished without the advice and participation of a trained counselor. The lesson learned is that leaders lead by inviting others to lead during difficult situations, especially emotionally charged situations, and not pressure themselves to feel they need to know it all.

Violations of Professional Norms — Case Study

This last category is perhaps the toughest of all in leadership. These are situations in which you are confronted with violations of professional norms, such as harassment, research misconduct, and intentional project mismanagement (fraud, waste, and abuse). These are violations, and by definition they do not feel good. This is when a staff member and colleague does something that is against the organization's code of conduct or interferes with other people's rights.

Research Misconduct — Case Study

Our example is about research misconduct. Research misconduct occurs more often than one might guess. It is something that we as leaders and researchers must take very seriously. Since confronting these situations and taking action can be so unpleasant, it is tempting to do nothing and stick your head in the sand, hoping the next guy will act. You may even be advised by respected colleagues to do nothing. We argue this is the wrong advice.

The situation started off quite subtly through accusations of what appeared to be self-plagiarism. In this example, the process required that one of my direct reports be responsible for gathering the facts, interfacing with other colleagues, and directly confronting the individual in question. I needed to assume the role of leader-coach to help my staff member navigate the process, and remind them that they are neither judge nor jury.

As the process unfolded, the incident escalated into something more serious than "inadvertent self-plagiarism." This increased the pressure on the individual required to gather the evidence. Word about the incident was quietly spreading. Many colleagues felt we were not doing enough, or we were not doing it fast enough, or we were doing the wrong things. There was grave concern that we were going to damage reputations — both for the individual being implicated and for the organization as a whole. Here, the leader-coach must be strong and supportive. It was important to follow the organizational process to gather the facts. I had to be supportive of my direct report who had the unfortunate burden of leading the development of the case; I had to be supportive of the faculty who were enraged and yet make them stand back so due process could unfold. Most importantly, I had to withhold my personal feelings and judgments. All the facts were going to come to me for review, and at that point, I would need to recommend a decision.

These situations are very, very challenging. I have been confronted with situations like this at every institution in which I have held a senior leadership position. I always remind myself that even though these are long, hard, and ugly situations, I know that in the end, sound scientific practices will prevail. This is a great example where a balance between being a sympathetic enabler versus creating an environment of collaborative problem solving is paramount.

Trust and Fairness

The above case studies focused on unforeseen events that required strong leadership applying balanced principles. This style of leadership must be built on a bedrock of trust and fairness prior to a crisis. The lessons learned from the case studies apply to the more subtle events that occur day to day. The balances we have noted create an environment enabling trust between the leader-coach and her or his staff. This is a different type of rapport with staff than many leaders currently enjoy. This type of rapport is based on mutual trust and integrity, and on fairness and openness. There are a number of techniques that aid building rapport between a leader and her or his staff. One technique is the deep listening we referred to earlier. It is important to demonstrate that one genuinely cares about her or his staff and what concerns them. Listening to understand is a powerful tool to help with this.

One area in which the leader needs to use deep listening is to understand the aspirations of her or his staff. The leader-coach must maintain a broad enough perspective to balance the needs of each supervised individual with the needs of the organization. This includes being honest about staff performance and by bringing the organization's needs into the conversation. This can include encouraging the staff person to get outside help, for example, when her or his performance is below par and weakening the organization.

There is a degree of personal honesty that is required by the leader when viewing her or his staff. Being honest with one's own beliefs, assumptions, and tendencies to avoid situations, such as conflict, is important. Being judgmental, being overly sympathetic, and jumping to conclusions only gets one into trouble, leading to a more complex situation to manage.

To build trust, the leader-coach must understand herself or himself. She or he must understand her or his own value system in order to not impose it on others. The leader-coach cannot hold her or his values and beliefs so tightly that she or he cannot listen to understand. Deep listening shows caring; it shows emotional involvement in what is being said with a focus on the present and future of the speaker.

Another important ingredient in building trust is that your staff has to judge you as being sincere, meaning what you say, and being reliable, that you will do what you say. This is particularly important with professional development issues. For example, you may choose to support someone's professional development goals by giving them time to complete several research papers related to great work that she or he has done. However, situations change, and you may need to make decisions that work against somebody moving towards, her or his aspirations, at least temporarily. A project issue may come up for which her or his participation is critical. It is important to acknowledge that this decision is taking her or him off track, and there are plans to get her or him back on track as soon as possible.

Allowing problems within an organization to surface is key to a healthy work environment. Fostering an atmosphere of fairness includes having a practice that encourages people to raise problems and bringing them forward for discussion. Whether you are an organization leader or research project manager, listening to all problems is important. Concurrently, however, one should have the expectation that along with the information about the problem comes a possible implementable solution. Finding a resolution together engages staff in collaborative problem-solving.

In Closing

These are lessons we have learned. Now we expect the unexpected. Whether it is an out-of-control wildfire, a devastating hurricane, death in the workplace, or research misconduct, the lessons we have learned lay a foundation of trust and practices between a leader and her or his staff that eases the process of dealing with catastrophic events. They prepare leaders to encounter and manage through the "unexpected."

15

Leadership and the Legal System

Mary W. Gray

American University

Introduction

Lean In is the eponymous advice of a recent book by Facebook Chief Operating Officer Sheryl Sandberg (Sandberg, 2014). Perhaps good advice, but sometimes even the most talented woman needs a little push or even a hand up if "leaning in" has resulted in a fall. There are a multitude of sources of assistance in the achievement of leadership for women in statistics — innate talent, excellent education and training, coaching, support of family and friends. But when all of these still do not provide the boost desired, there is always the legal system. The enforcement of its provisions can help, by deterring and redressing discrimination, to ensure that women have access to the resources that they need to flourish. The specifics in this chapter treat the prospect for help built into the legal structures within the United States. Situations vary from country to country; in some cases, they are better, and some cases worse than in the US, both in the on-the-books laws and in their enforcement. However, nearly all countries (not including the United States) have ratified the United Nations Convention on the Elimination of All Forms of Discrimination Against Women (CEDAW), which provides protections similar to United States laws in several areas[1].

Clearly, laws in the United States, even now, have a way to go — we still do not have an Equal Rights Amendment to the United States Constitution, and in 2014, Congress is unwilling to pass the Paycheck Fairness Act (S. 84, Paycheck Fairness Act, 113th Congress, 2013–2014). This bill would, if enacted,

---

[1]The United States and Palau are signatories of CEDAW, but have not taken the step of ratification needed for the provisions to have domestic effect. UN member states that have not signed the treaty are the Holy See, Iran, Somalia, South Sudan, Sudan, and Tonga.

amend the Equal Pay Act of 1963[2] to clarify its meaning and to make it easier for women to seek its enforcement. Equal pay is an important consideration in leadership, because in society, money often represents an evaluation of one's worth, translated into one's ability to lead. Moreover, unequal pay is crippling to self-confidence in exercising leadership as well as more broadly. Discussion surrounding the recent firing of *The New York Times* editor Jill Abramson focused on a number of issues, in particular on whether leadership characteristics found positive in men were viewed negatively in women (*The New York Times*, May 15, 2014). That she reportedly was paid less than her male predecessor and the lesser status that would represent to many may have been a factor in undermining the authority exercised in her leadership role, not to mention constituting possible illegal discrimination (*The New York Times*, May 15, 2014).

Review of Women's Rights in the United States

This chapter will provide a brief review of the history of women's rights in the United States, with some discussion of the situation elsewhere, in particular the specifics of CEDAW. Rights that strengthen women's ability to gain and effectively carry out leadership roles include the right to full citizenship and protection of the government and the right to vote. Laws prohibit sex discrimination (including sexual harassment) in education, employment (including compensation), public accommodation, and health care. While these constitutional provisions and laws ban all discrimination on the basis of sex, not only discrimination against women, there are other laws that are sex-neutral, but because of the way society is structured, provide protection disproportionately to women, such as the Family and Medical Leave Act. Note that "sex" is used here rather than "gender," because for the most part, that is the term used legislatively and judicially.

Legal support for the rights of women, whether to attain and maintain leadership or more generally, depends in the United States on three branches of government: legislative, executive and judicial; and two levels of government: state and federal. Legislatures, in the federal case the US Congress, pass laws; the executive implements (or fails to implement) laws. Courts have a dual role: to decide whether the law has been followed or broken, and whether the law is constitutional, that is, permissible under the state or federal Constitution. Here, the focus is on the federal level, since there is not uniformity on the state level. There are tensions between state and national power on the issue of women's rights as in other areas. In 2013, the US Supreme Court declared

---

[2]The Equal Pay Act mandates equal pay for jobs requiring equal skill, effect, and responsibility performed under similar working conditions, except where such payment is made pursuant to (i) a seniority system; (ii) a merit system; (iii) a system that measures earnings by quantity or quality of production; or (iv) a differential based on any other factor other than sex. Among other provisions, the Paycheck Fairness Act would limit the "any other factor" to those relevant to the employment and would make it easier for women to learn whether they are in fact underpaid.

the federal Defense of Marriage Act (DOMA) to be unconstitutional under the US Constitution, because it did not permit the grant of marital status for federal purposes (in the case before the court, estate tax law) to those married under state laws that did not prohibit conferring such status on same-sex couples (*United States v. Windsor*, 133 S. Ct. 2675 (2013)). Subsequently, federal and state judges in some states have held that prohibitions against single-sex marriages are inherently unconstitutional, while others have ruled that under the "full faith and credit" provision of the 14th Amendment to the US Constitution, those married under the laws of one state must be considered married in all other states. The other side of the federal/state dichotomy, that is, the constitutionality of state laws, is represented by the ongoing series of decisions about whether state abortion restrictions violate women's US Constitutional rights established by the US Supreme Court in *Roe v. Wade* (Roe v. Wade, 410 U.S. 113, 1973).

The United States, at its founding, was, like most countries at the time, governed by prosperous white men. Although there were other human beings not part of the ruling class whose inferior status was embodied in religious or secular law elsewhere[3], in the US Constitution, that evaluation was reserved for the slave population, counted only as 3/5 of its total in calculating the size of a state's delegation in the House of Representatives. Women's rights were not, in spite of the admonition of Abigail Adams[4], mentioned in the Constitution, even though at that time their rights were certainly inferior to those of free men. The beginning of organized feminist activity in the US is usually traced to the Seneca Falls Convention of 1848, the first conference devoted specifically to women's rights. However, while many women's rights, particularly property rights, were on the agenda for it and subsequent women's rights conventions of the last half of the nineteenth century, not all of the feminists of that era believed that women should have the right to vote[5]. The early feminist stirrings had their roots in religious, primarily Quaker, communities and the abolitionist movement. The ability to make contracts, own property, to have a profession,[6] or at least to have some protection from abuse was the major focus. Today, in theory, these handicaps have been overcome, although

---

[3]There are biblical references to valuing a female at 3 and a male at 5 shekels; in some courts under Sharia law, the testimony of two women is equated to that of one man.

[4]"[R]emember the ladies, and be more generous and favorable to them than your ancestors," Abigail Adams, in a letter to her husband, John Adams, March 31, 1776.

[5]In *Minor v. Happersett*, 88 U.S. 162 (1875), the ruling that women did not have a federally established right to vote was based on an interpretation of the Privileges and Immunities Clause of the 14th Amendment. The court held that while Minor was a citizen of the United States, the constitutionally protected privileges of citizenship did not include the right to vote. The 14th Amendment specifically guarantees all male citizens the right to vote.

[6]In *Bradwell v. Illinois*, 83 U.S. 130 (1873), the US Supreme Court upheld the right of the state of Illinois to prohibit women from becoming lawyers, in the majority opinion, on the grounds that married women could not be bound by a contract without the consent of their husbands (and single women wanting to be lawyers would be very rare), and in a concurrence because, "The paramount destiny and mission of women are to fulfill the noble and benign offices of wife and mother."

spousal abuse certainly remains a problem. Although some states had enacted laws to provide for the right to female suffrage, it was not until 1920 that the Nineteenth Amendment to the US Constitution[7] guaranteed this right for all women. Traces of *de jure* discrimination remained[8] for some time, and today we are still living with *de facto* discrimination in many walks of life, including leadership in business, government, and education. An Equal Rights Amendment (ERA) to the US Constitution was written in 1923 and passed by both houses of Congress in 1972[9]. Because an insufficient number of states ratified the ERA before the ratification period ended ten years later, it failed to be adopted.

Legal Recourse

In addition to specific federal legislation prohibiting discrimination on the basis of race, color, sex, age[10], disability[11], religion, or national origin[12], the Equal Protection Clause of the 14th Amendment to the US Constitution forms the basis for such decisions as *Brown v. Board of Education* (Brown v. Board of Education, 347 U.S. 483 (1954)), which overturned *Plessy v. Ferguson* (Plessy v. Ferguson, 163 U.S. 537, 1896) by establishing that "separate but equal" facilities, being inherently unequal, could not meet the standards for equal protection on the basis of race. Constitutional protection is tricky; the standard for cases involving race is "strict scrutiny," which requires that any "suspect classification" involving race must serve a narrowly defined, compelling state interest that cannot be met by less restrictive means. Classifications involving sex are subject only to an "intermediate" standard that requires that there be an important government interest and that the classification be substantially related to achieving that interest[13]. The still lower standard of review is the "rational basis" test.

---

[7]The Nineteenth Amendment provides: "The right of citizens of the United States to vote shall not be denied or abridged by the United States or by any State on account of sex."

[8]*Taylor v. Louisiana*, 419 U.S. 522 (1975), established that men and women must be treated in the same way with respect to service on juries.

[9]The Equal Rights Amendment: "Equality of rights under the law shall not be denied or abridged by the United States or by any State on account of sex."

[10]The Age Discrimination in Employment Act of 1967 provisions apply to discrimination against those of 40 or more years of age and has exemptions for executives, public safety employees, and certain bona fide occupational qualifications.

[11]The Americans with Disabilities Act of 1990 is a comprehensive civil rights act, including employment. The crucial provision requires that "reasonable accommodations" be made for those with disabilities.

[12]Some states and municipalities also ban discrimination on the basis of sexuality, student status, marital status, and other characteristics irrelevant to the employment.

[13]In *Craig v. Boren*, 429 U.S. 190 (1976), an Oklahoma law that established 21 as the legal drinking age for men and 18 for women based on the hypothesis that young men were more likely to be driving while drunk was declared unconstitutional, noting that the provision was overly broad and its age classification was not narrowly tailored to achieve its announced purpose.

Education

Constitutional prohibitions restrain only governments (national, state, local) or entities acting as government agents. However, the US Congress has extended protections through legislation based upon powers granted to it by the Constitution. Congress may regulate interstate commerce, so enterprises engaged in interstate commerce — in most cases, this means nearly all business having 15 or more employees — are covered by the Civil Rights Act of 1964[14]. Also Title IX of the Education Act of 1972, prohibiting discrimination on the basis of sex in education, covers all institutions receiving federal funding, while carving out exemptions for traditionally same-sex schools or certain sports[15]. The success of American women athletes in recent Olympics has highlighted the success of Title IX, as has the fact that approximately 50% of medical students and law students in the US are now women, up from single digits 50 years ago. Less is heard about the sciences, although in mathematics, the percent of PhDs earned by women is 25%–30%; it is higher in statistics, but lower (a reversal after several decades of progress) in computer science and physics. In social sciences (other than economics) and humanities, gender balance is essentially reversed. Unlike the case in sports, there has been little higher education litigation concerning academic issues. It is interesting to note that, applying the intermediate standard, the US Supreme Court found "separate but equal" facilities constitutionally acceptable in the Philadelphia public schools[16]. Subsequently, the schools' policy was found to violate the Pennsylvania state constitution's equal protection provision. One feature of the separate Central (boys) High School was that it offered more calculus, whereas Girls High offered Italian.

Several US Supreme Court decisions found discrimination against men to violate the Constitution. For example, in *Mississippi University for Women v. Hogan* (Mississippi University for Women v. Hogan, 458 U.S. 718, 1982), the exclusion of men from a nursing school was found unconstitutional. In *Frontiero v. Richardson*,[17] the court held that it was not permissible to provide

---

[14]Civil Rights Act of 1964: Title I–some voting rights protection, Title II–public accommodations (except for "private" clubs), Title III–public facilities, Title IV–public schools (race only), Title VII–employment (the Equal Pay Act of 1963 also covers employment), Title VIII–housing.

[15]Equivalency in terms of number of participants, expenditures, and level of interest has been permitted in lieu of opening all sports to both sexes, but restricted through a series of Title IX decisions, e.g., *Cohen v. Brown University*, 101 F.3d 155 (1st Cir. 1996).

[16]*Vorchheimer v. School District of Philadelphia*, 430 U.S. 703 (1977). Actually the Supreme Court affirmed a Third Circuit Court of Appeals decision in an equally divided (4-4) vote, thus leaving it standing but without precedential effect.

[17]*Frontiero v. Richardson*, 411 U.S. 766 (1973). Of course, one could consider that the female officers were actually the victims of the discrimination. A plurality, but not a majority, applied the strict scrutiny standard. Relying on Frontiero, in *Weinberger v. Wiesenfeld*, 420 U.S. 636 (1975), a unanimous court found that providing Social Security benefits only for female surviving caregiver spouses both discriminated against the women who had contributed to Social Security and the potential male recipients. In arguing for Wiesenfeld, Ruth Ginsburg (subsequently Justice Ginsburg) additionally argued that the Social

lesser benefits to male spouses of female Army officers than to female spouses of male officers. Eventually, separate sex-segregated public colleges also failed to pass constitutional muster[18].

Health

Access to professional education is not required for women's leadership, but it is certainly helpful, although not the exclusive right to be considered. Health is another important factor. Years ago, clinical trials excluded women on the grounds that the volatility of their hormone cycles made study design difficult and expensive. But now for more than twenty years, the National Institutes of Health (NIH) has required that women be included so that we know what helps or hinders the health of both women and men. Just recently, NIH has informed grantees that they must also include female animals in preclinical testing. Health insurance (unless as an employment benefit) was formerly more expensive for women (and hence sometimes avoided), the premiums being based on higher claims costs for young women due to childbirth and preventive care in general. After about age forty, the cost advantage shifted, but most insurers continued to charge women more. The Affordable Care Act (Patient Protection and Affordable Care Act, Public Law 111148, 111th United States Congress, 2010) now prohibits sex-based discrimination in health insurance. It also has other useful provisions, such as the Well Woman and other preventive health care coverage, as well as the controversial "contraceptive mandate" that obliges employers to provide contraception coverage in health insurance plans. Religious institutions are exempted, but a recent Supreme Court decision has broadened the exemption to closely held private corporations whose owners object to the mandate on religious grounds (Burwell v. Hobby Lobby Stores, Inc., U.S. Supreme Court, No. 13-354, June 30, 2014).

The right to control her own body is basic to a successful life for a woman, including leadership roles. The legal protection of reproductive rights upon which women can rely has had a tangled career; while its legal history focuses on women's rights, obviously, women are not the only beneficiaries. Only in 1972 did the Supreme Court find that forced sterilization violates the fundamental right to choose to have children (Skinner v. Oklaoma, 316 U.S. 535, 1942). In 1965, the court struck down the state of Connecticut's prohibition on the use of contraceptives by married couples as an intrusion on fundamental marriage rights (Griswold v. Connecticut, 381 U.S. 479, 1965). The landmark 1973 *Roe v. Wade* (Roe v. Wade, 410 U.S. 113, 1973) established the right of a woman and her doctor to decide whether to have an abortion, based on the constitutional penumbra privacy right. However, subsequently, the court

---

Security regulation in question fostered a stereotype that only women were primary care-givers or dependent on spouses.

[18] *United States v. Virginia*, 518 U.S. 515 (1996). The result forced Virginia Military Institute to admit women.

has permitted incursions on the right to choose in such cases as *Gonzales v. Carhart*[19].

The road to requiring that health insurance cover pregnancy has also been rocky. In *Geduldig v. Aiello* (Geduldig v. Aiello, 417 U.S. 484, 1974), the Supreme Court declared that excluding pregnancy coverage was not discrimination on the basis of sex, as it simply distinguished between pregnant and nonpregnant individuals; two years later, the court came to the same conclusion with respect to the nondiscrimination provisions of Title VII of the Civil Rights Act of 1964 (General Electric v. Gilbert, 429 U.S. 125, 1976). Congress reacted in 1978 by passing the Pregnancy Discrimination Act, defining discrimination "on the basis of pregnancy, childbirth or related medical conditions" to be sex discrimination for all employment-related purposes. Subsequently *Newport News Shipbuilding Co. v. EEOC* (Newport News Shipbuilding Co. v. EEOC, 462 U.S. 669, 1983), extended compulsory pregnancy coverage to spouses of employees.

Another important aspect of the legal protection of women's health in the process of acquiring a leadership role or more broadly is to be certain that pharmaceuticals and medical devices are safe. Although now these must be tested on women, should dangerous objects get through the review process, legal remedies are important. The most far-reaching illustration of the difficulties — in particular the difference between scientific proof and legal proof — is the case of the Dalkon shield. The Dalkon shield was an intrauterine device (IUD) developed and aggressively marketed in the 1970s, in use by an estimated 2.5 million women at its peak. 200,000 to 300,000 women claimed that their use of the device had caused serious injuries, including infertility and even death, and filed suit against the maker. Among the claims were that while the shield's effectiveness in preventing pregnancy was tested, its safety was not. Although pelvic inflammatory disease has other causes, and indeed, women using other IUDs suffered also, one study showed that Dalkon shield users were five times as likely to experience injury as were other users. Buried in the onslaught of litigation, the shield manufacturer, A.H. Robbins, filed for bankruptcy and was bought by American Home Products (Gina Kolata, 1987). Settlement of the claims resulted in the establishment in 1989 of a $2.4 billion trust to compensate victims, administered by American Home Products. After that, IUD use nearly disappeared until the recently new, allegedly much safer models became available; perhaps as a Dalkon legacy, IUDs are still more widely used in Europe than in the US (Business Week, 2014).

Publicity surrounding the Dalkon shield cases resulted in the passage of the 1990 Safe Medical Devices Act and the 1992 Medical Device Amendments to the Food, Drug and Cosmetic Act. Previously and subsequently, there has

---

[19] *Gonzales v. Carhart*, 550 US 124 (2007), overturned a circuit court decision that had declared the federal Partial-Birth Abortion Ban Act of 2003 to be unconstitutional. An amicus brief from a group of statisticians sought to negate the effect of an improper statistical analysis permitted at the lower court level, but the statistical evidence was not determinative in the Supreme Court decision.

been a great deal of other legislation offering protection. On the other hand, and of particular interest to statisticians, charges that drugs or devices have caused injuries led to a case that has established a new standard for scientific (including statistical) testimony, *Daubert v. Merrell Dow Pharmaceuticals*[20], believed by some to have eliminated "junk science" in the courtroom.

The effect of so-called protective legislation concerning types of jobs or working hours or conditions has often been to limit women's opportunities[21]. The requirement at American Cyanamid that women seeking certain jobs had to undergo forced sterilization was found to violate Title VII (Christman v. American Cyanamid, 578 F. Supp 63, W.D.W.Va, 1983). Potential child-bearing as a bar to employment was found to be illegal in *UAW v. Johnson Controls* (UAW v. Johnson Controls, 499 U.S. 187, 1991), and in *Cleveland Board of Education v. LaFleur* (Cleveland Board of Education v. LaFleur, 414 U.S. 632, 1974), the court ruled that the mandatory pregnancy leave policy was so arbitrary and unrelated to actual medical conditions as to violate the Due Process Clauses of the 5th and 14th Amendments.

Fortunately, it is relatively rare that violence or the threat of violence is a hindrance to women's leadership aspirations, but as it can happen, it is important that there be protections available. Although ordinary crime is customarily covered by state rather than federal law, the Violence Against Women Act (VAWA), Title V of the Violent Crime Control Law, was originally passed in 1994, reauthorized in 2000, 2005, and, after protracted haggling, in 2013. The act provides funding for the investigation and prosecution of violent crimes against women, for the support of victims, and for training for coordinated community response consistent through the country; it also imposes automatic and mandatory restitution on those convicted. Over the years, the VAWA's focus has expanded from domestic violence and sexual assault to include dating violence and stalking. In 2000, the Supreme Court held that parts of VAWA that permitted suits in federal courts by victims against their attackers were unconstitutional because they exceeded congressional power under the Commerce Clause and under section 5 of the 14th Amendment, intruding on states' rights (United States v. Morrison, 529 U.S. 598, 2000). However, other parts of the federal protection remain in effect.

Employment Law

Although leadership is important in politics at all levels and volunteer work in non-governmental organizations, for most women it is access to and success in leadership in their work that is most significant for them. Mention has

---

[20]In *Daubert v. Merrell Dow Pharmaceuticals*, 509 U.S. 570 (1993), the issue was whether animal studies, chemical structure analysis, and an unpublished "reanalysis" of published data were adequate to show that the drug Benedectin had caused birth defects.

[21]For example, in the past, women have been told that they cannot be astronomers, as the observatories have no women's restrooms.

been made of two lines of assistance, the constitutional standard that applies to government employment (including public universities) as it does to fundamental rights, like the right to vote and the right to privacy, and federal legislation, such as Title VII and the Equal Pay Act. State or municipal laws may provide additional assistance, but they vary greatly from place to place, so the focus here is just on how federal employment law works.

There are two kinds of sex discrimination that are prohibited: disparate *treatment*, which is treating an individual differently because of her/his sex, and disparate *impact*, which arises when the effect of a facially neutral policy is to disproportionately disadvantage members of one sex[22]. In a disparate impact case, the intention is irrelevant; in a disparate treatment case, it is critical. Statistical evidence can be useful in strengthening a disparate treatment case; in disparate impact cases, it is essential, first to determine whether the policy had an adverse impact, and then whether the impact was so disparate as to have been unlikely to have occurred by chance. In either case, what must be shown is that similarly situated women and men have suffered different outcomes. A possible, though rarely applicable, defense is that the factor is a business necessity. For example, in one case, a height requirement for pilots of 5 feet 10 inches had a disparate impact on women and men from certain ethnic groups. The airline claimed that this height was necessary in order for the pilot to reach all the controls in the cockpit. When a model of the workplace was brought into court, it was demonstrated that someone 5 feet 8 inches tall could very adequately reach the controls. The plaintiff succeeded in his suit, but the employer was allowed to retain a lessened height requirement for pilots.

Title VII did not become applicable to professional employment until 1972, primarily because it was recognized that an objective determination of the qualifications of individuals would be difficult — and it is. But statisticians know the benefits and pitfalls of using statistical techniques in multivariate situations and recognize the difficulties in making an objective evaluation of factors contributing to success. In hundreds of cases involving hiring, pay, promotions, and job loss, judges and juries have displayed varying degrees of sophistication in evaluating such evidence when statisticians bring it to them.

While this is not intended as a comprehensive manual for those considering making use of legal remedies for discrimination they believe has occurred, it is important to recognize both the advantages and disadvantages of reliance on the legal system. In particular, a legally mandated remedy simply puts the successful plaintiff in the position she would have been in but for the discrimination. There is no compensation for the pain and suffering that the victim of sex discrimination may feel or for the personal time and energy expended; attorney's fees are reimbursed only for successful suits.

---

[22]The seminal disparate impact Title VII case is *Griggs v. Duke Power Co.*, 401 U.S. 24 (1971), where the requirement of a high school education for power linemen was shown to have a disparate impact on African-Americans.

When someone believes that she has not been hired, promoted, or granted tenure, or is being paid less than similarly qualified colleagues doing the same work in the same place, because in whole or in part because of her sex, the employer may have internal grievance procedures that can be utilized in seeking a remedy. Obviously, it may be difficult — especially if the employer is a private entity — to assemble information to back up a claim of discrimination. If the internal procedure fails to produce a satisfactory remedy, litigation can be considered. Several important observations: first, it is important to ascertain whether the stated procedures were actually followed, as demonstrating the employer's failure to have done so may be the most promising avenue to redress. Second, if the complainant has suffered retaliation for filing a complaint internally or externally, she may be compensated for losses suffered through the retaliation, even if the original complaint is not sustained[23]. Finally, there are statute of limitation time limits on the filing of complaints that must be observed. Lilly Ledbetter had worked for Goodyear Tire & Rubber for many years when she discovered that she had always been paid less than men in the same job with the same or less experience. When she brought suit, she lost because she had not filed a claim shortly after her first paycheck many years earlier (Ledbetter v. Goodyear Tire & Rubber Co., 550 U.S. 618, 2007). As in the case of pregnancy benefits, Congress responded to this manifest injustice. The 2009 Lilly Ledbetter Fair Pay Act treats unequal pay as a continuing violation so that each paycheck initiates a filing period; however, careful attention must still be paid to time frames. Another law that has been helpful to women workers is the Family and Medical Leave Act, which mandates a minimum of 12 weeks *unpaid* leave for childbirth, childcare, or care of spouses or parents of the employees. This falls far short of what most industrialized countries provide, and while many employers have more generous benefits, there has also been some reluctance on the part of women workers to take advantage of such provisions because they feel that they will be perceived as less dedicated to their jobs and so suffer in pay and advancement (Chronicle, 2014). The Fair Labor Standards Act requires employers to provide adequate facilities for nursing mothers to express milk, and while federal law provides expressly only that breastfeeding may be conducted in federal buildings, most state laws at least exempt the practice from indecency statutes even if more positive support is not enshrined in law. Accessible and affordable child care is lacking in workplaces and communities in general, hampering achievement of an acceptable work–family life balance.

The advantage to litigation is that because of very liberal federal disclosure rules, it will be possible to acquire extensive detailed information about the qualifications and employment records of competitor colleagues or potential colleagues. The disadvantages are that legal challenges can be extremely expensive, adequate counsel may be hard to find, and since disclosure is available

---

[23]However, a recent Supreme Court decision, *Nassar v. University of Texas Southwestern Medical Center*, 133 S.Ct. 978 (2013), has made it more difficult to establish a retaliation claim.

to both sides, any personal blemishes of the plaintiff, no matter how slight, are likely to be exposed and exploited in the course of the litigation. A 2004 report by the American Association of University Professors (Dyer, 2004) provides a comprehensive discussion of the process of attempting to gain a legal remedy for perceived discrimination in granting tenure at academic institutions; much of the information is also useful for those in nonacademic employment. Not only is it a violation of Title VII to retaliate against a complainant by taking an adverse employment action, but this protection also extends to those who may assist a complainant. Similarly, the Whistleblower Protection Act of 1989[24] (as amended) protects federal government employees, government contract employees, and government grantees who file complaints that they believe reasonably evidence a violation of a law, rule, or regulation; gross mismanagement; gross waste of funds; an abuse of authority; or a substantial and specific danger to public health or safety. The most recent revisions to the law were designed to broaden the scope of protection after a Supreme Court decision[25] held that federal employees do not enjoy First Amendment free speech protection when they speak pursuant to their official duties. In October 2012, President Obama issued Presidential Policy Directive 19, providing additional whistleblower protections for intelligence agency employees; other legislative provisions cover specific areas, such as the SEC and IRS. Although anecdotal evidence of problems faced by whistleblowers abounds, on the other hand, there are whistleblower reward programs, and some agencies have threatened to go after employees who fail to report certain violations.

In addition to the specific statute noted above, the Department of Labor's Office of Safety and Health Administration enforces the more than twenty whistleblower protection statutes that protect employees who report violations of laws relating to health and safety issues, some of which relate directly to concerns of women workers, such as the provision of adequate restroom facilities.

For women in professions like law or accountancy, where leadership positions may require selection as a partner, it is comforting to know that they, too, may rely on Title VII in many circumstances (Hishon v. King & Spaulding, 467 U.S. 69, 1984).

That sexual harassment by an employer constitutes sex discrimination actionable under Title VII was determined in *Meritor Savings Bank v. Vinson*[26]. However, it may be difficult to determine who is the employer. In *Vance v.*

---

[24]The basic idea of encouraging reports of fraud and mismanagement originated in the False Claims Act of 1863, designed to root out fraud by suppliers of goods to the Union Army during the Civil War, but has expanded to include remedies for a variety of abuses, including, of course, discrimination.

[25]*Garcetti v. Ceballos*, 547 U.S. 410 (2006). In *Lane v. Franks* (no. 13-483, U.S. Supreme Court, decided June 19, 2014), the court narrowed the scope of Caretti to speech that is ordinarily within the scope of an employee's duties, not whether it merely concerns those duties, in particular, holding that sworn testimony on matters of public concern enjoys First Amendment protection.

[26]*Meritor Savings Bank v. Vinson*, 477 U.S. 57 (1986). The court held in *Oncale v. Sundowner Offshore Service, Inc.* 523 U.S. 75 (1998), that same-sex sexual harassment could form the basis for a valid claim under Title VII.

Ball State University (Vance v. Ball State University, 133 S.Ct. 2434, 2013) a food service worker was sexually harassed by the person to whom she reported and who made work assignments; however, since that person was not empowered by the employer to take tangible employment actions against her, he was not her "supervisor" for whose actions the employer could be held responsible. Harassment by a colleague is actionable under Title VII only if the employer was made aware of the harassment and failed to take action to remedy the situation. Additionally, at educational institutions, sexual harassment of employees, students, and potential students is prohibited under Title IX. Currently, the US Department of Education, motivated by the recent broad attention to campus rapes and sexual assaults, has opened investigations at dozens of colleges and universities. Many educational institutions (and, for that matter, other employers) have not established or reviewed their regulations on sexual harassment, have not adequately informed constituents of their content, or have failed to provide appropriate procedures for reporting and adjudicating complaints and assessing penalties. Thus, protection against sexual harassment may still fall short of what is needed.

But what if sex discrimination is only partly responsible for an adverse employment action? Or what if the reason for the lack of a woman's success was that she did not fit the sexual stereotype her employers had in mind? These, too, can produce a Title VII remedy (Price Waterhouse v. Hopkins, 490 U.S. 228, 1989).

A litigation technique that makes feasible the bringing of suits not otherwise possible because of lack of resources is that of "class action" suits[27]. The standards for certifying a class action are complicated, but essentially, the concept is that a group of victims has been subjected to the same sort of discrimination by the same employer, with a "named plaintiff" representing the claims. Obviously, it may be difficult to show that the jobs are the same and the discrimination is the same, but it may also be hard to show who actually is the employer for the purposes of the case. Lower courts had certified a class of one and a half million current and former employees of Walmart Stores nationwide who alleged that actions by their supervisors in pay and promotion decisions constituted discrimination against women. The Supreme Court dismissed the class certification because of the distributive decision-making process of the defendant (Wal-Mart Stores v. Dukes, 131 S.Ct. 2541, 2011). Other suits involving Walmart employees in a single state continue, but prospects for success are not good.

Benefits, including retirement benefits, are also covered by Title VII. The city of Los Angeles deducted a larger amount for pension benefits from the salaries of women employees than from men's under the theory that since women's life expectancy was longer, their pensions would cost the employer more and the women should pay for this. The Supreme Court found this

---

[27] *Shymala v. University of Minnesota*, Civil No. 4 -73-435, N. Minn, August 12, 1980, is an important class action case resulting in a broad-scale settlement involving employment practices at the defendant university.

discrimination in what is called a "defined benefit" plan to be illegal (City of Los Angeles v. Manhart, 435 U.S. 702, 1978). Other pensions, the most common now, are "defined contribution," where a fixed amount is contributed and accumulated and then may be paid out as an annuity. In one plan popular at universities, the men's monthly benefit had been 15% higher than that of women with the same accumulation of funds retiring at the same age; that is, women were charged the same for less in benefits rather than being charged more for the same benefits as in *Manhart*. The Supreme Court agreed that this variation also constituted sex discrimination (Arizona Governing Committee v. Norris, 463 1073, 1983). The pension company had claimed that their plan discriminated on the basis of longevity rather than on the basis of sex. However, a nice statistical argument shows that because of overlap in the distribution of life expectancies, 86% of the annuitants would die at the same age; that is, a death at one age of a woman is matched by a death of a man at that age, leaving 7% of the cohort as men who died early unmatched by early deaths of women, and 7% as women who lived longer unmatched by long-lived men. But 86% are "similarly situated" with respect to longevity. Since obviously it cannot be known in advance who is in this overlap group, all are entitled to equal protection.

CEDAW

With the limitations of the legal system as assistance for women who seek leadership positions in mind, would ratification of CEDAW (a very unlikely event) provide more assistance to women in the United States? The treaty has a provision that calls for equal pay for work of "equal value," much like the notion of "comparable work" that has been rejected by lawmakers in the US over the years. So long as job segregation continues in this country, not only at the professional level — look around at the faculty of any university — but even more so in skilled trades, actual application of the "equal value" standard might, if the work in disparate occupations were to be deemed to meet this standard, improve the compensation of large numbers of women and open up new leadership opportunities. More likely, however, any impact would be more limited, confined to the regularization of pay for more obvious cases, such as positions involving the same job but labeled differentially; for example, similar jobs could be labeled administrative assistant (female) or assistant manager (male). In a case against Northwest Airlines that dragged through the courts for fourteen years, one issue was that male flight attendants were labeled "pursers" and paid more than female "stewardesses," the difference in the jobs being that on international flights, the purser handed over the flight documents to the agent who opened the door of the plane when it landed. On domestic flights, there was not even that difference, and if there was no purser on an international flight, a female stewardess performed the task. The courts found for the plaintiff class, whom they awarded $52 million, but there was continued litigation over millions of dollars of attorney's fees (Laffey v. Northwest Airlines, 746 F.2d, D.C. Cir. 1984).

Of course, by looking at the world news, it is easy to see that not all nations that nominally adhere to CEDAW offer what might be considered appropriate opportunities and protections for women[28]. Some countries ratified the convention with exceptions, many of which have to do with agreeing to conform only to the extent that CEDAW does not conflict with their own national or religious laws. It is fair to say, however, that in some countries, the conditions for women are very grim (e.g., Afghanistan, Saudi Arabia), but on the other hand, many places where the effects of CEDAW would seem limited, some women achieve a degree of professional success. For example, in some countries, the majority of engineering students are women (e.g., Iran), or women have achieved leadership positions of president or prime minister, a pinnacle not yet reached by women in the United States (e.g., Argentina, Bangladesh, Brazil, Germany, India, Ireland, Liberia, Lithuania, UK, to name just a few).

The Role of Women Statisticians in the Legal System

It is clear that women statisticians have in the past and continue to have a special role in combating discrimination. The importance of statistical evidence has been noted — not only what to look for and how to analyze data are important, but also how to make clear, first of all to the complainant whom one is advising whether there is a case to be made, and, secondly, to the finder-of-fact, be it a judge or a jury, what the statistical evidence means, remembering always that the role of statistical expert is not to draw legal conclusions. Think of testimony as teaching an elementary statistics class!

Statistics are obviously important in insurance cases. At the time of the pension benefits litigation, there were also lobbying efforts by women's groups to eliminate sex discrimination in other kinds of insurance — including where, as in auto insurance, males suffer from discrimination. Since insurance is generally regulated by states, that meant traveling to a lot of state legislatures to make the case for nondiscrimination. Personally, I felt the need for more statistical know-how in the women's movement when I had to fly out of Montana in a snowstorm to testify in Oregon the next day. The opposition, primarily the insurance industry, always had plenty of technical backup, whereas the women's groups did not. The need for broader involvement in the statistical aspects of public policy provides another argument for interesting more women and girls in studying statistics; in fact, in the past thirty years, the

---

[28] Adherents to CEDAW are obliged to eliminate discriminatory laws, policies, and practice in the political (right to vote; hold office; participate in civil society; acquire, change, retain nationality for themselves and their children), social (education and training, health), economic (employment, family benefits, access to credit), and cultural fields, to modify or eliminate practices based on assumptions about the inferiority or superiority of either sex, and to take steps to suppress the exploitation of prostitution and trafficking in women.

actuarial profession has become much less segregated, and actuaries of both sexes are more concerned with equity and diversity issues.

The pension battle has been won; getting involved led me to become a lawyer. When the lawyers from TIAA-CREF (having previously explained that I did not understand statistics) allowed as how I might understand statistics but I did not understand the law, I determined to remove that argument as well. In the time it took the cases to move through the courts, I was able to qualify as a lawyer able to file an amicus brief in the Supreme Court.

Although there are a lot of women statisticians and male statisticians eager to help combat discrimination, not everyone who needs justice can afford to pay for help. As an attorney, I have an obligation to volunteer my help *pro bono* to help those who might otherwise not have the assistance they need; as a statistician, I feel the same obligation, especially in light of the scholarships and fellowships that have helped pay for my education. Many other statisticians also volunteer their services, through Statisticians without Borders, the On-Call Scientists program of the American Association for the Advancement of Science, the Innocence Project[29], university-based statistics clinics, or other means (Asher et al., 2008), but more help is needed. Frankly, it would be very helpful were indigent criminal defendants entitled to adequate statistical as well as legal assistance[30]. More broadly, one effective way to take on a leadership role in causes of concern — human rights, the environment, health and safety — is to volunteer to serve as treasurer of a nonprofit organization. The statistical skills that are needed are often fairly minimal, but very valuable. Most important is the ability to recognize what data are needed and how to analyze them, whether it is a balance sheet or the results of the latest fundraising effort that need attention.

Conclusion

But surely some women in the US, through "leaning in" or other tactics, do attain leadership positions, even though the numbers are small. A 2014 survey of salaries of presidents of 454 public universities in the United States revealed that the median salary of the 40 women in the group was higher than the salary of the men (CNN, May 2013). News coverage offered the explanation that it was a supply issue; very few women who were "qualified" could be found, and thus they had to be paid higher salaries. It was noted also that discrimination is more difficult to conceal at that level because the salaries are

---

[29] "The Innocence Project is a national litigation and public policy organization dedicated to exonerating wrongfully convicted individuals through DNA testing and reforming the criminal justice system to prevent future injustice." www.innocenceproject.org (retrieved June 10, 2014).

[30] In addition to discrimination and DNA litigation, statisticians have been involved in cases involving Sudden Infant Death Syndrome. See, e.g., *Wilson v. Maryland*, 803 A.2d 1034 (Maryland Court of Appeals, 2002), *R. v. Clark*, EWCA Crim 1020 (2003), in particular in refuting convictions relying on the so-called prosecutor's fallacy.

not only a matter of public record, as would be all salaries at public universities (thus in theory making access to legal remedies easier), but well publicized. When comparing salaries, a statistician might ask about years and kinds of experience, size of institution served, and other characteristics of the group. Moreover, what was not emphasized was that the salaries were just that — they did not include total compensation, which in most cases of university presidents is at least twice as much as base salary, thus possibly concealing inequities. The same story noted that at most of the universities, the highest paid employees were the football coaches, all of whom were men. But that there were 40 women among the group of presidents at least is evidence of progress in the last 30 years.

In summary, when other techniques have failed, when the desire and aptitude for leadership continue, legal remedies may, with great caution, be considered.

Part V

Leadership Development Platforms

16

Professional Organization Membership

Lee-Ann Collins Hayek

Smithsonian Institution

Introduction

No single approach will create a leader, and no other person can make you into a leader. Hard work, dedication to the field of study, and determination are the fundamentals; the rest comes as a trailer. Oftentimes, just sheer perseverance provides the impetus to the top. Many, if not most, scientists, mathematicians, and mathematical statisticians are solitary people, with difficulty and low interest in overcoming natural reticence. However, taking part in professional activities is inevitable for and important to a successful and satisfactory career. In particular, participating in a professional organization can (1) open your eyes to new ways of thinking (2) keep you on top of new developments and advances in the field, (3) help you to learn of unique experiences that you would otherwise not come upon, (4) over time provide and foster the confidence needed to participate and lead, and (5) provide actual leadership opportunities.

In mathematical statistics and statistics, we have a number of major, well-established and well-known associations. Among these are: The American Statistical Association (ASA), the Royal Statistical Society (RSS), and the Statistical Society of Canada (SSC) at the national levels; and the Institute of Mathematical Statistics (IMS), the International Statistics Institute (ISI), and the International Biometric Society (IBS) on the international level. There are, in addition, other national and international groups that promote our specialty field as a whole, e.g., the Mathematical Association of America (MAA), and the Society for Industrial and Applied Mathematics (SIAM). There are also those that foster interest in differing aspects, such as computational statistics, statistics education, computer graphics, Bayesian statistics, and others. In addition, there are the more general groups, such as the American Association of University Women (AAUW) and the Association for Women in Science (AWIS), the latter of which is also composed of international, national, and multiple local groups, but dedicated to all women in STEM (science, technology, engineering, and mathematics, including mathematical and applied statistics).

The local Washington, DC, chapter of AWIS, for example, arranges dinners, happy hours, and talks, and is extremely proactive in providing information on career searches, advancement, and especially on problems unique to women professionals. Any of these groups, and especially AWIS, in my opinion, can serve to increase your knowledge, exposure, and visibility and aid in confidence building.

A professional association is often loosely defined as a body of people engaged in the same profession seeking to further the interests of individuals within that profession and/or to maintain standards and to enhance public interest in that field. While that is all true for all of our professional groups, each of their websites states clearly the particular organization's selected purpose(s), along with any added advantages to membership. For example, on www.amstat.org, ASA is touted as the "world's largest community of statisticians," though I am in the dark about its being a "tent." The term "community of statisticians" is loosely defined, however, and includes those in many in other fields who are users of statistics or administrators of statistical programs. The list of membership advantages includes meetings, publications, membership services, education, and advocacy. On www.maa.org, the Mathematics Association of America, an association geared toward education more directly than ASA, makes its list of advantages for membership more exciting and suggestive of club membership: "a wealth of world-class resources, numerous opportunities to network and grow professionally, highly regarded outlets for research findings and expository articles, and a close-knit community of creative and caring mathematicians actively working together to improve and advance mathematical scholarship at the collegiate level." Finally, RSS, IMS, and ISI emphasize the scholarly nature and exclusivity of their organizations. On www.rss.org is an entire booklet on membership to download and a discussion of membership, fellowship, and professional qualifications necessary, while www.imstat.org states that the IMS is an "international, professional and scholarly society devoted to the development, dissemination, and application of statistics and probability." The national AWIS website, www.awis.org, lists its multiple goals, prominent among which is leadership and talent development. These specific goals are accomplished through a multimedia approach, and joining allows you access to all, including all web-based talks, discussions, news, and advocacy results and information especially pertinent to professional women's issues. With just these few examples, it should be clear that not all professional organizations even in a single field are created equal; examine the website to find what is best for you. Do you want a seriously scholarly group? Is a group with a large membership from among all ranks and levels and countries more to your liking? Do you prefer to focus in on a group that is devoted to your potential or chosen field of work, such as education, government, university, or research? The only caution: Do not choose out of insecurity, choose with goals in mind. If you only prefer or think you can only tolerate small groups, then widen your horizons and give a large organization a chance. For example, ASA is the largest, yet it has local sections and chapters that are small and an exceptional source of networking in your backyard environment. Are you still at the stage in which you worry about where you will get a job? Try a group more specifically focused on a potential work arena for you, e.g., MAA for teaching; ASA and RSS for college and other entry-level positions. Alternatively, with student discounts or free offers, try as many as you can and then make decisions.

So the question remains: Why should I join at all? Below, I hope to provide information for statisticians at each of the levels of their careers. I hope convincingly to show you that even for those who claim they are not "joiners," or those who think they save money by not joining, membership in a professional organization is a major advantage, even possibly an essential element, for any statistician, regardless of career level or employment field.

For Students and Entry-Level Career

My first thought when a professor in grad school mentioned joining a professional organization was money, and my lack thereof, but also the lack of time and ability to fight through technical articles. I could barely keep up with course work, outside readings, projects, and research, and had no free time as it was. How would I ever gain anything from the journals or the mailed material? However, these objections are easily overcome, especially today. Student membership is most often either free or greatly reduced and often comes with the newsletter and one broad interest journal. For example ASA has *Amstat News* and *Significance* (joint with RSS), while RSS (joint with Taylor Francis) has *Chance*. These new beginning-level, public interest journals are tantalizing in their topics and articles and mostly can be speed read.

In my opinion, this is the perfect time to join one or more professional organizations. Most such organizations welcome you, give you benefits that more established members do not get, and you enter into the organization on the "ground floor" of your career when it will assist you in multiple ways to work toward a strong professional foundation. By so doing, you can be provided with assistance throughout your career and end up in leadership positions at different and self-selected points along the way. This assistance is not just professional but can also be personal. Many early career women and men are too timid to give a talk when just starting out in a career. Listening to others helps, but talking to the speakers while in a relaxed setting may give you pointers; some presentations provide personalized fodder, some do not, but, you have access to all. Many beginners can never seem to put pen to paper after doing the research, and those who can, seem to have a finger incapable of hitting that send button for journal submission. Learning that seasoned professionals willingly tell you stories of the same fears, problems, and worse can give you the power for flight. Mathematics and statistical science are fields in which each of us has at some moment in history entertained the depressing hidden thought, "Everyone knows more than I." Joining a professional organization highlights the truth of the universality of this statement in our profession, which may seem impossible to believe without hearing it from other members.

So of what use is this organization going to be for me as a novice? First, there is a wealth of resources available free for student and entry-level professionals who are members. These resources can help you build your skill sets, obtain an internship or scholarship, find a mentor or a job, and get discounts. Importantly, the primary function of your membership at this stage in your career is to allow you to listen, to learn, and to interact when possible. For example, meetings, to which students often can get free travel, are the best source for courses that can provide information on writing up your research, getting published, getting ideas to publish, and other topics, aspects that you may have thought the senior members obtained innately. Tip: Walk in a bit early to a talk and look for a person who may seem interesting to you, or a known face from other meetings, and, before talks begin, say something ... do NOT act like this is a social situation and comment on appearance ... try a simple hello or ask about the topic of the talk. After the talk, speak to a person who was in the audience about the potential for the topic as research or where those who work on that subject are clustered, for example. Even if you are too timid to actually walk up to another statistician (or a "big gun" in the field) and talk or ask a question, by the time you garner your courage and decide to call or email, you can initiate the conversation with: "I saw you at the *** meeting last year and didn't have time to ask...." As an entry-level professor or industry worker, you can present your results in a variety of media, such as posters, presentations, and roundtable discussions.

As we all know, it is increasingly difficult to find anything suitable in the way of continuing education, almost a pejorative term. And, a new professor hardly can face letting anyone know she or he might be lacking in some aspect of our field, of course. So, professional organizations such as ASA offer continuing credit courses of particular importance; these are often based upon the newest developments in the field. At professional meetings, the range of courses is almost always wide, and you can attend in anonymity, if desired. You can sign up for a course in your specialty to learn where current thought can take you, or, you can attend a course in a new specialty area of interest for potential future work. Job opportunities and offerings abound for members to interview or learn about different fields they could investigate. Student competitions are frequently available, and even entering, not merely winning, can be a boost to your career. There are a myriad of ways to learn, even from losing a contest, since name recognition will result. A professional organization membership puts you in the midst of others in the same field from whom you get advice and learn, or get criticized, and you, in turn, not only learn but get stronger.

For those who hesitate to attend a meeting, know that members in most groups understand this common sentiment and try to make it simple and enjoyable to sign up. The size of a group can be intimidating, but there are many ways in which professional organizations try and overcome this. For example, there is often roommate-sharing, and after talks, there is most often someone interested in a meal or a snack, and you do not even have to initiate this ... just walk along and go (only in high school do you get ignored).

When I was president of the Caucus for Women in Statistics, I had young children and no one to watch them while I traveled and attended meetings. I initiated the movement to have babysitting services at the annual Joint Statistical Meetings (JSM). Selfish though it may sound, I knew I was not alone in this need, and it proved to be of major benefit to many women and men professionals, especially those just starting out. Being inspired to take action by the shared vision of other professionals, both men and women, who saw these services as essential, is a key element in leadership. The networking potential and the services are there, you must just take advantage of them.

A professional organization is one of the only places in which you are actually solicited to sign on for mentoring, and you can do so easily, without looking like a friendless dork. In fact, it is encouraged and advertised in many organizations, such as AWIS, and can be very useful to your career. Once again, the size of a group can be overwhelming, but by getting involved with their mentoring program, you meet in small groups or even one-on-one; if that doesn't work, request a different mentor or setting. Available from mentoring as well as the newsletters, journals, and meetings is insight into the intricacies of how the field of statistics actually works, what statisticians really do and how they do it. All of this gives you helpful baseline data on how to navigate the field and organize your career.

For Mid-Level

At this point in your career, it is advantageous to have been a member of a professional group for a number of years. You will have discovered that these groups actually are run by volunteers, at almost all levels. You will have watched the manner in which others attend meetings and speak out (and see that no one ridicules or criticizes legitimate questions); you will have seen instances that have proven to be of both positive and negative personal value, but from all of which you will have gained insight. Finally, you will now be positioned to further your career by volunteering for committee work, editorship, or meeting chairing or coordination. You can even be ready to organize a symposium in your area of expertise; this is the beginning of leadership.

Asking yourself questions at this point is as important as at any other stage. What are you hoping to accomplish? Is your goal to further your career, to start or grow your own business, or just to socialize with others in your field? Whatever the answer, you will have had enough experience to envision exactly the steps to take within the organization that will be most advantageous. If not, then you have the best audience from which to gain that information.

There are also, of course, advantages to joining for the first time if you have not done so prior to this time. If in need of career advice, a professional organization is a helpful resource ... other members, specific talks and career webinars, an online directory, and literature, among other sources, are all available. Member surveys are frequent, and you can find others in your specialty or potential specialties. Since it is true that mid-career is often a time of change, a member has access to a wealth of job ads and programs, including information and advice on getting a raise, getting promoted, and handling a career and a family or home life successfully. Remember that the association has something to gain as well: such a group strives not only to give value to the individual member, but to enhance her/his professional performance as well, it then has members "in high places." Therefore, do not hesitate to maximize the benefits available to you.

You will not gain anything from a membership if you merely people watch. The professionals in these associations number in the thousands, and no one can stop mid-stride to say hello or answer a personal career question. Alternatively, I would say that with high probability no new member even conceives of approaching others with a personal career or other question when thousands are hurrying by in the halls of a convention center. There are many more productive avenues. Very few members think to call the headquarters directly, but it would be an unusual time if you do not get an answer in a courteous and helpful manner. Try calling the main association's number and asking for the employment ads person or the education person, for example. What about asking for the membership person or using the online directory to find someone in your area and asking that professional or career question? What about contacting the local chapter and volunteering for anything they might have? What about attending the annual board meeting and just sitting and talking or saying "hello" before you sit? Our profession includes some exceptionally kind, interesting, and helpful people who may not be exactly gregarious but always open to a simple greeting or to providing an answer to a question. If you hit upon a person who is slow to respond or does not respond as you may like, try again.

Volunteering for committees or task forces, even setting up chairs at a meeting, is an important avenue for developing working relationships at mid-career level with those of similar interests and expertise. The range of contacts widens the more you branch out. This year, for example, the president of ASA sent out a general email asking for volunteers, making participation extraordinarily simple and welcoming.

For Senior-Level

It is most likely that those who have attained this status, either employed or retired and emeritus, are relying upon their professional organization to keep them advised on the latest journals, the policy issues, board decisions, research, and meeting issues. Many are involved in mentoring, education, statistical research, or their own or others' business, such as consulting or particularized research areas. For those retired, there is usually the advantage of a single payment of "lifetime membership" as well as tax-advantaged travel for attendance at professional meetings as additional benefits. Experienced senior-level members are always welcome on committees, task forces, and outreach groups. These members are, in turn, often the most successful at advocacy and policy measures after a long career and are, of course, superb mentors if they so choose. Sometimes, there are even free or discounted services or offers available. At this point in your career, you can look back as a known leader and realize that professional membership provided the impetus you required at each stage in your career.

Conclusions

The often cited "under-recognition" of women for scholarly contributions (e.g., see Popejoy and Leboy, 2012) is most often attributed to a variety of causes and discussed with statistics and figures on alleged under-representation. However, it is my opinion that an overlooked contributory factor is the innate hesitation of many women to assert themselves and offer their opinions. (I have as a goal to stop all my female students and colleagues from saying: "Sorry" multiple times and "This may be a stupid question, but . . . " . . . when males rarely, if ever, do this). I have often noted in meetings that a large proportion of the male component gets louder when challenged or when making an incorrect or outrageously self-promoting statement. Women have been seen to sit down sheepishly, turn red, become obviously flustered or embarrassed or other self-effacing behaviors in the face of even mild challenge. Taking notice of such gender-based characteristics and using the professional organizations as a platform both to leverage your developing strengths and to utilize them in ways new to you will provide you with all the skills needed to become a leader in the field.

I submit that the most powerful way to lead is by example. As you travel along in your career, I can only suggest to you that my long-term membership in multiple professional organizations has furthered my communication and collaborative skills, enhancing my ability to work more effectively alone and with others on both research and committee goals. Membership in ASA, RSS, AWIS, and other specialized groups in statistical science, as well as in application areas, has provided me with a platform for the application of strategic thinking and the development of winning strategies for advancing my chosen field of statistical science and my career as a mathematical statistician. Providing guidance for others in our field resulted from my skills developed and honed within the ASA structure. For example, membership on the board of the local Washington, DC, Statistical Society was my introduction, then enhanced by assuming national leadership roles, such as co-chair of the ad hoc ethics committee and chair of the first permanent professional ethics committee (wrote the original guidelines and received ASA board approval). Through the mechanism of professional association membership and accepting increasingly more responsibility within this framework, as I did, you will be provided with numerous opportunities to influence both people and policies and in turn become able to foster positive change within the statistical profession itself.

I hope that I have shown in this chapter that early career-stage membership in a professional organization that is geared toward your strengths or potential is vital for many aspects of your continued work and future success. Overcoming initial shyness, observing other well-known statisticians (Figures 16.1–16.4) delivering talks and fielding questions, socializing with other members, purposefully steering your socialization to those who have traits that you admire or those who are doing research in an area of interest to you, all provide a foundation for the future. Once you have advanced beyond entry level, whether by time or promotion, joining or maintaining membership at mid-career is important for providing new avenues of thought, research, potential for branching into an entirely new work or research path, and enlarging the arena in which you become known. By taking advantage of all the benefits of an international and national professional organization, at mid-point in your career, you should be well on your way to assuming leadership roles. Once you have established yourself as a solid supporter of the organization, your volunteer efforts will have paid off and your steps to leadership roles fall into place for you. If you have a plan concerning exactly which roles are of most interest to you, then this is the time to execute and act. Then, when you have attained senior level in your career, you will enjoy a large number of collegial interactions, sources of volunteer leadership roles, and chances to mentor those at other stages as well as recognition for your career-long efforts.

Bibliography

Popejoy, A.B. and Leboy, P.S. (2012). Is math still just a man's world? *J of Math and System Sciences*, 2:292–298.

(a)

Helen M. Walker, ASA, 1944

(b)

Gertrude Mary Cox, ASA, 1956

(c)

Elizabeth L. Scott, IMS, 1978

(d)

Margaret E. Martin, ASA, 1980

(e)

Barbara A. Bailar, ASA, 1987

FIGURE 16.1(a–e)
Women Presidents of the American Statistical Association and the Institute
of Mathematical Statistics.

(a)   (b)

Janet L. Norwood. ASA. 1989   Katherine K. Wallman. ASA. 1992

(d)

(c)

Lynne Billard. ASA. 1996   Nancy Reid. IMS. 1997

(e)

Sallie Keller-McNulty. ASA. 2006

FIGURE 16.2(a–e)
Women Presidents of the American Statistical Association and the Institute
of Mathematical Statistics.

(a)

Mary Ellen Bock, ASA, 2007

Sally C. Morton, ASA, 2009

(c)

Nanny Wermuth, IMS, 2009

Nancy L. Geller, ASA, 2011

FIGURE 16.3(a–d)
Women Presidents of the American Statistical Association and the Institute of Mathematical Statistics.

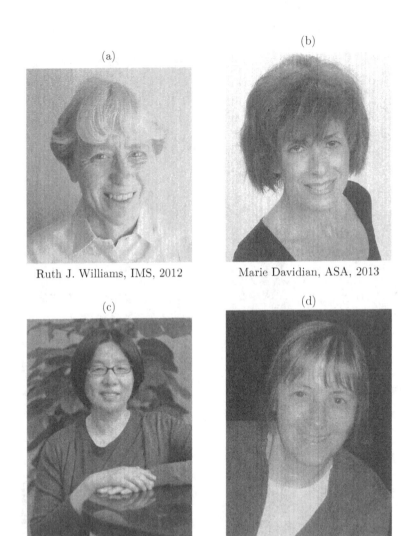

(a)
Ruth J. Williams, IMS, 2012

(b)
Marie Davidian, ASA, 2013

(c)
Bin Yu, IMS, 2014

(d)
Jessica Utts, ASA, 2015

FIGURE 16.4(a–d)
Women Presidents of the American Statistical Association and the Institute of Mathematical Sciences.

17

Leadership Development in the Workplace

Gary R. Sullivan
Eli Lilly and Company

Anyone who has seen the movie *Apollo 13* (Howard, 1995) is familiar with the famous "Failure is Not an Option" scene, that being the phrase uttered by Gene Kranz, the flight director (played by Ed Harris), at the end of a tense discussion. Prior to that quote, an engineer named John (who might as well be a statistician) is able to quickly convince a room full of other engineers, analysts, and managers to take a specific action that ends up helping save the lives of the astronauts in the damaged spacecraft.

When I talk to people about leadership or teach leadership-related courses to statisticians, I will often reference that scene from the movie, as it is an excellent depiction of what leadership for a statistician could look like. In the *Apollo 13* scene, John makes a provocative statement, puts himself (literally and figuratively) in the center of the discussion, makes a strong recommendation, takes on a barrage of questions and challenges, answers them through a keen understanding of the data and a grasp of the big picture and what's at stake, and ultimately convinces the decision-maker, Gene Kranz, to accept his recommendation.

The reason I share that scene is that if you want to develop your own leadership, you have to start with an understanding of what leadership is. That movie scene paints a great picture of what leadership looks like. Obviously, there are many ways to define leadership and many aspects of leadership that impact an individual's ability to inspire, influence, or convince others, and I will touch on those later in this chapter. But, I believe it is important to start with a foundation. As we begin the discussion on leadership development, I would encourage you to think about what leadership means and what your definition might be. I'll provide mine a little later but wanted to start your thinking on leadership.

I want to start with four critical points that I have learned about leadership in my years as a statistician and a manager that are important to understand and embrace before we get deeper into leadership development:

1. Leadership is not just for managers and supervisors.

2. Leadership is not for the weak.

3. Leadership makes a difference.

4. Developing leadership requires time, work and commitment.

I will spend most of the space on the final point and how one can go about developing her or his own leadership and/or the leadership of those around her or him. But let me start by saying a few words about the first three points.

Leadership is not just for managers and supervisors. Many believe that those who have direct responsibility over people or control of resources are the only ones who need to be leaders, or that leadership is much easier for those who have people or resource responsibilities. In fact, supervisors and managers of statisticians have the same challenges as individual contributors in developing and demonstrating leadership. They may have different opportunities to show leadership given their roles, but they still face the same challenges. For example, with their people responsibilities, supervisors need to inspire and reward their employees in a way that builds trust and optimizes productivity. How does one do that if wages are frozen and promotions are limited? Clearly, there are other ways, and it requires an understanding of what motivates each individual as well as a wide array of methods to recognize people and get them energized. Further, when a manager is asked to lead an administrative initiative with a team of their peers from other functions, she must rely on organizational skills, strategic thinking, negotiation, and good communication to get the job done. She can't simply tell people what to do and have her way. Analogously, a statistician trying to promote a new approach has to rely on influence skills, strategic thinking, negotiation, and good communication to get the job done.

Leadership is not for the weak. True leadership will require you to take chances and challenge yourself; make statements, recommendations or decisions that may make some people (including your peers or colleagues) very unhappy; find ways to work with people who may not share the same values or motivation that you do; compromise in order to achieve some gain or partial wins; and sometimes not get the credit you feel you deserve in the interest of the greater goal. The reward for a strong leader is the fulfillment of achieving a breakthrough that positively impacts the organization, the joy in seeing others grow and succeed, or the satisfaction of working with others through political and structural obstacles to complete a challenging task.

Leadership makes a difference. As I've learned more about leadership, I am able to assess the level of leadership in organizations both at work and outside of work. I see a clear relationship between the leadership abilities of those at high levels and the success (or failure) of the organization. Very few organizations succeed without strong leaders at the top and throughout the organization. I will provide specific examples I have observed as I talk more about how statisticians can develop their leadership.

Developing leadership requires time, work, and commitment. Most everyone has heard someone referred to as a "natural born leader." I do not believe in natural born leaders, but I do believe that some individuals are born with greater leadership potential than others. But anyone who wants to be a strong leader has to work at it, even those born with high potential. The primary focus of the remainder of this chapter will be ideas and guidance on how to develop into a strong leader.

Before I get into the leadership material, I want to set some expectations around the content of this chapter, as this is a book on leadership for women. Obviously, I am not able to share any first-hand wisdom about the challenges of leadership for women. I will allude to some strong women leaders I have had the pleasure of interacting with and will share some stories and experiences. Finally, the majority of my content is for general leadership development in the workforce and, in my opinion, is applicable for both men and women.

Developing Leadership: Starting Off

Let me go back to the groundwork I laid above around defining leadership. In the statistics group at Eli Lilly, we define leadership as the ability to inspire people to take a specific direction or action when they truly have the freedom, or choice, to do otherwise. As I said earlier, in order to develop leadership, you need to start by having an understanding of what leadership is, and a definition of leadership is a good first step to doing that. You can then start to look for the embodiment of that definition in people and situations that you encounter. I alluded to the scene from the movie *Apollo 13*, which is a good depiction of our definition at Eli Lilly. You might take some time to think about how you want to define leadership, whether it's a definition or a couple key words or phrases.

I cannot stress enough the importance of gaining awareness and understanding of leadership. Until you have a grasp of what leadership is and an appreciation of its importance, further developing your personal leadership could be challenging. In fact, much of the focus of our internal leadership development program is just that — understanding leadership and its impact on an organization. You can do this several ways:

- Listen to presentations on leadership or watch leaders in action and seek to understand the impact of their leadership.

- Talk to others about their perspective on leadership.

- Read about leaders paying special attention to their experiences. (Later in the chapter, I will discuss several books I have found helpful.)

As part of our leadership development program here at Eli Lilly, we have visited leaders in the Indianapolis community to provide a perspective on leadership outside the walls and buildings of our company. This is a great way to see and understand impactful leadership, identify skills and approaches that may be applicable to your situation, and observe different leadership styles, as not all leaders are alike. Let me share an experience with a leader from the local Indianapolis community whom I met through our leadership program at Eli Lilly.

In 2010, in the initial year of our leadership program, I accompanied a group of 25 of our statisticians to hear the leadership story of Ann Delaney, the executive director of the Julian Center in downtown Indianapolis. The Julian Center is an organization that supports victims of domestic violence and sexual abuse in Indianapolis. Its mission is to break the cycle of domestic violence. It is a well-funded and well-supported organization that has given great visibility to the problem of domestic violence, has helped more than 60,000 victims of domestic violence, and provided education to more than a 350,000 people on the impact of domestic violence to our community (*http://www.juliancenter.org/*). As Delaney shared with us, it was not always that way. She proceeded to share her story of leadership, taking the Julian Center from a poorly funded, unsecure facility that had to turn away victims due to limitations of space and support to a spacious facility, centrally located, properly funded that provides a safe haven for hundreds of victims and their children. Her leadership challenge told of risk-taking, negotiating, partnering, networking, teamwork, perseverance, effective communication, and influence with local politicians, businesses, law enforcement, and other community leaders to create the changes necessary to properly meet the needs of victims of domestic abuse. This was followed by a tour of the facility that provided tangible evidence of the difference that strong leadership can make (referenced from personal notes, March, 2010). Hearing and experiencing leaders like Ann Delaney not only provides examples and insights into leadership, but motivates you to become a stronger leader.

Focusing on Leadership Competencies

A corollary to my fourth point on leadership is that you cannot learn to be a leader by taking a class. This does not mean that leadership training cannot provide value, but that alone will not make you a leader. Once you grasp leadership and can recognize it, how do you develop it? I would suggest that it is analogous to becoming a better golfer. It's not as simple as going to the driving range and hitting a ball off the tee. It requires becoming consistent and expert in many areas: driving, hitting fairway irons and woods, hitting fades or draws, playing from the sand, chipping, pitching, putting, as well as course management. Any accomplished golfer will tell you that you will not get there in a week or even a year. You have to receive coaching, dedicate practice time to different aspects of the game, and then apply what you've learned on the course. Then repeat again and again. Leadership is no different.

What is the leadership equivalent of driving, pitching, and putting in golf? In other words, what are the competencies and skills that collectively make a strong leader? Consider individuals you believe to be strong leaders and think about some of their qualities and what they do particularly well. One of the strongest woman leaders I've had the pleasure of working with is Aarti Shah, a trained statistician and former vice president of our statistics function at Eli Lilly. Aarti's leadership truly elevated our function to a level of influence and company impact that would make any statistician proud to be a part of. Here are some of Aarti's qualities that make her such a strong leader:

- Communicates very effectively

- Makes good decisions

- Inspires others

- Is very convincing

- Holds people accountable

- Builds trust

- Understands the business

- Implements sound strategies

- Fosters strong teamwork

- Fully leverages the expertise of her organization

- Rewards and recognizes others

- Develops other leaders

That is a fairly long list, and I could add more. In fact, if you think about a comprehensive list of leadership attributes or competencies, the list can get very long. In their book *For Your Improvement* (Lombardo and Eichinger, 2005), Michael Lombardo and Michael Eichinger provide 67 different leadership competencies and share definitions, explanation, and relationships with other competencies. I do not share this to overwhelm you, but rather to ultimately allow you to focus on the attributes or competencies that you need or want to develop.

Whether you use Lombardo and Eichinger's list of competencies, or wish to compile your own list (perhaps with some help from some colleagues or mentors), I would suggest that you go through the following process:

1. Create a fairly exhaustive list of leadership competencies.

2. Identify the leadership competencies most critical for you to be a successful leader. (Note that this list may vary depending on the statistician, their role, the organization, and its goals or strategy.)

3. Assess your level of aptitude for the competencies you deem critical: strong, moderate, weak.

4. Focus on the 2–3 competencies most critical for you to strengthen.

Lombardo and Eichinger developed a set of cards and a method for identifying what competencies are most critical for the individual or group and the situation. It's a great exercise to go through with colleagues from your group or organization.

Putting Leadership into Practice

> "I hear and I forget. I see and I remember. I do and I understand."
>
> ――――――――――――――――――――――――――――――
> *Confucious (∼ 500 BC)*

Like anything, the most effective way to truly understand is actually by "doing" or applying what you have learned in practice. Leadership is no different. As we discussed previously, you can learn much about leadership by "seeing," including studying competencies, reading about how others apply them, and observing leaders practicing them. But the way to gain a true understanding and expertise about the competency is to take on a challenge where you will have to use the skill or skills you wish to develop. We can't always choose what we want to do in our work situations, but there are several ways that you can pursue the appropriate challenges:

- Review your current tasks and responsibilities and be mindful of opportunities to develop specific leadership competencies.

- Look to get involved or lead initiatives that would require the competency or skill you wish to develop or strengthen.

- Talk to your supervisor about your desire to develop a skill and ask that they help you find such an opportunity.

Let me give you two common experiences for statisticians along with the competencies these require. One common challenge for statisticians working in business or government is advancing the statistical practices of your customers or collaborators, and this challenge can vary widely. It may be a case where you are trying to convince a clinical researcher to adopt an adaptive design approach, promoting the use of control charts to a production manager, or advocating for missing data imputation to a survey researcher. Each of these situations requires a number of competencies to be successful. These include:

- *Technical Expertise:* Having a sound understanding of the methodology and how it can be applied

- *Business Acumen:* Understanding the costs and benefits the stakeholder must take into account

- *Persuasion:* Creating a convincing case as to why the change would create value for the stakeholder

- *Communication:* Effective articulation of your proposal, including responding to questions or concerns from the stakeholder

Obviously, these competencies overlap and are interrelated, a common theme you will find as you develop and improve your leadership.

Another leadership challenge a statistician may encounter as she or he gains experience is the opportunity to lead an initiative. Again, there are different competencies required, including:

- Organizational skills

- Motivating others

- Getting results through others

- Negotiating

- Facilitating

- Rewarding and recognizing their team

Depending on the responsibilities and interests of the statistician, the leadership competencies required will differ. Matching the competencies you wish to develop with the work challenge is a great way to hone your skills as a leader.

Although experience is the best way to develop your leadership, it cannot be the only way. You will not succeed in situations requiring leadership skills without having some knowledge or training in those skills. In the section on *Starting Off*, I discussed the value in learning about leadership by observing leaders, talking with others, and reading about leadership — all very important methods for developing specific leadership competencies. Let's refer back to these and discuss them in greater detail as they pertain to developing leadership skills.

Observing Leaders and Talking with Other Leaders: I will combine these first two, as they are clearly related. When you are faced with a leadership challenge or opportunity, there are undoubtedly other leaders in your workplace who have similar challenges, or other leaders facing challenges that involve utilizing some of the same competencies. You will oftentimes find yourself in situations where you can observe the other leaders practicing these similar competencies. If you are about to lead an initiative, you may be a team member on another project being led by someone else. You can learn by watching that leader or talking to her or him as she or he works through challenges and problems. For example, watch her or him facilitate a difficult discussion, or speak with her or him about how she or he motivated a team member from a different function to deliver on their step of the project, or simply observe as a senior leader communicates a significant change or new strategy to her or his organization. By observing leaders and talking with others, you can learn a lot about putting different leadership competencies into practice.

Reading about Leaders and Leadership: Another way to learn about leadership, as well as various leadership competencies is through reading. There are two approaches to selecting and reading books to improve your understanding of leadership competencies, and I use both of them.

The first approach is to read books about leaders and leadership written by leaders. Typically, these books share stories and experiences of proven leaders, the challenges they have faced, and how they have dealt with them. By having an awareness of critical leadership competencies, you can identify the practice and application of those competencies by the leaders as you read through their stories.

The second approach is to read books that are more focused on one or two specific competencies or aspects of leadership. This allows you to gain a deeper understanding of a single competency. Let me provide some examples of books and learning from each approach.

One book I would recommend is *Duty* by Robert Gates (2014), the former secretary of defense under Presidents George Bush and Barack Obama. In this book, Gates shares an accounting of his experiences as secretary of defense from 2006 into 2011. In reading the book, I was able to identify countless leadership competencies as they were practiced and applied by Gates as the administrative and operational leader of US defense forces. These include negotiating, decision-making, organizational and business acumen, political savvy, and communicating. The reader is able to gain amazing insights into these competencies and into Gates' thinking through the numerous experiences he shares in detail throughout the book. Some of these include:

- The hiring of key members to his staff throughout his tenure, as well as how he held many individuals accountable for their responsibilities, including asking some to leave their posts when there were serious missteps or failures.

- Numerous decisions, recommendations, and negotiations to move and withdraw troops, develop and deliver needed equipment, budget and plan, and develop and execute strategy — both military and organizational.

- Communications with the president, the president's staff, members of Congress, his own leadership team, officers and troops, as well as families, as he informs, negotiates, challenges, sells, motivates, reprimands, and consoles.

Another book with many great insights from a proven business leader is Sheryl Sandberg's book, *Lean In* (2013). Sandberg is the chief operating officer at Facebook and has held other leadership positions in both industry and government. She conveys leadership guidance through personal experiences gained in these roles, which provides great lessons and inspiration for women. Some of Sandberg's key points include:

- *Mentoring:* The importance of relationships and how to get a great mentor without even asking.

- *Career planning:* Truly considering all your alternatives in balancing your work–life while still pursuing challenging goals, and appropriately selling the idea that a career is more like a "jungle-gym" than a "ladder."

- *Confronting tough issues:* Acknowledging the challenges women face in the workforce, taking on those challenges in her own career, and ultimately bringing greater attention to the challenges for all women.

You can view her TEDTalk (*http://www.ted.com*), which provides a nice overview from some key chapters in her book. Although Sandberg's target audience is women, I was able to personally relate to many of her feelings and experiences.

Other books I would recommend written by established leaders include *It's Your Ship* (D. Michael Abrashoff, 2002) and *It Doesn't Take A Hero* (Norman Schwarzkoff, 1992). (This last book was one I read early in my career that gave me my first true insights into leadership.)

With regard to books that focus on one or two competencies or what I consider specific elements of leadership, there are many I have read that have provided me insights into key competencies, including decision-making, creating change, communicating, motivating others, and understanding others. I will speak about three that I believe provide great insights for statisticians who want to improve and develop leadership:

The book *Blink* by Malcom Gladwell (2009) is subtitled *The Power of Thinking Without Thinking*, but I see the book as being more about decision-making, and I recommend it for statisticians. One struggle some statisticians experience is the need to have all the information in order that they can be very confident about the decisions or recommendations they are making. Do not get me wrong. When it comes to quantitative data and inference, like any statistician, I want to have the requisite data to perform the right analysis in order to make an appropriate inference. However, leaders do not always have the luxury of having all the data and must rely on wisdom, experience, intuition, and an ability to process information from many sources to make timely decisions. Gladwell's book emphasizes the importance of relying on these to make quality decisions. Many statisticians do not do this well because they are trained to rely on hard data. The key learning point for statisticians is that once they gain experience, their ability to process information really equips them to make good decisions where hard data is not always available. They just need to get over the "fear" of not being 95% confident. *Blink* provides some good examples and a great perspective on decision-making.

Susan Cain's book, *Quiet* (Cain, 2012) is one that most statisticians will find impactful for several reasons. The subtitle of the book is *The Power of Introverts in a World that Can't Stop Talking.* Since most statisticians are introverts, they will gain tremendous insights into themselves and their statistician colleagues and collaborators. There is simply too much good information in the book to be able to fairly summarize, but let me provide a couple pieces that intrigued me. Early in the book, Cain debunks the myth that strong leaders have to be extroverted and charismatic, and points out that the correlation between extroversion and accomplished leadership is a myth. In addition, Cain provides great insights into how introverts think and behave. Even if you are an extrovert, as a leader you have to understand the qualities, strengths, and tendencies of those you work with to get the most out of them.

Another book I would recommend is *Metamorphosis*, by John McConnell (1997). McConnell is a disciple of W. Edwards Deming and is an expert in quality management and variability reduction. *Metamorphosis* is about creating change to improve performance. It speaks almost directly to the complaint that most every statistician has made: "I have a better way of doing things, if I could only get people to listen." Essentially, the challenge has been how do we (statisticians) convince (or influence) people to adopt a new or different approach that will lead to better results. The leadership challenge for statisticians is how to create change. McConnell's book provides great insights into the challenge of creating or driving change.

Although all of these books mention many business and industry examples to see relevance to the workplace, let me provide one example from the pharmaceutical industry that relates to McConnell's content. One good example of understanding the importance of how to create change in the pharmaceutical industry is the adoption of adaptive designs in clinical trial design. Knowledgeable statisticians will tell you that, in many situations, adaptive designs present clear technical advantages over fixed designs (e.g., a phase 2 design followed by a phase 3 design), namely more flexibility and less cost. However, in trying to persuade physicians and team leaders to use adaptive designs, statisticians must understand that making such a change goes beyond a technical argument. Consideration must be given toward structures and systems that support traditional fixed designs (e.g., clinical product supply), as well as emotional barriers (e.g., resistance to a new method that threatens the authority or knowledge of an established expert). Statistical leaders must recognize and address these important pieces of creating change.

Other books I would highly recommend do the same and speak to many competencies that are critical to being a statistical leader. These include *Drive* by Daniel Pink (2005), *The Power of Communication* by Fred Garcia (2012), and *Emotional Intelligence* by Daniel Goleman (1995).

I share these books, as it is important to understand different perspectives and aspects of leadership as you work with others. These are by no means the only books one should read, but just a sampling of what I have found helpful.

My recommendation is that you seek out different perspectives on topics that impact your ability to be a more effective leader.

Keeping Yourself Motivated

An important part of being a leader and developing leadership is to "keep the fire burning." That is, identify ways that you can gain inspiration in order that you stay motivated to keep learning about leadership and to continuously improve as a leader. Here are several ideas for doing this:

1. Meet with others working to develop their leadership

   At the end of our leadership course at Eli Lilly, we encourage the attendees to form small groups and to meet on a quarterly basis to share their experiences, successes, failures, observations, and learning (books, articles, other discussions, etc.). This small group approach allows you to share the work. For example, you can rotate who is going to lead the discussion at each meeting. This allows you to get a free lesson in leadership by hearing another's story or benefit from what they read or observed. This also creates a shared accountability for each other's leadership development. By not participating, you are not only hurting your own development but letting down your colleagues. Like everything else, a group like this will also have a finite lifespan (probably 2–3 years), so you may have to move on to another group or another approach after a period of time.

2. Leverage your mentors

   Typically, mentors are individuals with more experience who typically have a higher role (whether administrative or technical) in an organization. Accordingly, these individuals can speak to many personal experiences as a leader, or have had more opportunities to interact with and observe other leaders. One mentor I meet with often shares his experiences and perspectives as a way to improve my leadership competencies and understanding. For example, since he leads a large organization (several hundred people), I am able to gain insights from his interactions with more senior leaders and high-level governance teams into strategy discussions, business/financial decisions, and company direction.

3. Work with colleagues who are passionate about leadership development

   This is perhaps the activity that keeps me most motivated, and that is talking and working with others who are passionate about leadership development. One of my experiences in this area was leading the development of our leadership program in statistics at Eli Lilly. I enlisted several individuals across our function and our training group, and we would meet on a regular basis to brainstorm, plan, and share ideas on various leadership topics. As all of these individuals were very engaged and passionate about leadership, our meetings were filled with energy, ideas, and creativity. I came away from most every session more excited about leadership and motivated to keep working on improving it.

4. Use media resources

For introverts, it's great to gain inspiration privately, and there are many ways to do this. These include reading books, watching movies, and leveraging the Internet. I provided my perspective on the importance of reading in the previous section and also shared several books. As I mentioned above, I do not tend to gravitate solely toward books on leadership but would rather read books about leaders or on topics of leadership competencies. Watching movies is another way to gain insights into leadership. There are countless movies that provide examples of leaders and leadership. Biographies, historical achievements, innovations, and athletic achievements are often captured in film and provide great examples of leaders and leadership competencies. One of the local business schools we visited actually uses movies as a teaching tool in its MBA program, focusing on different scenes throughout certain films to highlight specific aspects of leadership.

A recent favorite of mine that exemplifies superb leadership by a woman is the movie *Secretariat* (Wallace, 2010). Secretariat was one of the last Triple Crown winners in horse racing that still holds the record for the fastest time in the Belmont Stakes, the longest of the Triple Crown horse races. The movie is based on the true story of how the lead character, Penny Chenery (played by Diane Lane), cleverly acquired the rights to the horse, identified and hired the appropriate trainer and jockey, and fought against the insistence of her husband and brother to not sell the horse in a dire financial situation. Chenery demonstrates courage, passion, risk-taking, keen negotiation skills, toughness, humility, and perseverance standing for her belief in the horse and its potential greatness.

Finally, the Internet is a fabulous resource for everything from papers, to insights from leadership institutes, to TED talks. TED talks are a favorite of mine, as they always provide strong examples of oral communication and often touch on specific competencies of leadership. Two very inspiring TED talks on these topics are by Simon Sinek describing the "Golden Circle" and Nancy Duarte on communicating ideas and inspiring others.

Critical Leadership Competencies for Statisticians

I have alluded to many leadership competencies through the course of this chapter but want to provide for you what I believe to be the most critical five based on my experiences as a statistician and manager working in industry.

Communication: As leadership is largely about influence, inspiration, and creating change, it is extremely difficult to be successful in any one of these without being a strong communicator. As I mentioned, this includes speaking, listening, and observing.

Business Acumen: Whether you work in industry or elsewhere, having a grasp of the big picture is critical to being a leader. Understanding strategy, company/organizational goals, bottom-line financials, and decision-making processes will go a long way in strengthening your leadership.

Negotiation: It is difficult for those not trained in statistics to make that quantum leap of fully adopting a new statistical approach for their part of the business. Hence, there must often be smaller steps that are negotiated to get to full implementation.

Decision-Making: As mentioned, one of the biggest challenges for statisticians is to make decisions or recommendations in situations where there is little or no hard data.

Integrity: Having integrity allows you to more easily build trust with peers, reports, supervisors, and senior management. Without trust, accomplishing anything with or through others becomes difficult.

Summary

I will close with a personal story about another strong woman leader at Eli Lilly, Carmel Egan. Carmel gave me my first management opportunity in 2002. She, too, has many outstanding leadership qualities, and I will share a story that speaks to some of those qualities, specifically building trust and inspiring others. Shortly after I was in this role, my first as a supervisor, I wanted to promote one of the statisticians in my group whom I felt had performed at a high level for several years and was very deserving. The process for doing this required a brief document sharing some of the statistician's key contributions and an explanation as to how this statistician was performing at the next level. The final step was signature approval by Carmel. I prepared the document and also prepared copious notes and testimony from her customers in support of the promotion. At our next meeting, I nervously presented the document to Carmel and made a statement in support of the statistician's promotion. I was prepared for Carmel to read the document, pepper me with questions, and ask for details before she would send me away and later render a decision. What happened was much the opposite. Carmel took the document, provided her signature in approval, and handed it back to me without reading a word. I was stunned and walked away from the encounter a bit confused at first, but realized the powerful message she was sending me: I trust your judgment, and I support your decisions (personal conversation, December, 2010).

Leaders aren't born they are made. And they are made just like anything else, through hard work. And that's the price we'll have to pay to achieve that goal, or any goal.

Vince Lombardi

Nothing would be done at all if we waited until we could do it so well that no one could find fault with it.

John Henry Newman

I conclude with two quotes that aptly summarize two important points as you move forward: (i) commit yourself and work hard at developing your leadership, and (ii) take what you learn and observe, and put it into practice. Good luck in your leadership journey!

Bibliography

Abrashoff, D.M. (2002). *It's Your Ship*. Business Plus, New York, NY, USA.

Cain, S. (2012). *Quiet*. Crown Publishers, New York, NY, USA.

Garcia, H.F. (2012). *The Power of Communication*. Pearson Education, Inc., Upper Saddle River, NJ, USA.

Gates, R.M. (2014). *Duty*. Alfred A. Knopf, New York, NY, USA.

Gladwell, M. (2009). *Blink*. Back Bay Books, New York, NY, USA.

Goleman, D. (1995). *Emotional Intelligence*. Bantam Books, New York, NY, USA.

Howard, R. (Director). (1995). *Apollo 13* (Film). Universal Pictures.

Lombardo, M.M. and Eichinger, R.W. (2005). *FYI: For Your Improvement*. Lominger Limited Inc., Minneapolis, MN, USA.

McConnell, J.S. (1997). *Metamorphosis*. Wysowl Pty Ltd, Cashmere, Australia.

Personal conversation, December, 2010.

Pink, D.H. (2005). *Drive*. Penguin Group, Inc., New York, NY, USA.

Referenced from personal notes, March, 2010.

Retrieved May 16, 2014, from http://www.juliancenter.org/.

Sandberg, S. (2013). *Lean In*. Alfred A. Knopf, New York, NY, USA.

Schwarzkopf, H.N. (1992). *It Doesn't Take A Hero*. Bantam Press, London, UK.

Wallace, R. (2010). (Director). *Secretariat* (Film). Walt Disney Pictures.

18

Research Team Experience as a Platform for Leadership Development

William A. Sollecito and Lori A. Evarts

University of North Carolina Chapel Hill

> "There is no substitute for teamwork and good leaders of teams to bring consistency of effort with knowledge."
>
> *W. Edwards Deming, statistician, leader, and pioneer of a worldwide movement in quality improvement (1986, p. 19)*

The increasing emphasis on the needs and opportunities for leadership development among professionals who have achieved technical success and prominence in scientific fields (Institute of Medicine, 1988, 2003) has included a greater recognition of the need for leadership development among biostatisticians (Rodriguez, 2012a). This need for greater leadership development among statisticians is being met in various ways, including texts, such as this one and via formal courses that are customized for delivery to statisticians; for example, two recent examples, specifically designed for biostatisticians, have been offered in schools of public health (Buchanich, 2012; LaVange, Sollecito, Steffen, Evarts, and Kosorok, 2012).

How Do You Become a Leader?

A couple of key questions that many statistical leaders, including the authors of this chapter, reflect on as we review our leadership journey include: how did we learn to become leaders, or more specifically, what skills did we have to master to exert our leadership abilities even before we were assigned formal leadership positions? In discussing these questions with fellow statisticians who have attained leadership positions, we found several common themes, but also observed there are a variety of ways to address these questions. One common theme we found is that although formal education is required to build and reinforce baseline knowledge, "much of the skill essential for effective leadership is learned from experience rather than formal training programs" (Yukl, 2002, p. 378). As noted by Kotter, "on the job" experiences provide valuable lifelong learning opportunities essential for leadership development (1996). This experiential learning provides an important and fundamental opportunity to develop leadership knowledge and related skills among scientists, including statisticians. In the same way that statistical expertise cannot be developed in the classroom alone, leadership skills require ongoing practice at various stages in one's career to be mastered and continuously improved. Unlike clear technical questions and solutions with which many of us who are trained as statisticians are quite familiar, learning leadership occurs in stages that are coupled with changes as we progress in our careers; at each stage, there are critical developmental tasks that must be achieved before we can progress to the next stage. This involves what some authors call an adaptive process. Whereas technical problems can be solved by following established protocols, procedures, and instructions, adaptive process problems are more complex and often require changes in attitudes, behaviors, and values. Learning to be an effective leader requires a lifelong adaptive process (Fernandez and Steffen, 2014).

For example, in discussions with past presidents of the American Statistical Association, several of whom are women, Bob Rodriguez notes that a common denominator for starting their leadership journies was a willingness to serve and work with others, by volunteering their time, then followed by a series of stages of greater leadership development over time; first starting small and then progressing over time, with one opportunity leading to another (2012b).

In our experience, as leaders in the contract research industry and later in academia, a similar pattern was observed. Humble beginnings, primarily motivated by a desire to help others in our organizations, led first to opportunities to volunteer to find solutions, then to recognition as influential problem solvers and eventually to more formal assigned leadership roles. Eventually, this also led to a certain degree of charisma, a leadership characteristic in itself, where others trusted our opinions and expected us to lead (Katz and Kahn, 1978).

While there are several ways that this lifelong adaptive process can occur, a rich source of experience that we have found to be especially valuable for developing leadership skills is participation in team activities and, in particular, research team participation and leadership. This chapter summarizes our experiences and observations in developing leadership skills that were built initially on our technical skills as biostatisticians, and later as research team members and team leaders, culminating in organizational/administrative leadership roles in the contract clinical research industry, and most recently our experience as teachers of leadership skills to practitioners in academia. The concepts presented here are not only based on our personal experiences; they have been supplemented by "leadership stories" of other colleagues and statistical leaders, both men and women, and in varying environments, that represent an "informal leadership panel." Together with our own experiences, our leadership panel's stories have been blended into the leadership concepts presented here. All are biostatisticians, mostly women, who started out their leadership journey as members of research teams and, through a stepwise process, assumed greater leadership responsibility and now have achieved notable leadership positions, such as the following: director at the Food and Drug Administration, chair of a Biostatistics Department, director of a multinational academic research institute, and several senior leaders in the pharmaceutical industry, including the vice president of a center for statistics and drug development, and a president of clinical operations for a multinational contract research organization. Although anecdotal in nature, this aggregation of experiences is uniformly consistent in regard to the value of early and continued experiential learning on research teams as a key element of leadership development.

What Is Leadership?

Before discussing how to develop the leadership skills needed to become a leader, or why we believe that research teams are good platforms for leadership development, we must first discuss the definitions of leadership that will be the focus of this chapter. Throughout this book, you will see this question answered many ways, some similar to what we propose here, some different. As statistical scientists and seekers of truth, we may ask why there is so much variability in the response to this question. The answer begins with the fact that unlike statistics, which utilizes hard scientific principles rooted in mathematics, leadership is often described as a soft science (Sapienza, 2004). Various leadership definitions have some very consistent principles, but there are also differences of opinions among various experts, about the details of how leadership is defined and also the most important skills for a leader to master. This became very clear recently when a statistical leadership panel of the American Statistical Association, which included several authors of chapters of this book, attempted to reach consensus on the definition of leadership; the result was a wide range of skills that were considered to be most important for statisticians to master. The differences that are found throughout the literature can be directly attributed to the wide range of applications of leadership principles, from everyday life experiences, spanning family, sports, politics, religion, to military examples that are common in the news, to business and scientific endeavors, including the leadership of clinical research teams (CRTs). While it is not reasonable to explore all leadership concepts and associated theories here, a brief review of some of the key principles that relate to leading and building leadership skills for statisticians is informative. The emphasis here will not be on the characteristics of distinguished, recognized leaders, as is often done when teaching leadership, but rather we will share leadership skills that can be mastered at early stages in a statistician's career, and specifically through serving as a statistician in a clinical research team environment.

First and most important for this discussion is the distinction between assigned leadership and emergent leadership. "Some people are leaders because of their formal positions in an organization, whereas others are leaders because of the way other group members respond to them ... Leadership that is based on occupying a position in an organization is assigned leadership" (Northouse, 2013, p. 8). Although the ultimate goal of leadership development may be to achieve an assigned leadership position, our intention is not to describe the characteristics of assigned leaders; instead, we will focus on behaviors that allow your leadership skills to emerge through an adaptive process, specifically through your role as a statistician on a CRT. Since CRTs are made up of a variety of cross-functional experts, various informal leaders may emerge on the basis of who may have the knowledge, will, and motivation to step up and solve problems for the good of the group. Leadership development of statisticians or any of the other functional experts through CRT experience is a form of emergent leadership, which many leaders (including the authors of this chapter and those on our informal leadership panel) identify as the means by which they first developed their professional leadership skills. Emergent leadership develops over time through effective communication and problem-solving activities on teams. "Some of the positive communication behaviors that account for successful leader emergence include being verbally involved, being informed, seeking others' opinions, initiating new ideas and being firm but not rigid" (Fisher, 1974). This statement identifies key skills to master in leadership development, and these skills will be explored in greater detail later in the chapter.

Central to emergent leadership is the concept that leadership is not a trait or characteristic of an individual; instead, it is a process that an individual uses to influence others. "Leadership involves influence. It is concerned with how the leader affects followers. Influence is the sina qua non of leadership. Without influence leadership does not exist" (Northouse, 2013, p. 5). In order for emergent leadership to occur, there must be influence, and this opportunity is available to all members of a team.

A leadership definition that is consistent with the concepts of emergent leadership and the importance of leadership as behavior and influence was first presented in a classic work by Katz and Kahn. It is unique in its simplicity and also in its applicability to the topic of this chapter: how does one develop leadership skills? These authors describe leadership as influential behavior: "every act of influence on a matter of organizational relevance is in some degree an act of leadership. ... We believe that the essence of organizational leadership to be the influential increment over and above the mechanical compliance with the routine directives of the organization" (1978, p. 302). Critical to this definition is the question: How does one exert influence? Stated differently, this question is: How does one achieve a level of power sufficient to have influence over others? Two types of power that are relevant for leadership emergence of statisticians on teams are expert power and referent power (Katz and Kahn, 1978). Expert power derives from the critical and unique role that we as statisticians have based on our education and statistical science skills; with time and experiences this same expertise then extends to team leadership skills. Referent power flows directly from the camaraderie of team members working together to achieve common goals and helping each other in this unifying effort; "referent power refers to influence based on liking or identification with another person" (Katz and Kahn, 1978, p. 302). We will return to explain further how these power and influence concepts tie together on CRTs later in this chapter.

As applied to principles described in this chapter, the "influential increment" concept simply means that the first step in leadership emergence is to do more than is required by your job description or role on a team — to "go the extra mile." Using this definition, we (authors of this chapter) can reflect on many applications of this concept to create leadership opportunities that we used to improve the quality of decisions, and outcomes for our teams, organizations and, most importantly for our customers; by volunteering to go beyond mechanical compliance and offer new ideas, innovations, and sometimes just working harder than others to help ensure success of our teams to the benefit of all. These ideas and actions were often in response to unanticipated challenges. The ability to have an influential increment and convert challenges into opportunities was often tied to risks but also led to substantial rewards both in regard to new scientific solutions and our own personal leadership development.

Team Participation as a Leadership Learning Experience

Many statisticians that we have worked with have spent a good deal of time working in some form of team environment. This is not surprising because of the specialization that our statistical training provides, but also because of broader trends that have occurred in recent years. As predicted by several notable authors, working in teams has become an important feature of organizational life that has grown and expanded, becoming a way of accomplishing challenging goals; furthermore, it has been well established that teams provide an excellent learning modality for all involved in team activities (Marshall, 1995; Katzenbach and Smith, 1999; Yukl, 2002; Fried, Topping, and Edmondson, 2006; Fried and Carpenter, 2013). Learning through team-work is a form of experiential learning, the value of which is well documented from experiential learning theory (ELT),which defines learning as "the process whereby knowledge is created through the transformation of experience; knowledge results from the combination of grasping and transforming experience" (Kolb, 1984, p. 41). Team learning is also a prominent feature of Senge's model of learning organizations, i.e., organizations that constantly obtain and transfer new information to improve performance. He describes teams as microcosms for the larger organization in fostering ongoing learning throughout an organization and introduces three critical dimensions of learning that are directly applicable to leadership development. These are the need to think insightfully about complex issues; the need for innovative, coordinated action; and the ability to teach others; and each of these dimensions can be accomplished through the role of team members on multiple teams (Senge, 1990). These dimensions provide guidance to statisticians in identifying skills that are needed to enhance their own leadership development. In this manner, not only will you become a better leader, but you will have the opportunity to go beyond the "mechanical requirements" of your role on the team to help ensure overall team success. This interaction of individual and team learning and success can also accelerate the leadership development process. For example, Senge specifically describes team learning as the fourth discipline of a learning organization, and notes that "when teams are truly learning not only are they producing extraordinary results but the individual members are growing more rapidly than they could have otherwise" (Senge, 1990, p. 10). You can grow as a leader by being involved in teamwork activities within a formal leadership program. There are many examples of formal leadership education programs where you can benefit from instruction that incorporates teamwork. An example of a program that places team-based learning at the core of instruction is the highly effective National Public Health Leadership Institute (Umble et al., 2005). An example that places team-based learning within graduate-level leadership courses is the Public Health Leadership Program at the University of North Carolina at Chapel Hill (Umble, Shay, and Sollecito, 2003). These programs and others demonstrate that learning leadership principles and practices can be greatly enhanced by incorporating teamwork activities.

Learning Leadership on Clinical Research Teams

While the leadership definitions introduced above apply to a variety of learning modalities and organizations, large and small, for the remainder of this chapter, we will focus on leadership stages, experiences, and skills that you can develop though participation in CRTs. The principles that we describe here are applicable to any type of research teams that involve groups of professionals working together to achieve a common goal. For example, in her book on leadership in science, Sapienza refers primarily to laboratory research environments but points out the generalizability of the leadership concepts that she presents to a broad spectrum of scientists and researchers (2004). To capitalize on our previous experience in contract research organizations (CROs), for illustrative purposes, we will focus primarily on clinical trial research examples for learning leadership, since that is what we had the greatest experience with while working for a multinational CRO. These types of studies are common and are well described by Pocock in his classical text: _Clinical Trials — A Practical Approach_ (1993). Most important, for the purposes of this chapter, is the fact that such trials will always include a role for one or more statisticians, either as team members or team leaders, including being principal investigators. We will assume that, in addition to statisticians, the "typical CRT" may include clinicians (physicians and nurses), data management professionals, computer programmers, medical writers, epidemiologists, and scientific advisors or consultants, either internal or external to the primary team.

Also critical to this discussion is increasing the influence of statisticians on teams and organizations undertaking CRTs, which is also expressed by Pocock (1993). Therefore, we present these ideas as a starting point for statisticians assuming greater leadership by going beyond their traditional roles. "It is a common mistake to assume that the statistician need only be concerned with the analysis of results. ... An experienced statistician should be a collaborating scientist in ensuring that both protocol design and interpretation of trial findings conform to sound principles of scientific investigation. In addition, the statistician is in a good position to act as policeman in ensuring that satisfactory organizational standards are maintained throughout a trial. ... I hope that in the future an increasing number of trial organizers will recognize the important collaborative role of the statistician from their trial's inception to its completion" (Pocock, 1993, p. 35). This position by Pocock clearly supports the influential increment concept by asserting that statisticians can and should do more than simply completing statistical tasks on a research team, thereby improving their leadership knowledge and skills through the referent and expert power that they possess.

Does the Complexity of CRTs Help or Hinder Leadership Development?

By their very nature, CRTs are complex and require leadership approaches that are flexible but effective. Clinical research is characterized by the unpredictability of science, and by scientists who are frequently highly trained solo contributors who might be hard to lead well. "Striking the right balance between, first, the freedom, ambiguity, and challenge necessary to foster creativity and, second, the constraints necessary for producing results with time, cost and perhaps commercial objectives is fraught with problems"(Sapienza, 2004, p. 4). This complexity contributes to what makes research team leadership an ideal platform for learning leadership skills, where both teams and individuals learn experientially through informal leadership development opportunities and learning from the experience and input of other team members (Dilworth, 1998; Yukl, 2002). Both the positive and negative experiences–characteristic of team-work — are a form of action learning (Dilworth, 1998), and while beneficial to learning and leadership development, they also present challenges. The challenges presented by cross-disciplinary teams, such as CRTs, which often face unique problems to solve, difficult obstacles to overcome, and risky decision-making, are the types of team experiences that enhance action learning. Both the successes achieved and the failures that are faced are important determinants of future leadership development (Yukl, 2002, p. 379).

Also important for leadership emergence is the fact that research teams are usually somewhat diverse, because they are often cross-functional in nature, composed of members who have different scientific skills and educational backgrounds. They are inherently complex, requiring interdependence of team members and coordination of their activities to achieve a common goal. As such, research teams provide ideal team learning environments and are excellent opportunities for development of team leadership skills These leadership development opportunities and benefits accrue through the daily activities that are needed to accomplish project goals. Further, these opportunities and benefits extend beyond the development of your scientific skills to include the development of administrative leadership skills that are critical to the successful planning and implementation of your team's activities. The seminal textbook on leadership in public health explains it more formally this way: "Teams are small learning organizations that integrate performance and learning. . . . Team members also learn team-building and leadership skills. They often learn that each member is a leader or potential leader. The conjoining of performance and learning in teams is generally a plus, since their conjunction is often a prerequisite for the organization to increase its effectiveness" (Rowitz, 2001, p. 54).

How well does this work? The extent to which you can learn leadership skills and values by participating on a team will be directly influenced by the type and complexity of the team's structure and activity. By their inherent design, CRTs involve an established period of time, multiple disciplines working together, and many challenges to overcome (e.g., time, cost, and quality issues). Another variation that adds to complexity on research teams is the growing number of virtual teams; teams that may either include members from various geographic locations or may include subteams of members from other organizations, which provide complementary skills to the central team (Byrne, 1993; Yukl, 2002; Leatt, Baker, and Kimberly, 2006). These complexities represent both challenges to team leadership and opportunities for significant team learning. The common denominator and the guiding principle that is central to this chapter is that teams are a microcosm of larger organizations that rely on leadership to be successful. Further, teams face many of the same challenges and opportunities for leadership development as larger organizations, albeit on a smaller, more digestible scale.

The Need to Seize Opportunities

How do statisticians use their roles on a CRT to expand their influence and learn leadership skills? Leadership learning opportunities will depend on the type of assignment you are given within the team and what you do with that assignment to enhance your learning and contribute to the team's success. Studies "indicate that learning from experience is affected by the amount of challenge, variety of tasks or assignments and quality of feedback" (Yukl, 2002, p. 379). As statisticians, you cannot wait for leadership roles to be presented; you must look for team leadership opportunities. In this context, "team leadership refers to the ability of individuals to influence other members toward achievement of the team's goals. Note the use of the word individual rather than leader. ... An individual with no formal authority may emerge as a leader and influence member behavior"(Fried and Carpenter, 2013, p. 143). In research settings, leadership opportunities may occur for all team members depending on the phase of the project and specific activities to be completed at various time points of the research effort. Finding these opportunities will become easier and will have a greater impact through experience, first as a team member, and then a team leader on a small team, and then eventually as a team leader of larger more complex CRTs. This is a pathway that we have both experienced. We first had roles as biostatisticians on small teams that had responsibility for statistical and data management activities alone. Later, as our organization grew, we used our cumulative knowledge and leadership experience to become team leaders of larger multinational teams that spanned a full range of clinical and statistical activities. At each incremental stage in this adaptive process, we faced greater challenges for learning and leadership; but also greater opportunities to exert influence and develop power. This is an example of how we were able to develop referent power, based on liking and identification by other team members, which also leads to increased acceptance of suggestions and directives (Katz and Kahn, 1978). In addition to our own experience, this is the story that was told to us by several of the statistical leaders on our leadership panel. One member of our panel told of her early experience on a research team where a problem or challenge would emerge that did not have a clear, direct solution. At first, the team would discuss the matter and then wait for a volunteer to speak up. In these instances, she proposed a solution or volunteered to take the lead on finding a solution; later, as similar challenges occurred, the team would turn to her and expect her to propose a solution. This is a clear example of how referent power and expert power resulted in her emergent leadership.

Trust

What else can you learn from working in teams? One of the best things you can learn is how to promote trust among the members of the team. This is critical in order to exert influence, build referent power, and also adds to team efficiency. Trust develops through the experience of working together as a cohesive unit, including collective conflict resolution, problem solving, and decision making, which require the inclusion of all team members in those activities. Trust also develops based on the degree of delegation and empowerment, all of which improve motivation, to accomplish team goals and support each other. Most important, trust requires open communication to apply these skills, and that begins with an understanding and commitment to the vision of the team, its goals, and objectives that can be facilitated by documenting team norms and expectations. Guidelines on how to develop these skills will be presented later in this chapter.

Trust may take some time to develop as team members learn from each other and share both successes and failures; learning from failure is not easy at first, but working in teams (because they are smaller and members generally are supportive of each other) can help start that process by providing a climate of psychological safety, where other team members are accepting of those who raise concerns, take risks, or admit errors (Fried, Topping, and Edmondson, 2006). When there is a culture of trust within the team, then the diverse viewpoints of individual team members can be freely shared and all can benefit. The idea is to create a so-called "impactful team," where the contributions (thoughts, experiences, and inputs) of all members of the team can be employed toward achieving a common goal or solving an important problem.

Diversity

How does diversity impact teamwork and leadership development? Learning to respect and encourage diversity is an important leadership skill that can be learned and strengthened through working on CRTs, and especially on multinational CRTs. Diversity has many dimensions, including: gender; age; individual characteristics and thoughts; education and training; country of origin; and life experiences. Indeed, it has been argued that innovation is often the greatest when the team embodies and embraces diversity (Hülsheger, Anderson, Salgado, and Hall, 2002). A team that values diversity can be expected to boost analytical thinking and critical appraisal of a situation or issue. As a team leader or team member, you need to strive to build your emotional intelligence, as defined by Goleman (1995), and cultural awareness, because then you are well positioned to nurture and maximize this boost. "Diversity can also contribute to a healthy level of conflict that leads to better decision making" (Daft, 2008, p. 300). Thus, embracing greater diversity and cultural awareness can expand your own leadership skills and competence as well as the leadership skills of your fellow team members, and enhance your team's performance.

Enlightened leadership recognizes that supporting all forms of diversity not only ensures representation and fairness, but it also leads to higher quality science, making an inclusive approach an important leadership skill to master in order to ensure success in research endeavors (Campbell, Mehtani, Dozier, and Rinehart, 2013).

Emergent leadership will be supported by greater inclusivity and diversity in team leaders as well as team members. But to accomplish this, there are also challenges to overcome by team leaders and team members. For example, Northouse points out that leadership emergence may be influenced by gender bias, making it more difficult for women to demonstrate emergent leadership, but the good news is it was found that when women exert emergent leadership behaviors that they are equally influential to men in their groups (2013).

An important aspect of diversity has to do with multicultural representation on research teams, especially teams that are transnational in scope. Worldwide connectivity and integration among organizations has underscored the importance of gaining an understanding of other languages and cultures so as to be understood as well as to understand others. As a leader, it is important for one to become culturally aware and also be receptive to preferences and approaches that are unlike their own. Cross-cultural awareness can be enhanced through experience in teams and also through a variety of formal mechanisms. For example, one method to begin to understand oneself can be accomplished through the completion of Hofstede's Culture in the Workplace™ questionnaire that examines one's cultural preferences on four dimensions: individualism, power distance, certainty, and achievement (Northouse, 2013).

Statisticians and others engaged in research team activities, and especially team leaders, must embrace diversity in the workplace and maximize the value of these differences and varying perspectives without imposing their personal bias. By learning how to do this at the team level, leaders can better prepare themselves for organizational leadership opportunities and, most importantly, help to ensure greater success in the research teams that they have influence over.

The lessons and benefits of understanding and promoting diversity were made very clear to us in our CRO experience, as we progressed from small, homogeneous teams of statisticians and data managers in one geographic location (North Carolina) to being members and leaders of multinational clinical development teams and organizations later in our careers; the benefits described here were real and led to establishing the leadership of our organization in a very competitive research industry. But perhaps most important was the strength of relationships developed among teams and the emergence of leadership opportunities for our many diverse fellow team members, and especially the many women, including several on our leadership panel, who ascended through the ranks to organizational leadership roles by learning to lead in a multinational setting with a high level of emotional intelligence.

Culture

Another important question is how does the culture of an organization support or hinder leadership development. And closely related is the question of whether teams have a culture of their own and how does that impact leadership development. As you work on various teams and in multiple organizations, you learn that there are many forces that influence how teams operate and how team members interact. These forces are critical for you to understand if you are to exert influence and become a better leader. In our experience, we have found that one of the most influential factors to team success, but one that is hard to grasp, is culture, both organizational and team culture. Organizational culture is defined by the basic assumptions and beliefs of an organization that shaped the way an organization views itself and sets standards for how tasks are accomplished and people treat each other (Schein, 2010). Culture is centered on the organization's core ideology; "key beliefs about those fundamental qualities that embody, for members, the organization's very reason for being" (Sapienza, 2004, p. 197). The culture of a team will reflect the overall organizational culture but will also be the product of the leader's and team members' beliefs and assumptions. These team characteristics are derived from their collective experience and include the functional groups that they represent. Since, by definition, multidisciplinary teams are drawn from a variety of functional disciplines, it is important to realize that, "most organizations are multicultural'... and different units or different professions may vary in their beliefs and behaviors" (Leatt, Baker, and Kimberly, 2006, p. 325). As part of this multicultural environment, team members within each team may bring different cultural beliefs and behaviors to the team; consequently each team may have its own culture. A simpler definition of culture, that is especially applicable to CRTs is that culture is the learned product of group experience (Kotter, 1996). As team members work together on a long-term research project and then later on multiple projects, culture will form and evolve, based on past team experiences, requiring ongoing learning and adaptation by team leaders and members. Also important to discussions about science and research is that culture can either inhibit or foster scientific creativity. Likewise, team and organizational culture can interact to impede or enhance team productivity, and your ability to succeed and lead. It is important for statisticians and other team members to understand the culture of your organization and your team.

How do you understand a team or organization's culture? What is most difficult, especially in the early stages of your career or your tenure with a new organization, is that culture is not easily visible; only its manifestations are observed (Schein, 2010). Similar to trust, understanding culture takes time and the collective experience of working with team members. Each individual should think about the type of culture that she or he is most comfortable working with and make sure that she or he also understands whether the culture of her or his organization or team will impede or further personal leadership development and the overall success of the team in being able to complete the endeavor. For example, it is our belief that leadership development opportunities will be more readily available in an open culture where empowerment of individuals is part of the core ideology. Both coauthors worked in a CRO that had a very strong culture that influenced our everyday activities and provided strong support to our personal leadership goals. A significant aspect of this culture was that team experience regularly provided leadership opportunities that first led to team leader roles, and later to administrative leadership roles, for those who wanted to move in that direction. This form of organic growth from within not only capitalized on the leadership strength of the statisticians in our organization, it helped to define and reinforce the culture as one where those who demonstrated leadership within teams had a clear pathway for promotion and greater leadership. The culture was also a stabilizing force during times of rapid growth. Furthermore, since the culture was a strong part of our personal leadership style, we were able to transition the positive characteristics of our CRO culture to an academic culture in a large university and succeed as leaders in both. We found that being able to articulate the culture to new employees and to understand how our culture changed as our organizations grew had a direct impact on our motivation and our ability to lead. Culture can enhance not only your leadership opportunities, but also your overall job satisfaction. So it is critical that statisticians and team members invest the time to understand culture, and this starts with open communication, which will be discussed in greater detail later in this chapter.

What Type of Leader Do You Want to Be?

In industry, politics, and science, there are many examples of both good and bad leaders; both effective and ineffective leaders. As you strive to master leadership skills that are needed to develop your leadership potential, you should define a target in order to define what type of leader you want to be. To begin, it would be useful to know of the characteristics that have been found to be associated with effective leadership. This will then allow you to begin to define the specific skills that you need to master in order to become the leader that you want to be. Kouzes and Posner, in their seminal work, *The Leadership Challenge*, summarize research based on a large sample of leaders who were asked to describe their "personal best" in leading others (1995). Their findings support the leadership concepts that we illustrated by the ability to develop as leaders through teamwork, such as experience on CRTs. Further, they emphasize practices "that are available to everyone and are not reserved for those with special abilities. The model is not about personality: It is about practice" (Northouse, 2013, p. 199). An alternative approach that is more "bottom up" is to ask the question from the "follower perspective," i.e., what are the characteristics you respect in a leader; or what are the characteristics that others have identified in "effective leaders." Taken together, the answers to such questions provide a sound basis for introducing specific leadership skills for statisticians to pursue, and which can be learned and applied in a research team setting. One criterion for defining these skills is to identify what constitutes effective team leadership among scientists. Derived from responses to a qualitative survey of 147 scientists, of whom two-thirds were PhDs, Alice Sapienza found that effective scientific leaders were identified as those who had the following characteristics:

- *Caring, compassionate, supportive, enthusiastic, motivating* (31% of responses)

- *Possessing managerial skills, such as communicating effectively and listening well, resolving conflicts, being organized, holding informative meetings* (26% of responses)

- *Being a good role model, mentor and coach* (17% of responses)

- *Being technically accomplished to lead a scientific effort* (15% of responses) (Sapienza, 2004, pp. 6–8).

Further comments included descriptions of leaders who expressed an "overall feeling of appreciation of the work done" (Sapienza, 2004, p. 6), which ties directly to one of the key leadership characteristics that Kouzes and Posner found in their study: "encouraging the heart" (1995, p. 9). Also informative in defining effective leadership are characteristics that apply across cultural dimensions, based on the GLOBE research project, which spanned 62 countries with more than 17,000 respondents. "GLOBE found that some aspects of leadership are culturally dependent, while charismatic and team-oriented leadership are universally desirable styles" (House, Hanges, Javidian, Dorfman, and Gupta, 2004). The finding of an exceptional leader is also summarized by Northouse: "That portrait is of a leader who has high integrity, is charisma/value based, and has interpersonal skills. ... The portrait of an ineffective leader is someone who is asocial, malevolent and self-focused" (2013, p. 403). The findings from these various surveys confirm our observations in a variety of settings, including our experience on CRTs, about the leadership skills that we mastered to become the type of leaders that we and our panel of leaders have become.

What Are The Skills That You Should Master to Become an Effective Leader?

The emergent leadership and the influential increment principles, together with the characteristics of effective leaders, presented above, provide a starting point to define some critical leadership skills that each statistician can learn, and can be practiced and applied while working in a CRT environment. The specific list that we present here is based on our experience and that of our leadership panel. This list is by no means exhaustive, but rather is a starting point for your reflection and lifelong leadership development.

Vision

Almost all discussions of leadership include the importance of vision (Kouzes and Posner, 1995; Kotter, 1996; Rowitz, 2001; Yukl, 2002; Daft, 2008). Central to the role and importance of vision is that it should be a reflection of the core values of an organization and present an ideal image of the future. "The vision should appeal to the values, hopes and ideals of organization members and other stakeholders whose support is needed. ... The vision should address basic assumptions about what is important to the organization, how it should relate to the environment, and how people should be treated" (Yukl, 2002, p.

283). These organizational concepts are equally applicable to research teams and also reflect the linkage to organizational and team culture, thus influencing leadership of teams. "Team leaders have to articulate a clear and compelling vision so that everyone is moving in the same direction. Moreover, good leaders help people feel that their work is meaningful and important" (Daft, 2008, p. 307).

A critical goal of team leaders is to foster commitment to team objectives. "Leadership behaviors that are especially relevant for increasing this commitment include articulating an appealing vision of what can be accomplished by the team, relating the task objective to member values and ideals, building member confidence in the ability of the team to accomplish these objectives, and celebrating progress made in attaining the objectives" (Yukl, 2002, p. 307). Vision is created by leaders and followers together, and its development is a critical skill that can be learned through experience, such as through leadership of or participation in a CRT. As you develop your leadership skills, you need to start with a personal vision for your career, your role in the organization, and the team(s) that you are engaged in. You also need to take the time and the effort to understand and have true commitment to the vision of your organization and team. As with culture, you need to understand whether the vision of your organization or team will impede or support your ability to achieve your personal vision and, most specifically, your leadership development goals. If not, then decisions may have to be made to change your personal vision, or join in efforts to change the organizational vision, or even perhaps find a new organization that supports your personal vision. The importance of vision during our experience in a rapidly growing CRO cannot be overstated. During high stress times in our organization, we found that dedication to the vision is what kept us grounded, and helped us to support change, innovation, and rapid growth, and further develop and exert our own leadership throughout. It also helped to keep the multiple teams and parallel projects within our organization moving efficiently toward successful completion, as was defined by our vision.

Further details on vision development are beyond the scope of this chapter and can be found elsewhere. (Recommended sources are: Bennis and Nanus, 1985; Kouzes and Posner, 1995; Kotter, 1996; Rowitz, 2001; Yukl, 2002. Additionally, a step-by-step summary of vision and vision statement development that incorporates total quality management concepts is presented by Melum and Sinioris, 1993).

Communication

A critical question that you must be able to answer to have an influential increment on teams is: How do I communicate most effectively with my fellow team members? Without effective communication, you cannot have influence. Effective communication skills are recognized as key components of emergent leadership (Northouse, 2013). Communication is also a critical step in ensuring commitment to a vision.

A CRT is a rich environment for team members and team leaders alike to learn, practice, and master various communication skills. Without appropriate and effective communication the vision, goals, timelines, scientific requirements, and specific tasks to be carried out by any CRT will not be effectively accomplished. Communication responsibilities for both team leaders and members start with ensuring that the content of any communication is accurate and well presented, with consideration of the perceptions and knowledge of those who will receive the communications. More is not always better, both in terms of length and complexity of communications; this is especially true for a diverse cross-disciplinary team with members who have a range of educational and functional backgrounds. It is critical as a sender of information to always be mindful of the receivers (Sapienza, 2004). Other needs to consider, which sometimes compete when making choices, are timing and the most appropriate format of communications. Finally, both organizational and geographical cultures of team members are important determinants of effective communications. While there are not always simple right and wrong ways to address each of these communication considerations, experience developed at the team level is invaluable in understanding how to undertake and promote effective personal communication.

One of the most critical individual communication skills that both team members and team leaders must learn in a CRT, or any level within an organization, is the ability to listen well (Covey, 2004; Sapienza, 2004). Team leaders and members need to be able to actively listen to each other as well as to external stakeholders; especially in communications directed at problem solving and decision making, both team leaders and members need to listen carefully to input from each other. To ensure this, team leaders must facilitate such discussions with a goal of drawing out the opinions of all team members; and team members must voice their opinions, whether they are in conflict with others or not. This is a critical skill and a requirement of effective team performance. How to do this? Active listening starts with a mental stance that is deliberate, open, and nonevaluative. It also includes utilizing of skills to reduce distortions that prevent mutual understanding; these can include paraphrasing what has been said and asking clarifying questions, as well as using simple nonverbal skills (Sapienza, 2004). Motivation and willingness to participate actively in team discussions is greatly enhanced when one perceives that team leaders and other team members are engaged listeners. Active listening techniques also stimulate better learning by all, and help to determine better solutions and decisions. Active listening and engagement also helps to overcome gender and other biases concerning perceived norms about who can contribute to solutions and decisions, thereby enhancing emergent leadership in teams (Northouse, 2013).

Effective communications can be enhanced by formal training in communication skills. However, team experiences provide a psychologically safe environment to learn and reinforce key principles that can be extrapolated to broader leadership settings and improved upon over time (Fried, Topping, and Edmondson, 2006).

In CRTs, effective team communication is not top down, but rather requires interaction among all levels of a team, using an open star communication network (Sollecito and Mele Dotson, 1995), also known as an "all-channel" network (Longest and Young, 2006), in which all team members communicate with each other as well as with the team leader. Consequently, the responsibility for effective communication is shared by all, both senders and receivers of information (Sapienza, 2004; Sollecito and Mele Dotson, 1995; Longest and Young, 2006).

Daily team interactions will require both ad hoc communications and those that are planned. Because of the complexity of clinical research activities, requiring interdependence among team members and the coordination and timing of events according to predefined schedules and protocols, most important communications must be planned in advance. An effective team leader must lead the development of a project communication plan with the team at the start of a research project (Project Management Institute, 2013). This plan should be updated and modified as necessary during the various phases of a project. The plan will define, at a minimum, how and how often communications will occur and what will be communicated and by whom and in what manner, among team members as well as other key stakeholders, e.g. sponsors or funders. Other aspects of a communication plan include the mode by which team members will be included in problem solving, decision making, issue escalation, and the documentation and approval path for changes to a team endeavor. The ability to develop and implement communication plans provides excellent learning opportunities for broader leadership responsibilities. Discussion and debates in team settings provide good learning opportunities for statisticians and other team members concerning how to expand their influence through effective communication about team issues, including active participation in problem solving and decision making.

Another important component of a communication plan and ongoing team communications is agreement on the mode of communication, which must consider the needs of both senders and receivers of information. The mode of communication must be chosen carefully to fit the occasion; for example, start up meetings and key interactions with sponsors/funders or regulatory agencies may require face-to-face interactions in order to be most effective, whereas more routine interactions, such as regular weekly status updates on completion percentages, can take advantage of technology to ensure efficiency. It is always important when choosing modes of communication to be flexible and consider both the needs of team leaders, team members, and other stakeholders, such that the communication mode chosen is available and accessible to all who participate and thus will enhance, not inhibit, communications.

Meetings

Meetings are a critical mode of communication in any organization, and especially for interdisciplinary teams. The value and need for meetings are another example of "soft" topics about which there are some differences of opinion in the literature. "A meeting is nothing less than the medium through which managerial work is performed. That means we should not be fighting their very existence, but rather using the time spent in them as efficiently as possible" (Grove 1995, p. 71). Efficient meetings are a critical component of any successful clinical research project. Research team meetings are a key activity to ensure communications within teams and across organizations, including all stakeholders. With proper planning and execution by team leaders and team members, meetings can be an efficient way to assess project status, share research findings, solve problems, and make decisions. Team leaders and team members together must ensure efficiency of meetings using established procedures. Many authors provide guidance on how to conduct efficient meetings (Grove, 1995; Rowitz, 2001; Lencioni, 2004). A couple of the points that are commonly emphasized in regard to efficient meetings are particularly critical to master for leadership development. First, team leaders and members have a responsibility to adequately prepare for meetings. A first step is the knowledge of whether the meeting is a process-oriented versus mission-oriented meeting, as described by Grove. Process-oriented meetings are meetings where knowledge is shared and information is exchanged. Mission-oriented meetings are sometimes described as ad hoc meetings, that is, meetings that occur on the basis of special need either to solve a problem or to make a decision. Decision making is one of the primary reasons for having mission-oriented meetings (Grove, 1995). Despite the type of meeting, ensuring that an agenda is developed and that team members contribute to the agenda for team meetings is important so that the interaction is beneficial to all. Team leaders must encourage all team members to participate by obtaining input from team members when evaluating progress, brainstorming possibilities, and concerning decisions and solutions to problems. Active participation in team meetings by statisticians and other team members provides important opportunities to demonstrate the influential increment principle and to strengthen referent power, both of which are critical to developing leadership skills and taking full advantage of the leadership potential that exists within a team. Our experience is that meetings provide numerous opportunities for team members to support each other, whether or not the task at hand is directly part of their responsibility, and thereby exert influence by going beyond routine participation. In summary, effective communication is an important leadership development skill that requires hands-on experience to master. Team leaders and team members can gain invaluable experience developing communication skills while serving on CRTs.

Problem Solving/Decision Making/Conflict Resolution

Effective teams require a foundation of trust, an ability to undertake positive conflict, commitment to decisions made, holding members accountable, and then team results can be achieved (Lencioni, 2002). Some of the best examples of experiential learning in teams come from group problem solving, decision making, and the resolution of potential conflicts that may arise during team processes. By their very nature — as a high-stakes, complex endeavor — CRTs will face numerous problems to resolve and decisions to make over the life cycle of a research project. Issues will naturally emanate from considerations related to deadlines and deliverables, but also those that are due to unpredictable scientific elements that are inherent in the research process. Problems solved and decisions made jointly by team leaders and members will result in better solutions and decisions, and will improve the overall effectiveness of the team. Such participation is also a source of greater motivation, since "people have greater feelings of commitments to decisions in which they have a part" (Katz and Kahn, 1978, p. 302). Although challenging, these interactions are a necessary part of successful clinical research, and they also enhance the ability of team members to learn leadership skills through active participation in these processes.

Problem Solving

Leadership effectiveness requires and is improved by experience with problem solving.

> Within this perspective, leadership behavior is seen as team based problem-solving, in which the leader attempts to achieve team goals by analyzing the internal and external situation and then selecting and implementing the appropriate behaviors to ensure team effectiveness. ... Leaders must use discretion about which problems need intervention and make choices about which solutions are the most appropriate. The appropriate solution varies by circumstances and focuses on what should be done to make the team more effective. Effective leaders have the ability to determine what leadership interventions are needed, if any, to solve team problems (Northouse, 2013, p. 290).

How do you solve problems on a CRT? When problems that need to be solved are identified, an efficient research team will take action quickly to address problems by first clearly stating the problem, identifying causes and potential (alternative) solutions, evaluating potential risks or unintended consequences of solutions, and then developing a plan of action with clear assignments (Grove, 1995). Team leaders must balance the pressures of time and cost to properly assess problem causality using team member expertise to find the true causes and to develop and implement appropriate solutions. Poten-

tial tools and models for causality assessment are found in the literature of
continuous quality improvement, where problems that lead to poor quality are
assessed along a spectrum that spans poor system design to individual performance issues, each of which is carefully evaluated with input from all team
members before proposing solutions (Deming, 1986; Sollecito and Johnson,
2013).

Although the entire team is responsible for finding solutions to problems,
team leaders have the direct accountability for ensuring that proper procedures are applied to find optimal solutions. Team leaders are also accountable
for clear assignments to team members who must take needed actions and for
appropriate follow-up of assigned tasks, using processes similar to the follow-
up of delegated tasks, discussed later in the chapter. In some cases, experts
are added to teams, and occasionally facilitators may be brought in, to ensure
that problem solving is carried out in the most efficient manner.

Volunteering to help others in problem solving is an important way to exert
influence and go beyond the routine requirements of a statistician's role on a
CRT — a way to develop leadership skills and referent power by creating the
expectation that you can be counted on to assist or take the lead in solving
future problems as well. This process was described in specific stories provided
by statisticians on our leadership panels as a way that they developed from
team members to team leaders, especially in their early careers.

Decision Making

Who makes decisions on a CRT? "Decision making is a task for all teams, and
the manner in which decisions are made is critical to team success. Teams may
reach decisions by consensus or voting, or team members may simply advise
the team leader, leaving the final decision to the leader. There is no single
best way for a team to make decisions, and the mode of decision making is
often dependent on the circumstances and goals of the team" (Fried and Carpenter, 2013, p. 150). In general, there are two major types of decisions that
are made in research team settings. First, there are forward-looking decisions
that often relate to planning activities, and second, there are decisions in response to problems or changes in procedures. In general, better decisions will
be arrived at through group processes. Using a group to make or at least participate in either type of decision can provide advantages over decisions made
solely by an individual leader. "Groups have more relevant knowledge and
ideas that can be pooled to improve decision quality, and active participation
will increase member understanding of decisions and member commitment to
implement them" (Yukl, 2002, p. 326). Due to the broad range of activities
and skills required on research teams, there may be some decisions that are
most efficiently carried out by subgroups that have the most relevant expertise; however, all team members should be informed and the opportunity to
participate should be made available for members if they feel that they have
the needed expertise. "Regardless of the mode of decision making, it is critical

that team members understand their role in decision making. Assuming an advisory rather than decision-making role is acceptable to team members as long as they do not expect to be in the position of making decisions. The decisional authority of the team and that of team members needs to be specific and clear" (Fried and Carpenter, 2013, pp. 150–151).

When team decisions are needed an ideal process for making the decision, with input from all team members, should include the following three steps. First, there should always be free discussion of the issue to be decided by all (relevant) team members. It is imperative during this process that team leaders draw out all important points of view from team members who have been asked to be part of the decision. Next, after sufficient discussion, a clear decision should be made and clearly stated to the group. Finally, all team members should agree to support the decision of the team, even if an individual member may not have been fully supportive during the initial discussion process. If a team member cannot pledge their full support, then the team leader may choose to reopen discussions, which may lead to modifications or compromises to the decision (Grove, 1995). In all cases, this process should emphasize the freedom of all team members to express their opinions with respect from all fellow team members.

Impediments to group problem solving and decision making can arise from various sources that leaders must learn to identify and resolve. These can include group dynamic issues, such as peer group syndrome, where individuals are reluctant to challenge their peers (Grove, 1995), or group think, which occurs when team members' desire for harmony override their ability to objectively appraise alternative decisions (Fried, Topping, and Edmondson, 2006). Effective team leadership requires the leader to overcome such impediments.

In summary, decision making has two components that provide valuable leadership learning opportunities for team leaders and team members. First is learning to take responsibility to make needed decisions, and second is the value of participating in decision making activities. The latter component will provide experience to foster better decision-making skills by all participants and also ties directly to the important concept of ensuring appropriate statistical influence on decisions made by research teams. In clinical research, it is important for statisticians to ensure that their expertise is included wherever they feel it is appropriate, either in an advisory role or in a direct decision-making role. The degree of statistical influence may improve the quality of decisions being made and is relevant for leadership development of statisticians with benefits accrued at later stages in their careers, as they increase their leadership responsibilities (Berg, 2013). This is consistent with Pocock's advice that statisticians should broaden their roles on CRTs (1993); this is also an example of how statisticians use expert power, which is an important component of the influential increment concept (Katz and Kahn, 1978).

Conflict Resolution

Is conflict always to be avoided for a CRT to be effective? Conflict on CRTs may not always be avoided; it may result from problem solving, decision making, or other team activities. On a research team, differences of opinion about scientific matters may abound and, in fact, should be encouraged as way to achieve the highest level of scientific excellence. Differences may also occur about how to resolve issues related to time, cost, and quality considerations facing the team. In all such cases, the team leader has a responsibility to allow open discussion on such issues and take steps to balance all opinions. Team leaders must take steps to prevent differences of opinions from escalating into personal conflicts and, if they do, to take careful steps to resolve the conflict. One of the key characteristics for effective research leaders, identified in the survey conducted by Sapienza, is the ability to resolve conflicts; and conversely, the inability to resolve conflicts was cited as a characteristic of an ineffective leader by almost 20% of respondents on her panel (2004).

Task conflict can be an important mechanism to expand clarity, increase team learning, and promote innovation. A useful recommendation for leaders that focuses on the advantages of conflict and diversity of opinions is as follows: "Shifting the orientation from looking for uniformly high 'team work skills' to managing team human resources as a constellation of talents should help to enable teams to be successful throughout the changing pressures of performance episodes" (Marks, Mathieu, and Zaccaro, 2001).

Keys to resolving conflicts include actively listening and promoting full engagement by all members in order to build a culture of trust and mutual respect among the team. Another way for leaders to resolve conflict or minimize unnecessary conflict is through a well-defined vision, which is clearly communicated and deployed. "A compelling vision can pull people together. A vision for the whole team cannot be attained by one person. Its achievement requires the cooperation of conflicting parties. To the extent that leaders can focus on a larger team or organizational vision, conflict will decrease because the people involved see the big picture and realize they must work together to achieve it" (Daft, 2008, p. 315). Once conflicts arise, it is critical that team members and leaders resolve them in an efficient and respectful manner. Although description of specific processes that can be used to resolve conflicts are beyond the scope of this chapter, it should be emphasized that conflict management and conflict resolution are important leadership skills that leaders must master. Experience in resolving conflicts at the team level will contribute to further development of conflict resolution skills at the organizational level. Whether they are team leaders or not, an important way for statisticians to demonstrate leadership and influence is through volunteering to participate in the resolution of team conflicts that are not directly statistical in nature and can be accomplished by providing an unbiased analytical view. Further guidance on conflict resolution can be obtained from Rowitz, 2001; Sapienza, 2004; and Ringer, 2006.

People Skills

The importance of developing people skills by leaders is underscored by the quote from Larry Bossidy, former CEO of Allied Signal: "At the end of the day you bet on people not strategies" (Tichy, 1997, p. 13). People skills are important skills to master by both research team leaders and team members. The same can be said of project managers in general. "Since leadership is essentially an interaction with people ... the most important skills a project manager must have are 'people skills'." (Lewis, 2003, p. 11).

The reasons for these beliefs are many, starting with the fact that leaders/project managers of CRTs rely on the collective efforts of team members to be successful, and team members are jointly responsible for the success or failure of a research project. It is the interdependence of team members that produces the synergy needed to achieve the full benefits of a team-based project or organization. Leaders must ensure peak performance by employees of an organization or members of a team and know when and how to exert influence and build relationships to create an efficient and effective team environment that also allows team learning by all. This relates directly to the influential increment concept and the need for team members to help each other by giving their fullest efforts and going beyond their usual role when necessary. We witnessed this many times in our CRO experience, where tight budgets and deadlines, coupled with the need for the highest quality, required team synergy, which could only be accomplished by all team members going that extra mile to support each other.

In a research environment, what are the people skills that are most critical to master and how is this accomplished?

Motivation

Grove asserts that, along with training, peak performance requires motivation, and further, that although motivation must come from within, it is a manager's job "to create an environment in which motivated people can flourish" (1995, p. 158). Based on Sapienza's survey of research workers, motivation is specifically identified as an important factor in creating a research environment that is characterized by energetic and enthusiastic action; and an important characteristic of effective research leaders is the ability to understand what motivates others and ensure an environment that leads to high levels of motivation (2004). As a team member who aspires to a higher leadership position or who simply wants to ensure that your research project is completed in the most efficient manner, you need to understand what motivates others on your team and, most important, what motivates you, recognizing and accepting that there may be differences between you and fellow team members in this important concept.

An effective leader "engages with others and creates a connection that raises the level of motivation and morality in both the leader and the follower.

This type of leader is attentive to the needs and motives of followers and tries to help followers reach their fullest potential" (Northouse, 2013, p. 186). Also directly relevant is the previously introduced definition of leadership as influence. A leader must influence the behavior of those around her or him, and nowhere is this truer than for team leaders who very often do not have direct administrative authority over team members, but do have the responsibility to elicit peak performance from them. This is a critical skill for statisticians to learn in order to become effective leaders. To accomplish this goal, the first question to address here is: How do I determine what motivates me and others?

A brief review of motivation theory indicates that research team settings can provide an environment for ensuring high levels of motivation. Using Maslow's theory of motivation as a model, each person's needs can be described by a triangle with physiological needs at the base (lowest level needs) and self-actualization at the peak (highest level needs) (Grove, 1995). Assuming that the basic physiological needs (e.g., the things that money can buy) are met in most research environments, then by its creative scientific nature, clinical research can be seen as an ideal activity for meeting self-actualization needs, e.g., the desire to carry out the highest level of science in a research endeavor, which is a reasonable assumption when describing why scientists, including statisticians, choose research careers. Also, in between these extremes, Maslow's hierarchy includes other levels of need that can also be met in an effectively led research team environment. These include social affiliation needs that can be met through membership in highly skilled research teams and esteem/recognition needs that can be met through team recognition factors, such as grants awarded and through scientific presentations and publications, as well as the ultimate reward of contributing to new scientific discoveries. Not everyone will be motivated in the same way. It must be recognized that every team is made up of diverse individuals with some similar characteristics, but also some that are different, especially in regard to role and position within a team's structure. Also very important is to be aware that what motivates a team leader may not motivate others on the team. Various authors describe individual work-related needs that provide a different, but complementary, dimension to the Maslow model. These needs can include power, a desire to have an impact on people; achievement, a desire to do things better or surpass standards of excellence; and affiliation, a need to maintain positive relationships with another person or group of people (Sapienza, 2004). Clearly, leadership or membership on a research team can be a way to meet some or all of these needs. The complexities of motivation are much more tractable on a research team than in a large organization. Experience in creating or working on a highly motivated CRT provides opportunities for learning leadership and how to extend this knowledge to the organizational level as well.

Empowerment

Closely related to motivation is empowerment as a way of improving the performance of individuals and teams. "Psychological empowerment probably has the same type of consequences as high intrinsic motivation and self-efficacy. The beneficial consequences include: stronger task commitment ... greater initiative in carrying out role responsibilities ... more innovation and learning ... higher job satisfaction" (Yukl, 2002, p. 107). Some of the same leadership factors that were previously discussed in this chapter are important to promote empowerment; these include a culture that encourages flexibility and learning, a decentralized organizational structure, and environments where there is mutual trust between leaders and followers (Yukl, 2002). Whether or not these other factors are present, some very specific principles define a leader's role in ensuring empowerment. In describing the role and responsibility of public health leaders in empowerment, Rowitz includes the following principles that should be applied to empower others: "Tell people what their responsibilities are. Give them authority equal to the responsibilities assigned to them. Give them permission to fail. Treat them with dignity and respect" (2001, p. 66).

A review of these concepts makes it very clear that empowerment is directly applicable to research team leadership and is a critical component for ensuring success. "Any leadership practice that increases another's sense of self-confidence, self-determination and personal effectiveness makes that person more powerful and greatly enhances the possibility of success" (Kouzes and Posner, 1995, p. 184). By practicing empowerment strategies, research team leaders will not only improve the chances of successful research team outcomes, they will learn the art of empowering others and how to become empowered themselves as they continue on their own leadership journey. By seizing empowerment opportunities, statisticians can build their expert power, leading to greater influence over team processes and learn the art of empowerment as part of their own leadership development. Two keys to this process are preparation, by learning broader skills that may go beyond traditional statistical education, e.g., leadership development; and willingness to take risks — by stretching beyond their usual roles and volunteering to do more than is required of their traditional roles on a CRT.

Delegation

Successful delegation leads to empowerment and corresponding higher levels of motivation when done properly. Especially in team settings, the ability to effectively delegate is a leadership skill to master. As with empowerment, effective delegation requires transferring, not only task completion responsibility, but also the appropriate level of authority that would otherwise be retained by the team leader. This is a form of power sharing that has numerous benefits to both the delegator and the delegatee, the person to whom a task is delegated. In addition to time management and other personal advantages for the delegator, the quality of decisions can also be improved, when the person to whom the issue is delegated possesses specialized skills, as will occur quite often on a research team. The most often cited benefit to delegatees is the development of new skills and confidence, making the job more interesting and increasing commitment (Yukl, 2002). Each of these benefits closely aligns with and promotes effective motivation and empowerment as well.

Knowing how to delegate requires time and experience to master, but two guiding principles that are critical for team leaders and team members to understand are as follows. First, the delegator and delegatee must share a common information base and a common set of operational ideas for how to solve problems, particularly if the advantage of achieving better decisions is to be met. Second is the importance of follow through by the delegator on all delegated tasks. "Delegation without follow through is abdication. You can never wash your hands of a task. Even after you delegate it, you are still responsible for its accomplishment, and monitoring the delegated task is the practical way for you to ensure a result." (Grove, 1995, pp. 59–60)

Other people skills that are complementary to both delegation and empowerment have to do with the ability to apply situational leadership concepts (Northouse, 2013; Yukl, 2002; Rowitz, 2001). Situational leadership facilitates both the training of the subordinate (team member) and the leader; it is a broad concept that has been described as a "credible model for training people to become effective leaders. ...Situational leadership is easy to understand, intuitively sensible, and easily applied in a variety of settings. ...Situational leadership stresses that leaders need to find out about their subordinates' needs and then adapt their leadership style accordingly" (Northouse, 2013, pp. 105–106). Closely related is the concept of task relevant maturity (TRM) of subordinates (Grove, 1995). Each of these concepts relate to a subordinate's ability (based on training and experience) and willingness (based on their degree of achievement orientation and readiness to take responsibility) to effectively respond to a task assignment (Grove, 1995). Situational leadership and TRM require the ability of a leader to understand when and what level of delegation or empowerment is most appropriate for each subordinate or team member relative to a specific task. Situational leadership requires an understanding of each employee's TRM level to address an effort and a comprehension of the seriousness of each task including time demands, as well as its applicability as a learning opportunity, including what Tichy describes as teachable moments, situations when people are most open to learning and accepting the guidance of others (Tichy, 1997, p. 55). Situational leadership and TRM are directly applicable to research team learning. "Situational leadership applies during the initial stages of a project, when idea formation is important and during the various subsequent phases of a project when implementation issues are important. The fluid nature of situational leadership makes it ideal for applying to subordinates as they move forward or go backward (regress) on various projects. Because situational leadership stresses adapting to followers, it is ideal for use with followers whose commitment and competence change over the course of a project." (Northouse, 2013, p. 109) As a statistician on a CRT, it is important to be receptive to guidance provided during teachable moments and most important to improve your TRM so that you will be in a position to take full advantage of empowerment opportunities. This may involve higher levels of communication on your part, e.g., asking questions or practicing active listening techniques; it may require you to seek additional training, beyond your statistical skills, e.g., project management skills; or it may mean that you need to ensure that you are motivated to take on additional responsibilities, an absolute requirement if you are to emerge from a team member role to an assigned leader role.

Motivation, empowerment, and delegation are critical concepts to apply and master to develop your leadership skills. Team leaders must create an environment where motivated team members are empowered to share responsibility and accountability by accepting team assignments and carrying them out with a sense of urgency. It is our experience that motivation, empowerment, proper delegation, as well as building our own TRM levels were important drivers of success in our careers, especially in the CRO industry, where rapid growth led to many opportunities to learn and take on greater leadership responsibilities. And the important learning of how to master these important people skills requiring the expansion of our emotional intelligence often made the difference between success and failure on our CRTs.

Further guidance on these important people skills can be found in: Grove, 1995; Kotter, 1996; Rowitz, 2001; Yukl, 2002; Sapienza, 2004; Goleman, 1995; and Daft, 2008.

Other Skills

There are many other leadership skills that can be developed thorough CRT participation, too many to cover in detail here. Among these are two that we found to be of particular importance in building our careers, but are not generally considered leadership skills, per se. These are quality improvement (QI) and project management, both of which have grown in recent years, as critical skill components of effective CRTs. In our experience, these are critical for statistical and other research leaders to master and take responsibility for learning, not only to improve their leadership abilities, but also to ensure the overall success of their research endeavors. We will briefly summarize these two areas before closing this chapter.

Scientific Integrity and Quality

The scientific integrity of a research endeavor is the direct responsibility of each team member and requires technical/scientific knowledge and leadership from each team member. Scientific integrity for research projects, teams, and organizations also includes ensuring quality in all aspects of research processes and outcomes. Although somewhat daunting at the organizational level, the development of project-specific processes and procedures to ensure quality on clinical research projects is more a manageable task and provides valuable learning experiences that can then be extrapolated to larger projects and organizations. Team settings are natural environments for ensuring quality (Fried and Carpenter, 2013). Ensuring quality is both a top down and bottom up activity, and as such is a responsibility for all members of a research team, with each team member most responsible for the scientific component that he or she contributes to the multidisciplinary team effort. For example, statisticians may have the most direct responsibility for defining quality assurance and improvement procedures for data collection, data management and programming activities, and statistical analyses. In addition, they may also take the lead on suggesting quality improvement strategies relative to other nonstatistical activities; for example, changes to patient recruitment efforts. Once again, this is an example of how statisticians can expand their influence on the research process. Ensuring quality at every research interface is an important leadership responsibility and is an area where statistical influence can be critical. Because of its role in problem prevention and solutions, as well as ensuring the integrity of research findings, knowledge of QI techniques is important for leadership development; its reliance on statistical reasoning skills also makes it a natural area for statisticians to become directly involved. Furthermore, any of the leaders who developed and promoted QI processes were statisticians, such as W. Edwards Deming (Sollecito and Johnson, 2013). Our experience in the contract research industry and in academia strongly reinforces our belief that as part of their leadership roles, statisticians should learn and apply QI processes and exert influence as needed on CRTs to ensure that quality and scientific integrity are well understood and never sacrificed for the sake of expediency or due to a lack of understanding.

The scope of procedures that can be applied to ensure the highest quality in clinical research is fairly extensive and includes, but is not limited to, ensuring sound randomization techniques to data integrity. Further guidance is provided by: Deming, 1986; Pocock, 1993; Sollecito and Fendt, 2006; Bialek, Duffy, and Moran, 2009; Loshin, 2011; and Sollecito and Johnson, 2013.

Project Management

Team efficiency, effectiveness, and synergy depends, heavily on the leadership skills outlined above, but also requires knowledge of a variety of other skills that are more traditionally defined as project management skills. At a minimum, these include team building, planning and monitoring of project budgets and timelines, and how to use traditional project management tools, such as the development of a project charter, scope, work breakdown structure, network diagram, Gantt chart, risk identification and management tools, change order documentation, and budget versus actual reports, to name a few. Development of these skills and appropriate application of these activities can be first steps in improving the business acumen of statistical leaders, especially as part of a stepwise leadership development process, moving from the individual project to the organizational level of leadership. The experience to develop and master these skills can be obtained through formal education and demonstrated competence via project management professional certification as offered by the Project Management Institute. In addition, as with other leadership skills described here, mastery of project management skills requires hands-on applications on actual research projects. Each of us and several of the members of our leadership panel honed our leadership skills at the same time that we learned project management skills, first on small local projects and later on multinational clinical research programs. Together, these prepared us for broader leadership roles.

The value of project management expertise was demonstrated many times in the contract research industry, and this was an area where statisticians in particular were able to demonstrate their influence on the larger organization by serving as project managers, as a first step toward greater leadership roles later in their careers. Further information on project management skills can be obtained from the Project Management Institute via its published guidelines in the PMBOK® Guide — Fifth Edition, 2013, and traditional project management texts such as Kerzner, 2001; Lientz and Rea, 2002; and Wysocki, 2014.

Summary

We have only touched on a small sample of leadership skills that statisticians need to learn, develop, and master to become better leaders, using CRTs as our focus. Although we emphasize research team-based experiential learning in this chapter, we also recognize that, as with statistical skills, there are leadership skills that are more amenable to formal academic courses and professional development workshops. These can be used, either as a starting point, or later as checkpoints, to better meet one's leadership needs, or to sharpen skills developed experientially. Like most specialized skills, time is required to achieve mastery; for leadership development, lifelong learning is required. This progressive learning relies on the use of team experiences supplemented by traditional academic approaches, professional development opportunities, and observation of the behaviors and outcomes of mentors and other leaders, and much more. Kotter points out that every leader in the 21st century must be a lifelong learner, and this requires risk taking to get outside our comfort zones, humble self-reflection, solicitation of the opinions of others, careful listening, and openness to new ideas (1996).

We believe lifelong leadership learning will supplement our training as statisticians to not only be better technical leaders, proficient in our statistical skills, but also will enhance our abilities as leaders of teams and organizations who stretch beyond our technical, cultural, and organizational boundaries to influence all around us.

Bibliography

Bennis, W.G. and Nanus, B. (1985). *Leaders: The Strategies for Taking Charge.* New York: Harper & Row.

Berg, P. (2013 — September). Influence: Essential for success as a statistician. *Amstat News,* 18–19.

Bialek, R., Duffy, G.L., and Moran, J.W. (2009). *The Public Health Quality Improvement Handbook.* Milwaukee: ASQ Quality Press.

Buchanich, J.M. (2012 — February). Scientific course strengthens students' communication skills. *Amstat News,* 7.

Burns, C., Barton, K., and Kerby, S. (2012). The state of diversity in today's workforce. Retrieved March 4, 2014, from http://www.americanprogress.org/issues/labor/report/ 2012/07/12/11938/the-state-of-diversity-in-todays-workforce/.

Byrne, J. (1993). The virtual corporation. *Business Week,* 3304: 98–102.

Campbell, L.G., Mehtani, S., Dozier, M.E., and Rinehart, J. (2013) Gender-heterogeneous working groups produce higher quality science. PLOS ONE 8(10)e79147 dol:10.1371/journal.pone.0079147.

Covey, S.R. (2004). *The Seven Habits of Highly Effective People.* New York: Free Press.

Daft, R.L. (2008). *The Leadership Experience* (4th ed.). Mason OH: Cengage Learning.

Deming, W.E. (1986). *Out of the Crisis.* Cambridge: Massachusetts Institute of Technology Center for Advanced Engineering Study.

Dilworth, R.L. (1998). Action learning in a nutshell. *Performance Improvement Quarterly,* 11(1), 28–43.

Fernandez, C.S.P. and Steffen, D. (2014). Leadership in public health. In L. Shei and J.A. Johnson (Eds.), *Novick and Morrow's Public Health Administration — Principles for Population Based Management* (3rd ed.). Burlington MA: Jones & Bartlett.

Fisher, B.A. (1974). *Small Group Decision Making: Communications and the Group Process.* New York: McGraw-Hill.

Fried, B.J., Topping, S., and Edmondson, A.C. (2006). Groups and teams. In S. Shortell, and A.D. Kaluzny, (Eds.), *Health Care Management — Organization Design and Behavior* (5th ed.). Clifton Park, NY: Thompson Delmar Learning.

Fried, B.R. and Carpenter, W.R. (2013). Understanding and improving team effectiveness in quality improvement. In W.A. Sollecito and J.K. Johnson (Eds.), *McLaughlin and Kaluzny's Continuous Quality Improvement in Health Care* (4th ed.). Sudbury, MA: Jones and Bartlett Publishers.

Goleman, D (1995). *Emotional Intelligence* (10th ed.). New York: Bantam Dell.

Grove, A.S. (1995). *High Output Management* (2nd ed.). New York: Vintage Books.

House, R.J., Hanges, P.J., Javidian, M., Dorfman, P.W., and Gupta, V. (Eds.) (2004). *Culture, Leadership and Organizations: The Globe Study of 62 Societies*. Thousand Oaks, CA: Sage.

Hülsheger, U.R., Anderson, N., Salgado J.F., and Hall, R.H. (2002). *Organizations: Structures, Processes, and Outcomes* (8th ed.). Upper Saddle River, NJ: Prentice-Hall.

Institute of Medicine. (1988). *The Future of Public Health*. Washington, DC: National Academies Press.

Institute of Medicine. (2003). *The Future of the Public's Health in the 21st Century*. Washington, DC: National Academies Press.

Katz, D. and Kahn, R.L. (1978). *The Social Psychology of Organizations* (2nd ed.). New York: Wiley.

Katzenbach J.R., and Smith, D.K. (1993). *The Wisdom of Teams — Creating the High-Performance Organization*. Boston: Harvard Business School Press.

Kerzner, H. (2001). *Project Management – A systems Approach to Planning, Scheduling, and Controlling* (7th ed.). New York: John Wiley and Sons, Inc.

Kolb, D. (1984). *Experiential learning: Experience As a Source of Learning and Development*. Upper Saddle River, NJ: Prentice Hall.

Kotter, J.P. (1996). *Leading Change*. Boston: Harvard Business School Press.

Kouzes, J.M., and Posner, B.Z. (1995). *The Leadership Challenge*. San Francisco: Jossey-Bass.

LaVange, L., Sollecito, W., Steffen, D., Evarts, L., and Kosorok, M. (2012 — February). Preparing biostatisticians for leadership opportunities. *Amstat News*, 5–6.

Leatt, P., Baker, G.R., and Kimberly, J.R. (2006). Organizational Design. In S. Shortell and A.D. Kaluzny, (Eds.), *Health Care Management — Organization Design and Behavior* (5th ed.). Clifton Park, NY: Thompson Delmar Learning.

Lencioni, P.M. (2002). *The Five Dysfunctions of a Team: A Leadership Fable.* San Francisco: Jossey-Bass.

Lencioni, P.M. (2004). *Death by meeting: A leadership Fable about Solving the Most Painful Problem in Business.* San Francisco: Jossey-Bass.

Lewis, J.P. (2003). *Project Leadership.* New York: Mc Graw Hill.

Lientz, B.P., and Rea, K.P. (2002). *Project Management for the 21st Century* (3rd ed.). San Diego: Academic Press.

Longest, B.B. and Young, G.T. (2006). Coordination and communication. In S. Shortell and A.D. Kaluzny, (Eds.), *Health Care Management — Organization Design and Behavior* (5th ed.). Clifton Park, NY: Thompson Delmar Learning.

Loshin, D.L. (2011). *The Practitioners Guide to Data Quality Improvement.* NY: Morgan Kaufman OMG Press.

Marks, M.A., Mathieu, J.E., and Zaccaro, S.J. (2001). A temporally based framework and taxonomy of team processes. *The Academy of Management Review,* 26(3), 356–376. Retrieved March 31, 2014, from http://www.jstor.org/stable/259182.

Marshall, E.M. (1995). *Transforming the Way We Work — the Power of the Collaborative Workplace.* New York: AMACOM.

Melum, M.M. and Sinioris, M.K. (1993). *Total Quality Management,* Chicago IL: American Hospital Publishing.

Bialek, R., Duffy, G. L., & Moran, J. W. (2009). *The Public Health Quality Improvement Handbook.* Milwaukee: American Society for Quality, Quality Press.

Northouse, Peter G. (2013). *Leadership — Theory and Practice* (6th ed.). Thousand Oaks, CA: Sage.

Pocock, S.J. (1993). *Clinical Trials — A Practical Approach.* New York: John Wiley and Sons Ltd.

Project Management Institute. (2013). A Guide to the Project Management Body of Knowledge, (PMBOK®) (5th ed.). Newtown Square, PA: PMI Publications.

Ringer. J. (2006). *We have to talk: A step by step checklist for difficult conversations.* Retrieved. March 31, 2014, from http://www.mediate.com/articles/ringer1.cfm.

Rodriguez, R. (2012 — February). Statistical leadership - preparing our future leaders. *Amstat News*, 3–4.

Rodriguez, R. (2012 — April). Statistical leadership — perspectives of past presidents. *Amstat News*, 3–4.

Rowitz, L. (2001). *Public Health Leadership — Putting Principles into Practice.* Gaithersburg, MD: Aspen Publishers Inc.

Sapienza, A.M. (2004). *Managing Scientists — Leadership Strategies in Scientific Research* (2nd ed.). Hoboken, NJ: Wiley-Liss Inc.

Schein, E.H. (2010). *Organizational Culture and Leadership* (4th ed.). San Francisco: Jossey-Bass.

Senge, P. (1990). *The Fifth Discipline: The Art and Practice of a Learning Organization.* New York: Doubleday.

Sollecito, W.A. and Mele Dotson. M. (1995 — October). Communication guidelines and networks for drug development teams. *Proceedings of the Annual Meeting of the Project Management Institute.*

Sollecito W.A. and Fendt, K. (2006). Continuous quality improvement in contract research. In C.P. McLaughlin and Kaluzny, A.D. (eds.), *Continuous Quality Improvement in Health Care* (3rd ed.). Sudbury, MA: Jones and Bartlett Publishers.

Sollecito, W.A. and Johnson, J.K. (Eds.). (2013). *Mc Laughlin and Kaluzny's Continuous Quality Improvement in Health Care* (4th ed.). Sudbury, MA: Jones and Bartlett Publishers.

Tichy, N.M. (1997). *The Leadership Engine.* New York: Harper Business.

Umble, K.E., Shay, S., and Sollecito, W.A. (2003). An interdisciplinary MPH via distance learning: Meeting the educational needs of practitioners. *Journal of Public Health Management and Practice,* 9(2), 123–135.

Umble, K., Steffen, D., Porter, J., Miller, D., Hummer-McLaughlin, K., Lowman, A., and Zelt, S. (2005). The national public health leadership institute: Evaluation of a team-based approach to developing collaborative public health leaders. *American Journal of Public Health,* 95(4), 641–644.

Wysocki, R.K. (2014). *Effective Project Management: Traditional, Agile, Extreme* (7th ed.). Indianapolis, IN: Wiley Publishing, Inc.

Yukl, G. (2002). *Leadership in Organizations.* Upper Saddle River NJ: Prentice Hall.

19

How Statisticians Can Develop Leadership by Contributing Their Statistical Skills in the Service of Others

Sowmya Rao

University of Massachusetts

I was excited and humbled when I was invited to write this chapter. Writing this chapter caused me to reflect on my journey as a statistician. My foray into statistics started at the age of 17 when I was admitted to the undergraduate program in statistics at Madras Christian College in Chennai, India, and my formal education as a statistician continued as I proceeded to obtain a doctoral degree in biostatistics from Boston University several years later, after a short detour into software programming. Every position I have held during these years, be it as a research assistant, statistical analyst, post-doctoral candidate, faculty member, or consultant, has enabled me to enhance my existing skills and acquire new ones — both statistics-related and otherwise (e.g., team member, leadership, and negotiation skills). Although I had many experiences that helped shape my career and contributed toward my leadership skills, I describe in this chapter one particular experience, a trip to Sierra Leone on behalf of the volunteer consulting group Statistics Without Borders (SWB), that I believe effectively illustrates leadership in the context of service.

Leadership Development: My Trip To Sierra Leone

(SWB) is an outreach group of the American Statistical Association (ASA) comprised of statistician volunteers who provide pro-bono statistical help to nonprofit organizations and governmental agencies with limited resources for statistical support, working primarily in developing countries. SWB was founded by Steve Pierson, Fritz Scheuren, Gary Shapiro and Jim Cochran in response to statisticians' requests for volunteering opportunities. Gary and Jim were the first cochairs. SWB currently consists of more than 1,300 volunteers from all over the world, ranging from high school students to very experienced statisticians with expertise in different areas of statistics, such as study design, survey development, business analytics, and analysis of differ-

FIGURE 19.1
Dr. Sowmya R. Rao (with the cast on the left) and Dr. Theresa Diaz (UNICEF; on the right) with the team of statisticians and survey interviewers from Statistics, Sierra Leone who conducted the first survey in 2010.

ent types of data. I currently hold the office of secretary of SWB. I credit my experience on the project described below to be the catalyst in propelling me to this position.

Before joining SWB, I volunteered in an informal manner by helping family and friends make sense of data being reported in the media or in published studies, an experience to which I'm sure many other statisticians can relate. When I learned about SWB at one of the Joint Statistical Meetings a few years back, I decided to join and become a volunteer, because I was looking for opportunities to work in global health. Furthermore, I wanted to volunteer my time applying my statistics knowledge to a worthwhile cause. SWB seemed to be the perfect fit, combining both my interests. In January 2010, a few months after I had joined SWB, I responded to a call for volunteers to help design a United Nations International Children's Fund (UNICEF) led project to assess under-five mortality caused by malaria, diarrhea, and influenza in Sierra Leone, Africa. The goal of the project was to evaluate the effect of employing rural community health workers (the intervention), who would treat children under five who were diagnosed with malaria, diarrhea, or influenza, with the goal of eventually reducing related mortality. The study was comprised of two waves of data collection that were two years apart. The first wave was taken before the intervention, while the second wave was taken after the intervention. A total of four districts were selected — two for implementation of the intervention, and two where the intervention was not implemented, for comparison. The principal investigator of the study, Theresa Diaz, from UNICEF, is an MD and an epidemiologist with 20 years of experience with the Centers for Disease Control and Prevention, USA, and she brought the rigor of research methodologies to the study. She wrote and submitted a formal protocol that was approved by the Institutional Review Board of the government of Sierra Leone, Office of Science and Ethics Review Committee, Ministry of Health. She needed SWB's help with sample design and also to compute the required sample sizes to achieve sufficient power to detect statistically significant differences, if any, between the intervention and comparison districts.

As is typical for such an undertaking, a team consisting of two senior statisticians and a junior member (myself) was assembled. Previous to this, I had analyzed survey data for both my doctoral thesis, as well as for my postdoctoral fellowship, and I was excited to have the opportunity to learn from my senior colleagues how to overcome the challenges in designing a study that would use complex sampling methods in a developing country. Unfortunately, unforeseen circumstances caused one of the senior statisticians to withdraw from the project soon after we had started discussions. However, this challenge turned out to be a blessing in disguise for me, because it created an opportunity for me to take a more active and leadership role on the project. Fortunately for me, Gary Shapiro, who has years of survey research experience, graciously agreed to mentor me through the process.

As we finalized the study design and the sampling scheme, Theresa asked if one of us could travel to Sierra Leone to help her with the analysis once the first round of data were collected. I immediately accepted, since Gary was unable to go and, I have to admit, the idea of traveling to Africa was quite appealing. I had never visited any developing country other than India, and I looked forward to the trip with both excitement and trepidation. If you have travelled in the developing world, you know that some places are easier to get to than others. In August 2010, I flew from Boston, USA, to Brussels, Belgium, where I joined Theresa for the onward journey to Lungi International Airport in Sierra Leone. That wasn't the end of the trip, since we still needed to get to Freetown, which lies across a large body of water from the airport, a journey which must be undertaken by car, ferry, hovercraft, or helicopter. (According to the US State Department travel website for Sierra Leone, none of these options is without risk, but I am convinced that no one would travel at all if they took all of these warnings at face value.) We chose to travel to Freetown via the seven-minute helicopter ride with our luggage on our laps, a ride that was fortunately uneventful. One van ride from the heliport in Freetown later, and we were finally at our hotel. As if the length of the journey and the numerous modes of transportation weren't complicated enough, I made this trip while wearing a cast on my leg due to an ankle injury! To add further adventure to my physical challenges, the apartment we'd rented was on the third floor of a building with no elevator. We arrived quite late, and after the arduous journey, I slept quite well!

The next day, after reporting at the regional UNICEF offices, we traveled to the offices of Statistics Sierra Leone. It was interesting for me to learn how investigators had dealt with the challenges of sampling and surveying a population in a country that was still recovering from a civil war. One issue the investigators faced is that there was no good address list of all the households in Sierra Leone for us to draw a sample from. The investigators dealt with this by using an innovative method, whereby they sent interviewers to the districts to map every household using a GPS-enabled personal digital assistant (PDA). The PDA was also programmed to sample the required number of households from the mapped list of households. When it came time to do the survey, the GPS then guided the interviewers back to the sampled households based on the stored coordinates, by displaying a dot indicating the location of the house on an image of a compass on the PDA screen.

Another issue with conducting surveys is that the interviewers may take "shortcuts" to reduce their workload (and this is certainly an issue that is not limited to the developing world). In the case of this survey, the use of GPS-enabled PDAs helped alleviate this problem. When surveys are conducted manually with paper and pencil, the interviewer has the chance to tamper with the responses or cheat by filling out the answers themselves instead of interviewing the selected people.

The advantages of using a GPS-enabled PDA were: the interviewers did not have to carry huge loads of paper surveys around; the interviewers did not have to manually skip questions since the survey was programmed into the PDA with automatic skips; the GPS coordinates of the interview location could be compared with the actual location of the household that was mapped earlier, making it difficult for interviewers to cheat; and the data were easily downloadable for immediate analysis. The use of this technology in the study has been described in detail in our article in *Chance* magazine, a publication of the American Statistical Association (Rao, Shapiro, Diaz, 2012).

When it came time to download the data from the PDAs, the local statisticians had some problems with a few of the PDAs and Theresa stepped in to help. I was impressed by how knowledgeable she was on every aspect of the project, including the capabilities of the PDA. I was not sure what was expected of me on-site, but I was ready for anything and so I was happy to assist in the downloading and merging of the data sets prior to statistical analysis. Some statisticians may not have the programming experience required to conduct the analyses efficiently themselves, but fortunately for me, I have always had an interest in working with data, and I have done an extensive amount of programming (statistical and otherwise). Thus, I was able to assist Theresa in whatever was necessary to obtain the data for analysis. We spent the first few days at Statistics Sierra Leone extracting all the data from the PDAs.

At the end of each of our trips, Theresa had set up dates to present the results from the study. This did not allow us much time in which to carry out the analyses, and I had to work very efficiently. We conducted the analyses back at our hotel. Theresa would be on her computer at one end of the room preparing her presentation, trying to analyze the data for the tables in her slides, while I was on my computer at the other end merging and cleaning data sets and conducting the more statistically complex aspects of the analysis, while also assisting her with her code or slides. We worked almost nonstop; the only breaks we would take were a couple of hours for dinner at one of the nearby restaurants. Being a vegetarian, it was not very easy to find food to my liking. I survived mostly on the few nutrition bars I had picked up at the New York airport as I ran for my flight. I had a few packets of microwaveable food with me that I could have eaten, had I only noticed the microwave in the living room of our apartment. (It was pointed out to me on my return from Sierra Leone by my husband when I was showing him the pictures of where we stayed!) This is, perhaps, an indication of how absorbed we were in our work, since I had spent most of the two weeks working while sitting on a couch directly across from the microwave.

The local statisticians had requested that we train them to perform data analysis using the statistical software SPSS that was available to them. So after we completed our analysis, we ran a one-day workshop where we covered the basic steps needed to conduct analysis (e.g., data cleaning, obtaining descriptive measures, running regressions, creating graphs) with a combination of presentation and hands-on assistance. We found them to be very enthusiastic and engaged at the workshop. They also utilized the skills learned at the workshop to collaborate on a peer-reviewed manuscript (Diaz et al., 2013).

At the end of the workshop we presented the statisticians and interviewers with certificates of achievement from UNICEF for their data collection efforts, and shared with them the results of the study. They seemed very delighted to be presented the results, and many of them mentioned that this was the first time they had seen any results from studies they had participated in, since most other foreign investigators only used them to collect the data.

In addition to traveling to Sierra Leone at the end of the data collection for the first survey, I also went there two years later (August 2012) to assist with the data extraction and analysis of the second wave survey. In the two years since my first trip, they had grounded the helicopters as being unsafe. They instead introduced speed boats that one boarded from floating docks made of plastic blocks that would ride up and down with the waves. To make matters worse, I am quite afraid of the water, so needless to say, this was an adventure I would rather not have had! Thankfully, this time I was not in a leg cast.

The statisticians seemed to be happy with our first workshop, and so during my second trip, they asked us to conduct another workshop, this time using STATA. This group had some statistical analysis experience, and so the workshop was more interactive. I had to extend the workshop for a second day, since they had many questions about the analysis as well as the software.

Some of my family and friends were worried about my safety, since they had heard about the civil war in Sierra Leone (which had actually ended by 2002). In particular, my in-laws, having only heard negative stories in the media regarding Africa, and having little experience traveling in the developing world, were worried for my security and articulated their concerns to my husband. I understood their worries, but I am glad I did not miss this adventure. Both trips were memorable to me in many ways: I had an opportunity to be involved in a very important study for a country that is trying to rebuild itself after years of civil war, I was able to use my skills in the teaching and mentoring of others while also being mentored, and I had a chance to enhance my statistical skills and build my own leadership skills while also empowering others to build their own sustainable skills.

In addition to being a remarkable experience, the study has also been academically helpful to me, since it has already produced two publications and presentations with more in various stages of progress. I also became an integral part of SWB by first being selected as the secretary of the New Projects Committee that handles all requests for pro-bono work, and then being elected to serve as the secretary of SWB.

Statistical Skills in the 'Service' of Others

I would classify the areas of my service to the UNICEF project in four broad categories: teaching, mentoring, collaborating and consulting.

Teaching

Many statisticians teach formally as part of their profession as full-time faculty members of schools, colleges, and universities where students receive academic credits for courses, while others perform this service by offering workshops or accepting short-term teaching assignments. In this role one can effectively develop students' interest in statistics by making the subject engaging.

On both trips to Sierra Leone, I conducted workshops to educate the local statisticians on data analysis methods using SPSS and STATA. Both workshops were developed based on requests from the local statisticians on what they wanted to learn in the program. They had never analyzed data before and were very excited to have the chance to learn. Rather than create a presentation based on what we thought they should learn, the local statisticians selected the content for the workshops, and as a result, the workshops were very successful and fulfilling to me and the participants alike. This enabled them to collaborate effectively and make significant contributions to the manuscript as coauthors.

Mentoring

Similar to teaching, mentoring occurs in both formal (e.g., student advising) and informal ways (e.g., mentoring a colleague). Mentoring is a great responsibility, since the mentor has to understand the needs of the mentee and guide her or him effectively to achieve her or his goals and not just further the mentor's career. The mentee/mentor relationship does not always end at the conclusion of the formal mentoring. My postdoctoral mentor once told me what he had heard from his mentor: "Once a mentor, always a mentor." Some of us continue to work with our mentors throughout our lives developing from a mentee to a collaborator.

It is also possible to be mentored and be a mentor at the same time, even within a single project. For example, I was mentored by Gary when I first joined the Sierra Leone project, and I was in turn able to mentor a doctoral student who travelled to Sierra Leone to assist with the data collection and analysis of the second survey. I continued to collaborate with her as she used these data for her doctoral thesis.

Collaborating

Most applied statisticians spend much of their time collaborating with teams of interdisciplinary scientists providing guidance in study design and statistical analysis, interpreting the results in terms that nonstatisticians can understand, and writing the statistical methods for grant applications and manuscripts. It is not always clear what is required of a statistician as a collaborator. On some projects, I have been involved in most aspects of the study, from writing the grant proposal, to designing the study, to analyzing the data, to writing the manuscript. At other times, my contributions have been limited to a subset of project activities.

As described above, I started my collaboration with Theresa on the sampling design and quite unexpectedly became involved in other aspects of the project — data extraction, statistical analysis, and manuscript writing. I was more heavily involved with the project, including data analysis and manuscript writing, for the first survey, while my role was more limited on the second survey. I did assist with the data extraction and merging of the data for the second survey, but the doctoral candidate conducted all analyses of the data, while I was available to answer any questions she might have on the data or statistical analysis. I also continue to collaborate with her on the scientific manuscripts.

Consulting

With the growing demand to extract actionable information from massive data collections, statisticians have a vital role in making sense of the data. Since the supply of statisticians has not equaled demand, there has been an increased demand for both paid and volunteer consulting opportunities. While paid consulting can be very financially rewarding, unpaid consulting on a worthwhile project can be quite enriching and gratifying in other ways.

I have served as a consultant and a collaborator for investigators at academic institutions where I have been not only paid to conduct data analysis but have also been granted coauthorship on those manuscripts. Although the remuneration on these projects is not very lucrative, it is satisfying to be part of good research and publications.

I was an unpaid consultant in my involvement with the SWB project. Although I did not gain financially, I learned about the challenges of conducting studies in a developing country by using innovative methods such as (i) first, collecting GPS coordinates of every house in a particular village; (ii) second, sampling households from this list to survey. I was also exposed to the novel use of technologies for efficient data collection — the use of PDAs to not only map the households, but also to sample respondents and conduct the survey.

I view teaching, mentoring, collaborating, and consulting to be intertwined as opposed to being mutually exclusive. Most statisticians usually spend time performing most of these four services, albeit in different proportions of their time. So, how do these skills help one become a leader? In order to answer this question, one needs to first define "leadership."

What is Leadership?

Although it might seem simple at first to define leadership — everyone probably has an intuitive idea of the concept — there are actually several definitions of what a leader or leadership is: "something or someone who leads or guides other people" (Merriam-Webster's dictionary), "someone who inspires others to dream more, learn more, do more and become more" (paraphrased from John Quincy Adams, sixth US President), "unlocking people's potential to become better" (Bill Bradley, former US Senator and former NBA basketball player), "is about nurturing and enhancing" (Tom Peters, business author), "leadership lies in guiding others to success" (Bill Owens, 40th governor of Colorado), "someone who has followers" (Peter Drucker, management consultant), and "those who empower others" (Bill Gates, US business magnate).

Initially, I defined a leader as someone who is an expert in one's field (scientific or otherwise) and not as someone who just holds a position of power. I talked to a few other people to understand their perception. A number of people (mostly statisticians) concurred with my own definition, but then I talked to my family (all nonstatisticians) who surprised me by their perspectives. One of them defined it as, "Leadership is all about influencing others towards a shared purpose or goal," while another pointed me to a quote by Ronald Heifetz, a professor of public leadership at Harvard's Kennedy School of Government: "Leadership is not about you: it's about them" (NPR, 2013). I believe, as Professor Heifetz says, that a leader in this context should not just be one who finds the solution to a problem, but someone who also empowers and helps others by giving them the tools to solve their own problems.

Leadership is not simply about demonstrating technical expertise or competence in interpreting or working with data. Leadership in statistics occurs when we can guide investigators to design studies in a statistically sound fashion to answer their research questions, and communicate their findings in a manner that is understandable. In this sense, leadership in statistics is not only about methods, probability models, or elaborate analysis. It is also about helping people understand and visualize the results of the analysis in a form that has meaning to them and enables them to make use of that derived meaning.

In my opinion, serving as a statistician in any of the four categories listed above leads to developing one's own technical and leadership skills while also developing other leaders. My involvement in the Sierra Leone project helped me hone my teaching and mentoring skills, learn to program in STATA (a software I had never programmed in before), gain expertise in the application of survey methodology in real-life situations, consult and collaborate with international investigators, be elected to the executive committee of SWB, and, eventually, be invited to write this chapter.

If we consider leadership to be about the empowerment of others, then a leader is someone who strives to have a lasting impact not only in his/her area of expertise, but also on the community (s)he serves as well. So, how can statisticians utilize their skills and leadership for creating lasting impacts on society?

Impact

More countries are moving toward evidence-based policymaking and need statisticians to help design studies, collect, and interpret data in different sectors such as the social sector, the health sector, and the judicial/legal sector to name just a few.

Social Sector

Social policies affect the lives of all citizens, particularly the disadvantaged. The social sector requires effective policies in poverty-alleviation, nutrition and food sufficiency, education, employment-related issues (e.g., equal pay by gender, work–life balance). Statisticians can employ current methods or develop new ones to evaluate existing policies and suggest improvements considering changing demographics (e.g., an influx of refugees and immigrants, the aging population, and declining fertility rate) or propose new policies where none exist.

For example, World Health Organization/UNICEF field data on malnourishment among children below six years in India led to an early child care and development policy that necessitated changes to the budget, resource allocations, and programs for health, nutrition, and preschool education, making this a strong program since 1975 (UNICEF, 2014).

Health Care Sector

Winston Churchill once said, "Healthy citizens are the greatest asset any country can have." Health care has become one of the most important areas for research in which statisticians play a key role. Access to affordable care, effective and inexpensive treatments for diseases, reducing avoidable medically adverse events, and reducing infant mortality are areas that can be improved by evaluating the impact of prevailing policies and suggestion of new policies to replace ones that are not currently effective. Statisticians have a wonderful and unique opportunity to collaborate with investigators from other disciplines (e.g., economics, social sciences, medicine, and health services) in this quest for excellent health care for reasonable prices.

Our results from the Sierra Leone study indicated that the intervention was beneficial. We hope that this will lead to policies to employ and train more community health workers to provide health care in rural areas leading to a reduced under five mortality rate.

In addition to the Sierra Leone project, I consulted on a project conducted by the Ministry of Health of the Hashemite Kingdom of Jordan with support from the World Health Organization, International Advisory, Products and Systems, the Massachusetts General Hospital Center for Global Health, Harvard University and the Jordan University for Science and Technology. The goal of the project was to study the impact of the influx of Syrian refugees into Jordan on the Jordanian health facility capacity and utilization. We hope that results obtained from the survey of all the health facilities (clinics and hospitals) in Northern Jordan will aid policies to not only benefit the Syrian refugee population, but also the Jordanian health care system as a whole.

Judicial/Legal Sector

Collecting and analyzing data on criminal activity is vital to developing sound law enforcement policies, as well as to guide lawmakers in the development of policies to reduce crime. These policies and laws need to be adapted to changing times and crimes (e.g., cyber attacks). The examples below demonstrate the role of statisticians in the judicial/legal sector.

A New Delhi based nongovernmental organization, working to address the Delhi rape in 2013, conducted an analysis of data collected on the youth population in India — aged between 15–25 years and concluded that lack of youth education and life skill opportunities in India leads to the increase in violence and has also predicted more crime among the youth in the future.

The government was also advised on the specific areas to focus on regarding youth education and life skill opportunities (Academia, 2014).

In India, data generated on the increase in the female work force in information technology, business process outsourcing, and garment manufacturing sectors where night shifts are common, were linked to violence against women (FICCI, 2014), leading to amendments to labor laws as well as legislative support for women.

Other sectors in which statisticians contribute are the economic sector (e.g., business analytics, analysis of stock markets), environmental sector (e.g., mitigation of soil erosion, landslide recurrence, prediction of earthquakes and tsunamis), sports and recreation (e.g., creation of opportunities for the youth by studying return on investments of different sports), foreign policy (e.g., evaluation of the effect of policies on intercountry relationships), and national security (e.g., tracking of social media to identify potential terrorist threats).

Conclusion

The field of statistics has grown widely in the last two decades, impacting many areas directly relevant to our daily lives. With the development of an interconnected world and increase in the awareness of world events come a thirst for information and an appetite for reliable evidence. We are surrounded by statistics of all kinds from all different walks of life — education (e.g., proportion of children dropping out of school, ranking of different schools), entertainment (e.g., the top money-making movies every week, the suggestions of movies made by Netflix based on tracking movies that individual clients watch), wars (e.g., the number of civilian casualties in a war, the number of refugees crossing national borders), and national security (e.g., the number of violent crimes per year, data mining of social media to track terrorist plots). The health care industry has been an important catalyst for the increase in the information seeking behavior from leaders to consumers being interested in, among other things, tracking their calories consumed, optimal amount of exercise needed, and who will be susceptible to certain diseases.

Developing countries are lagging behind in adapting new statistical methodologies and applications compared to developed countries. Poorer countries do not have access to the modern research methodologies or analytical software. These countries are becoming interested in conducting studies to help propel their development in all areas. With this comes opportunities and challenges, since many of the modern statistical techniques (designs and analytical methods) have been developed in industrialized countries.

This information age has also led to some serious pitfalls, such as a massive misinterpretation of available data and statistics. For example, Statistics Canada released its official report on the 1999 crime statistics in their publication, *The Daily*, on July 18, 2000, showing a decrease in crime rate. Some journalists possibly did not understand the report and gave their own interpretations as the cause for the decline — with one headline saying, "Hot economy stalls murder rate." Statistics Canada had, in fact, not evaluated the effect of economy on the crime rate (StatCanada, 2014).

This is only one example of the misreporting one sees on a daily basis. Most people are overwhelmed by the numbers but do not want to appear ignorant either. So, they interpret results as they prefer. A vested interest in creating hype and focusing on their ratings and attractive sound bites causes media persons to, at times, report information outside the context in which the data were collected and analyzed.

As statisticians, we are in the forefront with a unique opportunity to be leaders in interpreting data, creating awareness, making possible simple ways of understanding by the lay public, and engaging with the community to see the analysis as toolkits to ask the right questions for a better world. We should not miss such an opportunity. It is difficult to say whether using statistical skills in the service of others develops leadership skills or whether having the leadership skills leads one to more opportunities to serve. One needs some skills to serve, and serving in various capacities in turn helps develop more skills, leading to more opportunities. It is a circular process. As described in my experience in Sierra Leone, I do not know if I would have had the opportunity to work or be successful on the project if I did not have some skills to bring to the table, and involvement in the project also helped me develop my skills and find other opportunities to lead.

I was willing to volunteer on the Sierra Leone project not with any expectations for personal gain, but because I wanted to help on an important project, and what I got out of it was serendipitous. I do believe in this verse from the Hindu scripture Bhagvad Gita — "Karmanye Vaadhikaraste, Ma phaleshu kadachana; Ma Karma Phala Hetur, Bhurmatey Sangostva Akarmani," which means, "One should perform their duties without expectations of fruits of the labor; By not making the fruits of labor the purpose of one's actions, one would not be attached to not performing their duty." I do hope that my experience will inspire others to take every opportunity that is presented to them to use their statistical skills in the service of others (whether by volunteering or for remuneration) and to not worry about the fruits of their labor, since the rewards follow passionate endeavor.

Acknowledgments

I would like to thank my husband, Dr. David S. Morgan, for his continued support and reviews of multiple revisions of the chapter, and my other family and friends, and colleagues for their invaluable help on this chapter.

Bibliography

Diaz, T., George, A.S., Rao, S.R., Bangura, P.S., Baimba, J.B., McMahon, S.A., and Kabano, A. (2013). Healthcare Seeking for Diarrhoea, Malaria and Pneumonia Among Children in Four Poor Rural Districts in Sierra Leone in the Context of Free Health Care: Results of a Cross-Sectional Survey. BioMed Central Public Health. 13:157.

Rao, S.R., Shapiro, G., and Diaz, T. (2012). Use of GPS-Enabled Mobile Devices to Conduct Health Surveys in Sierra Leone. Chance. 25, 3:38–42.

http://www.npr.org/2013/11/11/230841224/lessons-in-leadership-its-not-about-you-its-about-them. Accessed May 10, 2014.

http://www.statcan.gc.ca/edu/power-pouvoir/ch6/misinterpretation-mauvaiseinterpretation/5214805-eng.htm. Accessed May 13, 2014.

http://www.unicef.org/india/nutrition_1556.htm. Accessed May 13, 2014.

http://www.academia.edu/6088068/Rising_Rape_in_India_..._Why_it_is_happening_and_how_do_we_Fix_The_Problem. Accessed May 13, 2014.

http://www.ficci.com/SEdocument/20249/Safety-of-women-at-workplaces-Recommendations-for-Businesses.pdf. Accessed May 13, 2014.

Part VI

Individual Strategies

20

Lessons from an Accidental Leader

Arlene S. Ash

University of Massachusetts

In this chapter, I draw lessons and offer advice based on my own journey from bright student (who found math easy and writing papers hard) to division chief in a medical school and activist scientist for social justice.

Why 'Accidental' Leader?

As a female born in the US just after WWII, I grew up thinking that my education would help me run a household and teach my kids to appreciate and navigate the world. Dads worked; moms stayed home and raised kids. Today's young women enter a very different world, and most expect to have careers.

Even if I was not being groomed for a career, my childhood family supported intellectual growth and valued education. When I undertook graduate school, I felt family pressure to continue through to a PhD in mathematics and "not be a drop out." Still, my family did not know much about academia; so, even as I made honors throughout school, nobody sat me down and said, "You could achieve much more than you currently aspire to." Unfortunately, promising women are still far less likely to get such advice than their male peers.

I completed my PhD in 1977 and a post-doc in the Dartmouth Mathematics Department in 1978, then held faculty appointments at Boston University (in math, until 1984) and at the BU School of Medicine (in math within General Internal Medicine, until 2009). I have been a division chief for Biostatistics and Health Services Research in Quantitative Health Sciences at the University of Massachusetts Medical School since 2009. Along the way, I became an applied statistician (more on that in a moment).

Googling my name and a keyword or two will find some of my less-traditional endeavors. These include: teaching math at Mindanao State University in the Philippines through the Peace Corps, helping found women's self-help health care centers in Chicago, and a health care informatics company (DxCG, later absorbed into Verisk Health, Inc.) in the Boston area; expert witnessing in sex discrimination cases for the Massachusetts Teacher's Association and in a wide range of other public policy lawsuits; being a visiting scholar/delegate in Vietnam; advocating for electoral integrity (following my testimony in *Bush v. Gore*); and testifying before Congress on payment reform in the US Medicare Program. While many of these activities are not directly relevant to my career in biostatistics, they have increased my "name recognition" and have certainly made life more fun.

Advice from My Life Experience: Generalizing from an N of 1

Everyone has different lessons to learn. Being too cautious and being too bold are both dangers. I grew up with these mantras: Do not pay someone else to do what you can do yourself; make do with what you have; buy only after finding the best possible deal. Self-reliance and thrift were necessary for my family then — and they remain virtues — but new realities require new methods. As I have acquired more resources, I have had to unlearn some thrifty habits in order to invest in my own development and in the tools I need to be productive. Others may need to unlearn different lessons that — however successfully they have served in the past — are less well suited to present realities.

I had to let go of the illusion that I can, and thus must, perfectly allocate my time and money. For example, occasionally buying something expensive that does not add much value, or failing to make an investment that would have helped, is OK; making your own mistakes is a necessary part of finding a healthy balance.

Failures happen to everyone, and they do not mean that you are not good enough to succeed. When you screw up, reflect on what you might have done differently, learn the lessons you need to, take a deep breath, and move on.

Some years ago I set myself the task of reframing my language, to remind myself of my own agency. For example, saying (and, eventually, even thinking), "I did not make the time for that," rather than, "I did not have the time," or, "I was not able to get that idea across," rather than, "The student did not get it." Saying, "I am sorry; let me try to be clearer about what I need from you" is both a good way to move past a miscommunication and an acceptance of responsibility. Failed communication — regardless of who's at fault — is your problem. "Let us try to figure out how not to have this happen again," is another useful way to talk and think.

Family–Work Balance

Growing up, I always thought I would marry and have kids. That is what women did. I was well into my 30s before I realized that I had become "seduced" by my work, and that marriage might never be more attractive to me than the single scientist's life that I had stumbled upon. Having my own family could also have been wonderful, but I have never regretted how my life played out. I participate actively in other people's families including those of relatives and mentees, and have sustained deep, long-lasting friendships with my age peers, both men and women. I have loved and been loved by some extraordinary people.

Not having kids enabled me to work some pretty crazy hours during a lot of my life, but I soon realized that activities put off "until my career was more settled" would likely never happen. So, I committed to making time for travel, friends, and family, constructing a well-rounded life out of lopsided periods of intense, career-directed work, political activism, cultural immersion, and play. Good work usually requires intensity. Time management choices should be made actively and revisited at all major life changes — every few years at a minimum.

I was a visiting scientist in Hanoi, Vietnam, in early 1987. Women could be seen working in all kinds of jobs, for example, construction, which seemed kind of neat to me. However, the country's leaders were ashamed that Vietnam's poverty required women to do physical labor. In a conversation with Prime Minister Pham Van Dong, he spoke about the desire to get the country to a place where women could return to being mothers with time for their families. I said, "Maybe it is OK for a woman to just focus on work; I am not married and I have a good life." I realized that we were talking right past each other when he beamed at me kindly and said, "You are still young and nice-looking; there is still time for you." However, I did not marry, and I am still entirely OK with that.

Advancing in a Career

Careers evolve. Consider the following advice: "You just got promoted, that is great. Now forget all the good work habits that got you here." That may sound ridiculous, but different skills are needed to succeed at different points in your career.

You likely succeeded initially by concentrating meticulously on your own stuff. Later, you must learn not to do it all yourself. You should develop networks of collaborators and learn how to mentor and delegate responsibility. You must teach and empower other people to ensure that work products are trustworthy without your having to personally vouch for every single period, comma, and semicolon. You can not do that as the leader of a group of any but the smallest size. You must let go of many aspects of what previously made you valuable. For example, you may never have had anyone whose job it is to facilitate your work — and now you have (or share) an administrative assistant. A chief or chair needs different skills than someone more junior.

I am still learning how to delegate. Each of the following matters: finding good people, developing good relationships, providing the right training and feedback, and figuring out early whether there is a good fit. Not all successful managers use the same model. One woman I know regularly hires promising young people into low-level jobs and gives them as much responsibility as they seem able to take on, often "growing them" into researchers in their own right. A good manager recognizes and takes advantage of team members' different strengths and interests, and helps them work with the demands of their current realities.

Accomplishments

Each year, at least, review what you have accomplished and how that compares to the objectives you set last year. Spend time savoring your triumphs as well as critiquing your shortfalls. That is, be as encouraging to — as well as demanding of — yourself as you should be with those you supervise. Start with praise; criticize fully, but narrowly and specifically; end with encouragement and a plan for action.

Keep track of your accomplishments contemporaneously — partly to simplify promotion — but also to regularly check in with yourself and at least one mentor or peer. Find and use your institution's CV or résumé template. Learn how to read a résumé (your own included) to identify a healthy career trajectory. For example, ten abstracts might be nice, but if few lead to papers, they reflect too much time talking about preliminary work and too little in creating lasting products. Early career papers are likely to involve "sandwich" authorship, later ones, first or second authorship, and later yet, either senior (last-listed) authorship, or again, being back in the middle of the sandwich.

Periodic reviews are vital. Why did some things not get done? Were they important or could they just as well drop off the list? Where will the things you are doing now take you, and is that where you want to go? Set specific goals for the coming year and aspirations for the next five years. "Stretch goals" are good, especially in the longer time frames; if you never miss a goal your targets are probably not adequately ambitious. Use regular reviews to help set and achieve important milestones.

Have expansive conversations with people about their aspirations and frustrations. Perhaps one person wants to teach more, while another needs a teacher. The best leaders I know help people find not-so-obvious shifts in career (or merely in career emphasis) that play to their interests and strengths, and help put people with complementary skills together.

Always be encouraging, but, especially in formal, written evaluations, do not be overly nice. For one thing, writing only nice things will come back to bite you if the problems escalate and there is no previous record of the problems. For another, you help people most by spelling out the problems and how you expect them to be addressed. A few years ago, I worked with a very smart, young assistant professor who had an impressive research record, but needed to listen better and accept guidance in how to explain the significance of her complex methods to grant reviewers. Further, she was not attending to the consulting assignments that supported most of her salary. The previous year I had given her a glowing written review based on her successes and discussed the problems, but had not written them up. That year, I wrote them up. She had clearly worked very hard to be the best all her life; she was angry, upset, and ashamed to have a poor review go into her record. Wrestling over the review was painful for us both, and I wondered if trying to change someone so stubborn was a waste of time. However, she now participates much more effectively within the department, has been promoted, and has thanked me for pushing her. We continue to identify needed areas of growth for her, but far more collegially.

The above story has a happy ending (or middle, anyway), but being liked is not the goal — building a collegial, productive work environment is. Someone who is a bad fit for your job may be able to shine somewhere else. Helping people identify and grow into what they are best at (whether within your shop or elsewhere) is the mark of a good leader. While firing an employee is the least fun part of leadership, your first responsibility is to sustain the team spirit and cooperation essential to a satisfying workplace.

Finally, mistakes are a rich resource for learning. For example, I now probe more deeply with references before hiring and try to identify potential problems right away, to avoid having to wonder later whether — if I had been a better leader — things would have turned out better.

Ultimately, your success is measured by the breadth and depth of what you accomplish — not by the fraction of things tried that turned out well.

Skills

A first career task is to solidify your own knowledge base and become a "go-to" person for at least some of the skills needed in your field. Over time, you will also want a more broad-based understanding. What kinds of skills do others have that complement yours? What kinds of tools are needed to solve what kinds of problems? Work to ensure that you and your network of colleagues and vendor partners can provide "full service."

Seek a complete mental map. Readily acknowledge ignorance and ask for help filling in its uncharted regions. Learn not to nod "yes" when you are not really following. Ask again. Learn to say, "Let me see if I understand..." and then try to say it in your own words. The same insight applies when conveying knowledge to or requesting help from someone else. Have them repeat it back to you in their own words. Merely listening is too passive. It is hard to nail a new idea without making it our own: by using it, writing it up, or speaking about it.

Effective "scanning" is another important skill. My specialty is digging deeply into a thing until I understand it completely and can convey that understanding clearly. Yet leaders must rapidly evaluate and triage many possibilities. A first question to ask — that is often worth revisiting — is, "What is the main goal here?" A related skill is swiftly conveying your point. First, of course, you must know what your point is — an exercise best started early in the life of a project, and repeatedly revisited. Use informative subject lines for emails, titles, and opening and closing paragraphs for papers, proposals, and talks. Your work does not speak for itself — you must make it speak.

Investigators in my department often send around one-page project proposals to get a big-picture, high-level "is this exciting?" response. The goal is to engage a wide range of people, talents, and perspectives, and avoid wasting time on a "blah" idea. We also have research-in-progress seminars that people across the department, and the school, are encouraged to attend. Junior (and sometimes senior) investigators bring ideas to the group for feedback. Participating grows valuable connections and helps you identify colleagues whose skills complement your own.

Special Challenges for Women

Many women do not speak up, and when they do, they often speak too quietly or in an overly high-pitched voice. I always spoke loudly enough, but had to learn to speak more slowly and plainly, and, encouraged by an acting coach in high school, to speak at a lower pitch than my natural young woman's voice. The lower pitch makes people more willing to listen, and more likely to understand and remember what I say. Although communicating is particularly hard for people whose first language is not English, native English speakers must also work on this crucial skill. Clear speech helps you engage others and attract resources. Learn to convey key points briefly, and to shut your mouth when you finish a sentence, leaving room for others.

Women often put others' needs first, not making time for themselves. Years ago, a (male) colleague and friend did some thinking about why his female colleagues were not progressing in their careers as he thought they deserved. He had me make regular appointments with myself — creating blocks of time to do my own work. You can not advance your own projects while being limitlessly available for everybody else. If you must cancel a meeting with yourself, reschedule it, just as you would for anyone else. Respect your own time no less than that of others.

Strategic Volunteering

An early lesson in the value of volunteering came through my involvement in the women's self-help health movement in Chicago in the 1970s, in particular, in founding the Emma Goldman and Chicago Women's Health Centers. This work was enormously empowering; I learned the delicious, subversive lesson that a small group of women could decide on something that they wanted to make happen — something that they were not particularly trained or certified to do — and just do it! While the clinics' clients were grateful for the "pay what you can" woman-centered services offered, the act of creating the clinics transformed our lives. I discovered what I wanted to do with my math training as I moved from being a pure mathematician to becoming a biostatistician who tackled real-world problems.

Another important volunteer activity was testifying for the Seabrook Anti-Pollution League as it opposed licensing for a nuclear power plant in New Hampshire in 1978. This seemed crazy; I knew nothing about the subject matter. But I did have a mathematician colleague and close friend, John Lamperti, to help with the "Where to start?" phase. We pulled the top report from a huge stack, scanned it quickly, and kept going until we found something that we thought we could talk about. Having "the confidence to try" is important, as is having someone else willing to venture into new territory with you. Testifying at the Environmental Protection Agency hearings in this case gave me the experience and credibility to testify elsewhere. And expert witnessing helped me develop my skills and affect the world. John and I have collaborated on other issues, several relevant to "electoral integrity," such as: "Florida 2006: Can Statistics Tell Us Who Won Congressional District-13?" (Chance, 2008: 21(2): 18–24).

You will be asked to be on many committees. Seek out the volunteer and service activities that stretch you and extend your collegial networks. When asked to do other things, you will already be doing your share. In my first tenure track job (at Boston University), I volunteered for the faculty compensation committee, where I got to use and extend my statistical skills — by examining salary data for equity.

Seek entry-level jobs in professional groups — such as secretary or webmaster. That is how I started with the Medical Care Section of the American Public Health Association — an activist group of more than 1,200 people who care about the intersection of health care and public health. Now, I am its chair.

Indeed, I volunteered my way into expert witnessing, which eventually included consulting in a range of content areas, including environmental protection, jury selection methods, insurance rate-setting hearings, employment discrimination, and electoral integrity. While these experiences were "cool" and sometimes lucrative, I eventually put expert witnessing aside, in order to concentrate on health care payment reform.

When asked to take on a new responsibility, Miss Manners would tell you to start by saying, "Thank you for thinking of me"; you should also say, "I will get back to you." Then go talk the suggestion over with your "say no" person, who will encourage you to accept key opportunities (such as becoming an National Institutes of Health (NIH) study section reviewer) while saying no to others. If you do not yet have a "say no" colleague, enlist one; regardless, take the time to clarify (and possibly negotiate) expectations and responsibilities, and consider the consequences, before saying yes. For example, being an NIH study section peer reviewer is, on its face, insane. When I started, it took at least ten percent of my working life, effectively unpaid. Still my "say no" person was right to tell me to say yes. Study section is like super-graduate school; it teaches you state-of-the-art thinking and how to write successful grants. If you participate well, others will remember, and will look favorably on future grants in which you are involved.

Becoming a Manager

I naturally like people, and find it easy to be encouraging. I have loved mentoring someone and later hearing that she was happy in her career and that my encouragement — and my belief that she (or he) could make it — had mattered.

But, when hiring people, you also need to be critical; search and consult widely and probe deeply — even, and perhaps especially, if you are desperate to have someone on board "yesterday." Go beyond those offered as references to find others who are more willing or able to share criticisms as well as praise. Temperament and character matter as much as training and expertise. Share your group's mission and values with new employees, whether staff or faculty. Whatever the job description, its purpose is to advance that mission.

Further, you must continuously invest in human capital and make midcourse corrections to maintain a healthy work culture. Despite the advanced stage of my career, I have only recently become a supervisor, and know that I have room to improve.

It is easy to hold meetings in which people get excited about the idea of working together, but very hard to actually make that happen. Only one thing I know works: finding a specific project and getting started collaborating.

A leader needs followers. How do you recruit talented followers? Find work that excites you and get others excited about it. Help them find their place in it, knowing that their needs and abilities likely differ from yours. Help them grow their talent, and embrace the possibility that they may be more talented and successful than you. Even at this advanced stage of my career, I love learning, either from the direct mentorship of my talented peers, or by observing and emulating the highly effective behaviors of others, regardless of their age or status. I feel proud and energized to work in a department whose chair (Catarina Kiefe) writes on our website, "We believe in science that makes a difference, collective creativity, and social justice through improved health." I have learned from watching Catarina — a leader who nurtures "the success of others" — and from our department's vicechair, Jeroan Allison, whose passion for social justice was the driving force behind UMass' interdisciplinary Center for Health Equity Intervention Research (CHEIR). Notice the "I" in the name; the center's focus is on interventions to reduce inequities.

Nobody ever told my childhood self to work to achieve mastery. Rather, I absorbed the implicit goal of getting "As." But a leader must identify important problems and learn how to address them. People skills matter. These include: understanding the needs and capabilities of others, sharing praise, helping others to realize their full potential, and having the success of others as a goal.

Hone your "elevator speech" — that is, how you will explain your work during a short elevator ride. Briefly, why should anyone care about what you do? Like an artist whose portfolio serves to introduce the range and depth of her talent, develop striking examples that highlight your strengths.

You need several different elevator speeches. For example, when talking with a basic scientist at our school, I might say, "Our group was brought here to make research easier for people like you. We can help you collect, analyze, and archive data, plan efficient experiments, write up findings, respond to reviewers, and write grants." About my own work I might say, "I build models to predict important outcomes of health care. For example, when people in Medicare receive their care through an HMO, our government now uses such a model to pay the HMO more for taking on the care of someone with complex medical needs than for a healthy person." If the ride is long and my listener is still interested, I might add, "These models also help distinguish hospitals whose high quality improves patient survival from those whose patients survive better simply because they were healthier in the first place."

Listening skills are also essential, as is the ability to synthesize the perspectives of others. Mathematical training involves finding the structure of an argument. In meetings, you may be able to help people identify common ground, or to see how their visions differ. The unique circumstances of your nonprofessional life may also provide unique sources of strength. For instance, I grew up in a household of people with poor hearing, where I learned the art of quick summarizing while "simultaneously translating" from English to English. This can really help in meetings, where people often talk past each other.

If confronted with sexist behavior, try to develop a measured response, ideally leavened with humor. You will be judged on your ability to handle such problems with dignity. When I became the statistical leader of one section of the Eastern Cooperative Oncology Group, the medical leader, a surgeon, thought that he was flattering me by introducing me as his "fringe benefit." I was speechless, but was rescued by the male statistician whose role I was assuming. He said, "Actually, Dr. Ash is a very accomplished PhD statistician." What might I have said? Perhaps, "Let us not all be horrified by that introduction. Harvey is so 'old school' he thinks he is being charming!"

Pay attention to cultural norms, but do not fear to embrace "your authentic self." Here is an example. In the summer of 1999, I attended a final plenary lunch of the AcademyHealth (then AHSR) meeting in Nashville. I was packed to go home and having no other place to put it — my hat was on my head. Senator Bill Frist, MD, was the speaker. The budget for the "Agency for Health Care Policy and Research" — the federal entity that funded our kind of work — had nearly been abolished that year. Why? Agency-sponsored research had demonstrated little to no value for some aggressive medical interventions (e.g., for low back pain: medical hospitalizations (traction), invasive diagnostic tests (myelograms), and complex surgeries (spinal fusion). Further, the agency had published guidelines discouraging their use. Some doctors and other medical industry actors who resented researchers telling them what (not) to do had mobilized several Congressmen to defund AHCPR. Frist had helped broker a deal that preserved funding, under a new name: the Agency for Healthcare Research and Quality. Frist told us that the name change made sense because policy wasn't our domain. I felt impelled to stand up during the Q&A and push back, saying that while we appreciated his help, and I could understand the politics behind the name change, it most certainly was the domain of researchers to recommend policies. After lunch, many people came over to congratulate me for my comments. They could find me, of course, because of my hat. I quickly realized that Congresswoman and feminist Bella Abzug, another short woman, was definitely onto something with her signature flamboyant hats. Wearing a hat at professional meetings has become something of a signature for me. I am a little puzzled that others (especially short women) do not regularly exploit this highly functional tactic — but I like hats — and I am not going to stop wearing them just because most people do not. And here is a fun epilogue: Fifteen years later, when I told the story of Senator Frist and the agency's near death, one colleague said: No kidding? You were the lady in the black hat?

Medical School Careers

Medical schools are still male-dominated and hierarchical. In considering how to get a word in edgewise, draw confidence from your distinctive expertise. Even though medical doctors are used to being the experts, rarely deferring to other people's opinions, listening well and leading with your expertise can earn respect. Medical doctors know that they don't know what you know — and may even suspect that it might be useful.

To interact successfully with MDs and MD researchers, you must learn how to elicit what they actually need from you. When told, for example, "I need a sample size calculation (or a t-test)," you might say, "First, tell me what problem you are trying to solve." You could explain this by saying, "Suppose a patient came and asked for an MRI. You would not just order it, would you? If I do not understand the problem, I cannot help you find the right solution."

If possible, use examples from someone's own experience. For example, in explaining to a colleague who studies "Uptake of computer-assisted smoking cessation interventions," I translated the mantra that "Historical control is a weak design" into the question "If computer use is constantly rising, how would you know if increased uptake after an intervention is due to the intervention rather than merely what might have happened absent the intervention?" Likewise, a nonstatistical collaborator might not know the answer to the question, "Are these observations independent?" but might be able to tell you if the data come in groups that make it likely that they are similar to each other, such as being patients who may be attracted to a particular doctor based on her taste for conservative (or aggressive) treatment. Only a client/collaborator who understands what you need can provide the subject-matter insights that lead to the right statistical tools. If an explanation goes over your head, ask for clarification. When needed, ask for introductory reading or some Q&A time with your collaborator to help you get oriented.

Early on, start developing a document that describes the problem's key features, source data, and goals.

Performing well on assigned work is important. Learn to judge when to dig in harder yourself and when to ask for help. Scope out the work early on and decide whether your toolbox contains what you need. The worst outcome is arriving at the deadline without having done a good job.

Academic promotion usually also requires that you identify your own projects and goals. Keep your eyes open for opportunities to do this.

Developing Your Areas of Expertise

Becoming a principal investigator requires strategic thinking. How will you get there? Agencies and foundations don't just trust you to successfully manage something new because you are talented at something else. There are many ways to acquire new skills and credibility. These include: taking a course, teaching a course, reading and critiquing key papers in the area, and apprenticing yourself to a project that has, or needs, this expertise. Each contributes to your credibility as an expert.

Commitment to a content area is a fruitful path to methodological innovation. For example, improved statistical tools for looking at longitudinal data have emerged from the desire to facilitate better care for people with mental health issues. The statisticians involved did not start by saying, "We want to learn how to study longitudinal data." They said, "We are going to study a chronic disease in which treatments and outcomes for patients are constantly changing. We need longitudinal methods that work in this setting."

Communicating

Conducting a solid analysis is important, but you must also explain it. Learn how to tell your story, starting with being able to convey the gist of it in one or two sentences, and then digging down at the one-paragraph, one-page, and more expansive levels.

Know your audience and purpose. Focus on clarity. What story does the data tell and how do you know it? Do the raw cross-tabs tell the basic story that the sophisticated analyses refine and confirm? If not, what imbalances in the groups being compared contribute most to the changed perception? I have been very successful in court cases where the opposing witness used a lot of jargon. Implicitly he was saying, "This is too complicated for you guys to understand; let me assure you that I used the most sophisticated methods to arrive at my conclusions." In contrast, I tried to be clear, perhaps saying: Dr. X has told you that "after removing outliers, the men's and women's salaries looked very much the same." However, almost all the high outliers he removed were men, while almost all of the low outliers were women. [Then I would show a 3 by 2 table of numbers of outlier status (high-, versus non-, versus low-outlier) by sex.] So we should not be too surprised to find that salaries for the remaining men and women look pretty similar.

Writing papers is also important. The work needed to write a good paper usually shows me that until I sat down to write it, I was not thinking clearly enough. Writing a "simple" explanation is often hard. Getting the technical details right is necessary but not sufficient.

Another issue is making sure people can hear you, since many people have poor hearing. They will often sit there politely — not understanding — because they assume that everyone else can hear. I kick myself to this day for not shouting out, "We cannot hear," at an event where the speaker, a friend, was the victim of a poor sound system. Despite my good hearing, I could barely make her out. When she learned of the problem later, she was mortified.

A teacher I knew who was not a native English speaker insisted that the students in his class sit up front because (he told them) he had trouble understanding them and needed them to come closer. Asking people to assist you with your problem (e.g., in hearing them) is a good general strategy.

Beyond hearing is understanding. If I feel that a speaker is not being understood, I will often say, "Can I play this back for you? Do I understand you?" and then I will "translate" in a loud, clear voice. As a leader, you should help your junior faculty become better speakers. You may want to play the role I do in meetings, of intervening to ensure that the speaker communicates effectively.

Returning to My Story

As an assistant professor at Boston University, I testified against the school as a witness in the National Labor Relations Board hearings to decide whether BU's faculty could unionize. I knew that this would not help my (soon to be considered) tenure case at BU. And, indeed, despite securing the vote of my department, the school denied me tenure. But by then, I had other ideas about what I wanted to do. I have never regretted leaving.

From my experience testifying in discrimination cases, I never considered suing BU for tenure. The law is a slow and highly imperfect way to seek justice. If you feel unfairly treated, assemble your case and try to enlist your supervisor in helping to right the wrong. Do not assume malice; equity problems often develop inadvertently. Ask for help: "Will you work with me to develop a timeline? What milestones do I need to meet to make you want to push my case (for, say, an equity raise or a promotion)?" You will likely get satisfaction a lot sooner, while avoiding the considerable acrimony of a lawsuit, which can cut off your access to good references, and — regardless of the justice of your case — make potential future employers nervous.

When leaving the math department at BU, I received an offer to be an in-house statistician for an eminent cardiology group at Johns Hopkins. It looked like a great job, but I had strong roots in Boston. I thought of it as having to choose between an "A" job in a "B" location (Hopkins in Baltimore), or vice versa — a perhaps promising position that I could "cook up for myself" in Boston, where I had enormously important personal ties. I now think that the Hopkins job might not have been an "A" choice; being a lone statistician among clinician specialists is isolating.

In the end, I took a specially created position in the Division of General Internal Medicine at BU School of Medicine. Since they did not have the money for a full-time hire, they offered me a part-time salary. This worked for me because: 1) my financial needs were not great; 2) I had conducted a salary survey before negotiating with the Medical School that enabled me to ask for, and get, money that was similar to my full-time pay in the math department; and 3) I was also earning regular money as an expert witness.

Surprisingly, the job stuck. Although I had had many jobs and lived in many places since college, I remained in Boston, and with BU, from 1978 until 2009 (moving from the College of Liberal Arts to the School of Medicine in 1984). This was partly due to friendship and family ties to the Boston area (including aging parents) that led me to turn down feelers from other places. A mid-career move could have been energizing — working with new people, learning new things, and having my routines challenged. But my choices worked for me.

During this period, I started to use professional associations as a vehicle for growth. I have remained involved in one way or another with various professional groups ever since. I became very involved with the Caucus for Women in Statistics in 1985, and in a fluke, was elected president of the Caucus shortly after joining. I thought: a president should look for things that would interest members. That is how I learned of a Kovalevskaia travel grant, that I applied for and received, for a scientific exchange in Vietnam in 1986–7. Later, I became involved with other American Statistical Association (ASA) subgroups, notably the Scientific and Public Affairs (SPA) Committee. I led a SPA subcommittee focused on electoral integrity; my interest dated from my experience testifying in *Bush v. Gore* in Florida in 2000.

Professional organizations connect you with people, ideas, and opportunities to collaborate. Other professional service is also important: reviewing articles and grants teaches you the language and the norms of your field. Being a standing member of a grant review panel (say for NIH or AHRQ) that meets in person can lead to lifelong professional friendships.

Finding Your Niche

When I moved to BUSM in 1984, I inherited a project that became my signature work over the next 25 years; it was originally called the Average Annual

Per Capita Cost (AAPCC) project. Neither the field nor that work yet had its name; I now say that I work in health services (or health policy) research, and the AAPCC work eventually came to be known as "risk adjustment" or "predictive modeling." The problem was how to calculate capitation payments to HMOs that enrolled Medicare beneficiaries. Previously, Medicare had paid medical service providers on a fee-for-service basis — the more care given, the more money paid. In contrast, an HMO would receive a fixed amount of money per month to take care of all the medical needs of its enrollees – regardless of what services they used. Clearly, Medicare should provide more money to an HMO when it enrolled a sick person rather than a healthy one, but what data and models could be used to make these calculations?

This problem was going to take a while! Luckily, I did not realize then that it would be 16 years before the work was implemented. Such a long time frame might have seemed too daunting. However, it was well worth doing. Our models underpin the payment formulas for Medicare's capitated purchasing program (now called Medicare Advantage) to this day.

As I said, my first risk adjustment project was inherited. I was hired to do work that had been contracted for by others. After completing that work, I could have found other areas to focus on. So, before continuing, I asked people at Kaiser Permanente, in the government, and in academia, "How important is this problem? Would it become moot if the US adopted a single-payer system?"

My Kaiser friends said: "Think of Kaiser as the health ministry for a country with a single payer system and 8 million people. We collect revenues to pay for their health care and use it to staff clinics and hospitals in various locations. Our San Francisco location cares for many people with AIDS, Oakland deals with the health problems of an inner city, and many of Marin County's relatively well-off clients have substantial age-related cardiovascular problems. Of course, each thinks their patients are sickest. We need an objective way to allocate budgets."

Aha! Credible predictions of the resource needs of specific populations will always be useful, as well as models to predict other health care outcomes. Here is another example. Some people with heart attacks go to hospital A, and others to hospital B. How much of the difference in 30-day mortality at the two hospitals is due to differences in how sick their patients were when they came in as opposed to differences in the hospitals' quality of care?

To gain acceptance for my work, I had to demonstrate that these models were meaningfully better than the simple age–sex models in use at the time. The Kaiser anecdote became a good talking point. But people said, "Your models only explain 9% of the variance in total cost — is that useful? — what about the other 91%?" The low R^2 reflects the fact that such models cannot tell you, say, which individual will have a heart attack next year. But the models can predict how many more heart attacks will occur in a group of people with a lot of cardiovascular health problems than in a healthier group; this latter information is all that matters for getting resource allocations right for health care providers. So, I developed tests for comparing model performance for predicting expected outcomes in "relevant subgroups" that are in broad use today. I published papers, spoke at conferences, and talked to reporters. I was asked to advise state governments, the US federal government, and even the government of Germany, about this kind of predictive modeling and its role in health care payment reform. And when Congress considered new Medicare payment methods, I testified before the House Ways and Means Committee.

Maturing in Your Niche

My risk adjustment work led to my founding a company, DxCG, Inc., with colleagues, especially my long-term economist-collaborator at Boston University, Randy Ellis, in 1996. Within 8 years we had more than 60 employees and 300 national and international clients. However, we faced an increasingly consolidated world of health informatics suppliers. Even though I still consult with the successor company, Verisk Health, selling DxCG lessened the conflict of interest issues that arise when I use "my" models for academic work. Real-world consulting informs my academic research, since clients who use the software often ask interesting and important questions. Our work on predictive modeling was honored by AcademyHealth, a professional organization for health services and health policy research, with its 2008 Impact Award.

I continue to enjoy and learn from mentoring, supervising interns, and working with students. Sometimes, I reach out to communities that rarely become involved in statistics. For example, in the mid-2000s I met with students at a charter public high school in Dorchester, Massachusetts. One student asked, "How much money do you make?" and the room filled with an expectant buzz. I knew that my annual salary — which was substantial — would be compared with that of super athletes and thought of as paltry. Work in their community is typically paid in dollars per hour, so students have a better feel for that. I responded, "Let me just say that in coming here to give this talk, I put down work that I am being paid \$400 an hour to do." (It was a legal case.) And the room went utterly quiet. After a while, someone asked, "How long were you in school?" And when I said, "A looooong time," the room again went very quiet.

For about 10 years, I worked as an expert witness in many, varied cases. My most famous one was in December 2000: *Bush v. Gore*, the Martin County absentee ballot trial. That experience led me to work with the American Statistical Association on electoral integrity issues; it also led to consults with state legislators, scholarly papers, editorials, and conference invitations.

Partly, I went into math because it felt easier to solve math problems than to write term papers. But to affect the world you must write well, a skill I now embrace. Years ago, I participated in a health services research leadership seminar with a scientific writing workshop. I still rely heavily on a few key concepts I learned there. To improve my speaking, I envision, and when possible practice, talking to someone with a stake in my work (with health care, that is easy) who is entirely ignorant of statistical jargon; this readily reveals rough spots that need polishing.

I regret that after many years working in employment discrimination, I wrote very little on that subject and moved on without training anyone to follow me. I vowed not to repeat that mistake with my work in risk adjustment with Lisa Iezzoni (professor of Medicine at Harvard Medical School and director of the Mongan Institute for Health Policy at the Massachusetts General Hospital) and Michael Shwartz (professor in Management/academic lead of Health and Life Sciences Sector Operations and Technology Management at Boston University School of Management). Michael and I wrote several chapters for the four editions of the *Risk Adjustment* book that Lisa edited. Now when people ask for my advice on risk adjustment, I can say, "I put everything I know about that in the book! Go read it!"

The Present

In 2009, I was recruited by the University of Massachusetts Medical School to become professor and founding chief of their Division of Biostatistics and Health Services Research, Department of Quantitative Health Sciences. This was a big step for me, and at first I resisted it. Many people my age think about retiring, and I had been thinking about it. But it is energizing to work with good, new people and to help build a department.

I have some things I still want to accomplish — developing ideas with colleagues and working to influence Massachusetts' health care policies. Our department recently joined with others in the city of Worcester to secure a grant to make the city healthier. Our department's CHEIR provides a platform for interdisciplinary thinking to develop interventions that address (not just describe) health inequities. Earlier, I talked about choosing words and phrases carefully. Current language speaks of health "disparity," but "inequity" keeps the focus on social justice. I want to clarify the consequences of the analytic methods we choose on what we think we "find" in our data, as well as to make health care better in the real world. I want hospital quality reports to recognize that nonmedical factors (such as poverty) matter — and use the enterprise of hospital quality reporting to provide insight into when solutions to poor health care outcomes might best be sought within the larger community.

Finally, I view my professional identity as a magic cloak that has taken me all over the world, not as a tourist, but as a partner in problem solving. It connects me with wonderful colleagues and friends and has brought me financial independence. It also allows me to be heard on issues that matter and to advance social justice. I hope that my story and advice will help you find your own path to leadership as a mathematically trained scientist in a medical school or elsewhere.

21

Practical Suggestions for Developing as an Academic Leader

Charmaine B. Dean

Western University

Nancy Heckman

University of British Columbia

Nancy Reid

University of Toronto

University settings provide a variety of leadership roles. This discussion encompasses both research and organizational leadership roles, with a focus on key elements that are at the intersection of both. Through this discussion, we hope to encourage more women to step up to assume these roles.

Successful leadership in academia requires the ability to face challenges. Challenges can reflect one's personal internal drive to succeed in a competitive environment or one's organization's drive to stay ahead and transform. Challenges can come from external sources, such as changes in how the academic environment is viewed, or changes in priorities of governments and governmental agencies. These challenges coexist with more personal challenges, including family caregiving for children and aging parents, the increasing intrusion of digital communications, and the importance of the nurturing of the tri-instrument which allows us to serve — the body, mind and spirit synergy, the health of which provides the foundation for all other activities.

The challenges, and thus the workload associated with a leadership role, are not evenly distributed in time. Some aspects of work may gain higher priority at any one time or over short periods. However, some aspects, such as the nurturing of the family and the body–mind–spirit tri-instrument, respond best to continual attention. Half of the battle of managing workload is understanding that periods of exceptionally heavy workload are simply part of the normal cycle, even though these periods sometimes seem to be nearly out of control.

A university career provides many training grounds for those considering leadership positions. These can provide a sense of the challenges you might encounter and, importantly, a sense of whether or not leadership roles interest you. For example, departments are often in need of committee chairs, societies are always in need of volunteers for committees and for leading initiatives, and women are often in demand for various committees, as stipulated by equity guidelines. Even early on as a young academic, you can explore the leadership skills which you might bring to the table. You can see if you have a knack for bringing sense and clarity to a troublesome group conversation. You can begin to develop a leadership style and philosophy. You can explore some of the skills and elements of success as discussed below, seeing what suits your personality. By building strategy and confidence through your career in every aspect of your duties, you will have a measure of your success and personal satisfaction in these roles, allowing you to determine your desire for a greater involvement in leadership. Participating in these initial leadership roles will also mark you as someone willing to tackle difficult questions and offer you an opportunity to bring your personal style and viewpoint to bear on your institution.

This chapter discusses strategies that may be useful as you move into leadership roles as the head of a research group or as director of an institute, chair of your department or dean of your faculty. The discussion encompasses elements along the entire path to a leadership role. Some of these elements emerge at the very start of your career, where your work can be preparation for a leadership role.

Note that strategies for success are not constant in time, but change with your context and goals. Life strategies have to adapt to changing internal and external systems and need review periodically.

These guidelines are intended to apply to both men and women, but women may be in more need of harkening to the advice. Women may have less experience or confidence in envisioning leadership roles for themselves, especially if they have few role models at their institution. Broadly speaking, women seem to be less successful in negotiating terms of employment, for example, so may need to give these negotiations extra thought. Family care responsibilities still do generally fall more heavily on women's shoulders, although each case is unique. Women perhaps need more reminding of the importance of self-care — to pay particular attention to the tri-instrument of mind, body, and spirit, and to acknowledge that success in leadership may require aggressive prioritization of time and activities.

Vision

To be a successful leader, you must create a vision for what you are accomplishing through your leadership role and how this makes an impact. As a leader, the challenge is to go beyond the details of running your unit to look at the master plan and see how your work fits in. If you are an academic leader, how does your work fit into your university's success? If you are a research leader, how does your work achieve the goals of the research program being developed? How does it link with high-profile research groups working on related themes? Answering these questions allows you to shape your vision to more effectively reach your goals and your unit's goals. To answer these questions, you must understand both the internal and the external context, including where your unit's strengths and weaknesses lie, and at times including, for example, government or institutional priorities and constraints. You must come to terms with how your unit distinguishes itself.

Setting a vision, charting mechanisms for achievement of a vision, and developing metrics of success are all important tasks of any leader. The first step is listening: to your unit members, to other leaders, to stakeholders — students, industry partners of your unit, government and global partners. A careful and frank assessment of the strengths of the unit will help determine how those strengths can be brought into play in charting a course for the future. External reviews can help, as can specific metrics which reflect the implementation of the vision. As statisticians, we are generally comfortable with data gathering and data evaluation, but in a leadership role these are needed for a rigorous understanding of your unit's performance.

Your vision needs to be embraced by the unit, so it becomes the unit's vision and so you can count on champions on the ground to follow through and make it a reality. Key steps are to first draft a vision for the success of your unit that is in line with that of the leaders of your organization. You must articulate that vision to committee members, persuade them of the opportunities it provides, and offer them incentives to make it happen. And you must deliver on your vision through achieving or surpassing defined metrics of success. In this way, you establish yourself as one who delivers on goals, setting the stage for confidence in your unit's ability to succeed.

It is also helpful to develop and articulate privately to yourself your philosophy and principal priorities. Reviewing these regularly reminds you of your direction, and helps you avoid getting pushed and pulled every which way in the many activities which accompany a leadership role. It also assists with aligning your principal priorities and your unit's priorities with major projects, budget expenditures, and daily work.

Choosing the Team

As a leader, you need a team that will help you achieve your unit's vision. Team members also form part of a fan-out mechanism — a way to articulate and spread your vision. Hence, much thought and care should be put into creating the right team. The team should certainly be composed of those whom you believe can help you attain the goals you seek for your unit. Choose team members who bring talents which complement yours. Choose individuals with energy and drive, who understand and approve of the vision, and are ready to work with you wholeheartedly to achieve it. Choose individuals who have a reach into various parts of your unit and outside the unit so that the team as a whole exerts influence within and outside the unit. Don't ignore the naysayers; they help provide an alternate viewpoint. Build structures so that the team works in synergy; so, for example, undergraduate studies is not placing barriers to interdisciplinarity while the research side is promoting it. Use opportunities to train your team members in developing strengths or overcoming weaknesses so that the full team gains strength and knowledge from both experience and training.

Working with Your Organization

It is critical to understand the external environment and the priorities of your organization and how your unit and your leadership role play a part in its success. This is similar to understanding the playing field and the game and your role as a player in winning. Certainly, your unit's objectives need to be considered, but putting them in the context of the bigger picture assures your success. And there are situations where transformations will be required to achieve goals that fit into the needs of the organization. Showing that you are able to bring successes to the organization builds senior leadership's confidence in you. Senior leadership will see that your unit's goals are well in sync with the organization's goals. The senior leadership will be confident that when you are given an opportunity to pursue specific directions that meet your needs, you will be able at the same time to achieve university goals.

Focus

One of the most important characteristics of success in leadership is the ability to focus. Focus allows creative energy to flow and provides for a clarity of thought to develop clear directions and the vision for your leadership role and for your unit.

Focus is critical, but is not always easy to come by. It often eludes us as we are bombarded by information and digital means of accessibility to private space. One of the ways that focus is hampered is in the bustle of management tasks. This is particularly so when taking up a leadership position for the first time, when management actions such as email response and staff interruptions are seen as urgent, often incorrectly so.

First and foremost, it is important to create time for work beyond meetings. This requires, typically, working with your support personnel to identify work periods where you have protected time without interruptions, ideally your most productive times. For a senior leader, such as a dean or director of an institute, this can be an immense challenge; however, communicating your needs, and also articulating the major pieces of work to be accomplished and a timeline for accomplishing them, will create a framework for achieving your objectives.

Some simple strategies can easily bring more focus into the everyday routine. For example, the well-known divide and conquer strategy, where a major task is subdivided into small portions that can be managed within shorter periods of time, can assist in charting a way through a project by allowing focus on smaller portions daily or weekly. One strategy beyond divide and conquer is to alternate work periods of thirty or forty-five minutes between activities that are creative and require dedicated attention, and those that are easier to focus on, for example, handling the never-ending email inbox. In particular, when stuck on a challenging task, alternating a few work periods between the challenging task and one that is less challenging allows the mental processing time needed to "unstick" the challenging task. After that point, several work periods can be devoted continuously to the initially challenging work.

It can sometimes be difficult to settle into even an alternating work strategy; for example, there may be very many minor items with deadlines that need attention. In this case, it may be more useful to bite the bullet and "clear a thinking space" by prolonged and dedicated work to move through a block of the multitude of smaller items awaiting attention. Then you will have some breathing room to bring focus to bear on more challenging items. Creating such breathing space is important, not just for addressing large or complicated research problems, or for being creative and productive at challenging problems, but also for allowing focus on other elements of life, such as family and wellness activities. Keeping up with the minor items also ensures that you, as a leader, are not a stumbling block to smooth workflow for your unit. Often handling the many small managerial tasks in a leader's life requires this shotgun approach. However, in the normal workflow periods, settling into an alternating focus of challenging work, and more routine work, sets a better pace. Working too often to clear masses of smaller tasks away to create a clear thinking space denies the possibility of serious work that has greater impact. Tackle research, leadership, and difficult challenges with gusto; on the other hand, tackle managerial details and tasks with more constraint, keeping to the necessary level for completion of the tasks.

All senior leaders need to be highly organized and able to handle and quickly deal with a fire hose of activity coming in a steady stream interspersed occasionally with a torrential flow. This demands that you master the understanding of how to create time for focus and how to quickly dispense with incoming requests. Such mastery will only come by careful assessments of how you spend your time and what needs to be changed to allow more time for focus. As you develop as a leader, time spent on such assessments is invaluable, and assessments become quicker and less time intensive as you execute effective systems.

Challenge Yourself from the Start

From the beginning of your career, you can position yourself well for leadership roles at many levels by challenging yourself to broaden your perspective and to learn new skills.

The usual strategy immediately post PhD is to delve into many issues and sideconnections to the PhD research. This is a good strategy that helps to make sure that you've covered all aspects that immediately relate to the PhD topic, and it positions you well for your upcoming promotion and tenure evaluations. However, you are wise to also make connections to other areas where your research would be useful and to put your research out in a variety of fields, giving it a wider profile. Further widen your scope by opening yourself up to new areas that are linked to your expertise and by looking for opportunities to learn new areas of research that are hot topics and to which you can make a contribution.

Aim for research with impact. Just prior to CD's first sabbatical, a government health agency called for help with a problem. As a result, her entire sabbatical was redirected to learning in depth about the issues, learning new areas of research and new methodologies to provide a solution. The whole process charted a new direction that is still one of her passions. But this required openness to moving beyond an ad hoc solution, turning around a well-earned sabbatical to learn new methods and tools, and shifting research directions. It also required a passion for one's discipline and science broadly.

Importantly, rather than publishing smaller, isolated pieces of research, weave them together to tell a bigger and more complete story; look at competing approaches and make comparisons; look into limitations; be interested not just in the publication but in the problem and its solution, how the solution benefits the world in which we live, how it can make an impact. Continually assessing the impact of your work may offer you new ways to understand how your work fits into the bigger picture of the research universe and to raise your vision to a higher level.

Success requires understanding how work is evaluated. Careful preparation of grant proposals is, of course, crucial. Serving on grant review panels is extremely helpful in terms of getting an in-depth understanding of the review system — to improve your own proposals and to understand the funding landscape. Early in your career, talk with colleagues who have served in the review system to get advice on your proposal and research direction. Further in your career, ask senior colleagues to help provide opportunities for you to be involved on grant review committees. At all career stages, take advantage of opportunities to review grant applications and journal submissions that are of possible interest to you. Be bold in practicing leadership in evaluation at committee meetings, at society meetings and as a way of life, so others are inspired by your leadership, your integrity and evaluative skills, and see you as the right sort of person for significant professional roles, marking you as an emerging leader.

Your professional development training should include not only discipline-based methods and tools, but also so-called "soft skills"; grant writing, negotiation, consulting, and conflict resolution, for example. Challenge yourself also in your communications; being effective in communications is a key characteristic of a successful leader. Use opportunities that come to you as a young academic, for example, as an expert scientist helping the media, or in presentations at society meetings. Don't take these opportunities for granted; make your performance stand out through dedicated effort.

Sometimes challenging yourself involves exploration and taking risks that may not yield direct benefit, but even so, provide more world experience. In understanding your research and yourself, it's also important to know what directions not to pursue. Gaining this wisdom is best done through small, less risky ventures rather than through larger, high-stake ventures.

Much interesting science requires collaboration across disciplines, and there are often important sources of funding for team-based grants. Many universities have support for creating team grants; look to your research office for assistance and for guidance as to whether timing is right for you to take the lead. This also lets them know that you are ready to make an investment of time and energy to be a leader and will put you on their radar for future opportunities.

Be Diligent in Student Supervision and Training

Training is cornerstone to the life of an academic: from undergraduates in courses to graduate students and postdoctoral fellows. As a leader, you need to be diligent, innovative, and exemplary with regard to the levels of training that are part of your workload. Classroom instruction is changing by leaps and bounds, from inverted classrooms to global online learning environments. Invest in understanding what frameworks best suit what you are trying to teach and the learners you are reaching. Experiment with different tools and don't accept the status quo. Industry experts and guest speakers who use the tools you are teaching or who have problems that the students can solve add a great deal to the training environment.

Aim to provide the best possible environment for your graduate students; they are the lifeblood of your research. Help them find funding so that they can focus, teach them how to write grant applications, send them to other labs to develop complementary expertise, give them opportunities to interact with industry and to develop communication and presentation skills, help them to network with the best in their field of research, and spend time with them jointly to build a supportive learning environment. Give your students the opportunity to work in teams on small projects to set a foundation for collaboration skills. These are all important elements to help students become successful as academics and as future leaders. This support of graduate students is not completely selfless: building a strong research team can keep your own research moving along, particularly when you take on demanding leadership roles.

On Negotiation and Promotion

Throughout a career, negotiation may take place as you discuss a job offer or the changing conditions of an appointment. The negotiation periods are typically points in your career when you have the most power to set terms of your employment. Don't be in a rush to settle the negotiation. Take time to get information on parameters that are typical and that are at the boundary of possibilities, and why. This information helps you to negotiate from a position of strength. Look also to colleagues at other universities to hear of terms provided at a range of institutions. Be prepared in case your requests, or some of them, are turned down and an alternative put in place; that is you, yourself, should be ready to offer alternatives if initial requests are stumbling blocks. Understand the constraints that guide your employer so you are better able to offer suitable alternatives. More importantly, understand yourself and your needs: what would it take to have you accept this leadership position; what would it take to make you a success? Make specific comparisons with remuneration provided to other leaders, if you have these available, and bring items to the table that truly create the right environment to support your leadership role. Research benefits packages, teaching and service components of the position, research support, sabbaticals and administrative leaves, equipment, linkages to units with whom collaboration would be helpful — these negotiable items have a big impact on your work environment.

Gain an understanding of the environment and the challenges you might meet in your new leadership role. Talk with other leaders, senior colleagues, and new hires to hear their opinions on what might be changed at their institution, about challenges being faced and how they might be alleviated. Understand the philosophy that guides your employer, read the strategic plans for the university and the unit you are leading, and ask about the external and internal environment. Get information on budgets and what resources you have at your disposal, about support for what you would like to achieve, about metrics for success, and about other key deliverables. Discuss what team will be built to help you in your task.

Be creative in your requests, if the culture allows. When considering the headship, to accommodate time with her daughter, NH asked the dean if the monthly heads meetings could end earlier than 5 p.m. The dean had a family-friendly policy and readily agreed, changing end time to 3 p.m. NH also bargained for a six-month administrative leave halfway through her five-year headship, and this was granted.

Networking and mentorship are invaluable for raising your profile, for developing a broader view, and for finding out about possible opportunities and upcoming challenges. Use professional networks as well as university colleagues to build your professional profile, that is, to let others know your skill sets and your interests. If there are specific individuals at your institution or from other universities who could serve as good mentors, keep them engaged with you. Keep in mind that building relationships can often be easy through service to professional organizations.

Networking with key internal and external stakeholders gives you an awareness of how they view the work of your unit and lets them know that you value their opinions. It also assists in building a team around you to accelerate the achievement of your unit's vision. Networking with other leaders provides awareness of the variety of mechanisms being applied to advance the work of the university and keeps the whole leadership team engaged. Take the lead in bringing leaders together informally — informal chats are often a valuable way to gain information. For instance, informal chats with other heads can reveal upcoming important curriculum changes in other units or might provide insight into how to deal with certain new procedures requested by your institution.

Importantly, make sure to have a few trusted advisors who serve as sounding boards and who feel comfortable speaking frankly. When moving into a new leadership position, identify a few key individuals — perhaps a former leader — who can groom you in the position.

Build a Positive Work Culture

With the drive for excellence and a strong work ethic, it becomes critical to remember to take time to build relationships and a positive atmosphere in the workplace. Being compassionate, offering flexibility, and having a personal style that says "people before paperwork" sets an exemplary tone for your community. Sometimes, specifically adding items to your to-do list to enhance the workplace environment, even informally, can help in forming habits related to those actions.

Units are full of varied personalities, some with grievances and some with enthusiastic spirits. These offer a solid learning ground for leadership skills related to managing conflict, developing persuasive strategies for bringing a team together on a project, and creating change and transformation on a smaller scale. They help solidify your value systems — how you will operate with individuals who are challenged in working together; how you will speak out against elements that have crossed your lines of integrity; how you will create a positive environment for action and improvement in your unit. It is also important to recognize that different people have different strengths — assign roles that utilize an individual's strengths and passions.

Keeping the body, mind, and spirit healthy reduces stress and makes sure that you and your colleagues are fit to work hard and face challenges. Talking about wellness in the workplace is one way to make sure to give this attention — for example, you can build a team approach to wellness through walking meetings, or through an emphasis on fitness during the workday and weekends, showing by example that this is important.

Working toward the building of a positive work culture is the duty of everyone in a unit, and activities in this realm can become a learning ground for leadership from the start of your career. As a young investigator, establishing through your conduct that you have integrity and are objective, open, honest, and supportive lays a foundation to show yourself as a good colleague and a potential leader. This is not to encourage your involvement in difficult issues as soon as you are hired — often, it is better to listen and observe at first. Opportunity will arise from time to time, whereby you are able to slowly and surely display your values as you set a tone for the type of culture you would like to build in your unit. Remember that even as a junior member, the unit is also yours to shape; shaping does not belong only in the hands of senior colleagues.

Maternity and Other Leaves

To support a balance between the personal and the professional, the Canadian system allows for various leaves – maternity and parental leaves, sick leaves, personal leaves and, in academia, sabbaticals or study leaves. Careful planning pertaining to these leaves can maximize the benefits.

When each of CD's sons was born, she spent considerable time with them for many months. Her first son was born during graduate studies. She recalls preparing extra material to review with her supervisor, which she held back from him before her son was born so she would have a stockpile of new work and results to share with him after the birth. Since creating the right environment for work is important, she moved her office to home after her son was born so she could more easily meet the demands of both motherhood and career. This reduced her opportunity to network with other students and faculty but provided considerably in efficiency at a critical time, casting a net benefit that far outweighed the losses. On the other hand, when her second son was born, she deemed it important to continue to connect with her department and her students. To facilitate this, she moved her work office so that she would be in a noncentral location; in that way, she was able to return to work very shortly after delivery with baby in tow and not be disruptive to colleagues, yet be on campus to take advantage of the work environment. For others, it might make sense to connect using electronic media, much as when on sabbatical and physically far away from students.

NH became a parent later in her career — posttenure, as a full professor. While she already had an established research reputation, she knew she would need to "coast" for a few years. Therefore, before she took her leave, she spent a sabbatical year travelling to three different places to work with three different coauthors. She worked as in her graduate days — late at night, early in the morning — and so entered motherhood with a stack of papers on their way to publication, strong collaborators around the globe, and a full-out travel experience that would tide her over her stay-in-town years. During her leave, she worked from home during some of her daughter's naptimes. She sometimes asked students to meet her in her home, since commuting was challenging. She continues that meeting practice to this day, regularly seeing students off-campus at her home or in a coffee shop. She finds it an enjoyable change of pace, and finds meeting away from the worksite, with its constant demands, a good way to keep her meetings focused on the student.

In academia, we are fortunate to have regular study leaves to recharge our personal and research batteries. These leaves are a crucial opportunity to make breakthroughs, particularly after years of heavy administrative or parental duties, and thus should be used wisely. We must protect our leaves from excessive nonresearch intrusions of our home institutions — and moving to a new research location typically allows for that. For families, however, a study leave away from home can be logistically challenging, as such a leave requires coordination with a spouse and arranging new schooling, child care, and other support systems. For this reason, some academics consider a "stay-at-home" leave, much like the currently popular stay-cations. A stay-at-home leave, while perhaps not as invigorating as an away-leave, can be successful if one finds work space away from the work unit, arranges to have research collaborators visit for extended periods, or makes short research trips. A stay-at-home leave can also be used to develop valuable local research collaborations — collaborations that are easy to maintain during nonleave years.

NR also became a parent post-tenure and promotion, but if anything is certain about parenting and working, it is that every situation is unique. Certainly, there are many successful women academics with families, families started in a wide range of different circumstances. Many researchers at the beginning of their career, whether undergraduate students, graduate students, postdoctoral fellows, or junior faculty, wonder about the best time to start a family. Recently, NR heard of a panel discussion on careers, during which two members of the panel (both male) insisted that women should wait until they have tenure before starting a family. Categorical, if well-meaning, advice of this type should be ignored. The reality is that raising children while working at an interesting and challenging job is very demanding. When asked for advice from new parents, she usually advises accepting as much help as you are offered, and asking for as much help as you can. NR was department chair while her children were still quite young, but again, there is no one size fits all for leadership roles either. One advantage was that once home, dealing with small children was a refreshing break from the working day — completely absorbing, and leaving no time to obsess over problems in the department.

Parental leave, when available, is very helpful, although few academics will really stop research, it is simply too interesting! Most department chairs and senior administrators now work to reduce demands on new parents, whether or not there is an official leave policy, or beyond the length of the official leave policy. Don't hesitate to ask for some consideration in teaching, either a temporary reduction, or a repeat of courses already prepared, or something similar. Don't make the mistake of not asking for something on the assumption that it won't be possible. If it is truly not possible, your one-up will be the first to tell you.

While on leave, it may be wise to keep in touch with what is happening in your unit so you are not caught off guard by directions being set. Consider chatting with your department chair about major initiatives from time to time, as suits your schedule. This strategy assists in making reentry into the department a smooth one.

When returning from maternity and other leaves, a staged approach to reentry into the service components of workload may be useful. It might also be helpful to initially choose committees that do not require many face-to-face meetings, but rather permit advancement of the committee's goals through individual work.

Keep Personal Ties Strong

Having a supportive network of family and friends and taking the time to truly enjoy them are key to one's success. Throughout your career, try to structure your personal life to ensure enough time is available to nurture this network.

For families with children, consider encouraging everyone to be at home for the evening meal, even if this requires much working around of schedules and giving up some activities. Allow for times for conversation to flow, perhaps with that afternoon tea or midnight snack. Make time for holidays at all the usual celebration times in the year and in the summer. Building traditions through summer and other holidays helps to ensure there are special times for family and friends.

In the early childhood years of CD's sons, work was in the "out-of-control" state almost routinely for years, albeit with interludes where she was able to draw in the reins to put firmer boundaries on its intrusion into personal and research life. CD made a decision to spend time with children in the evenings from the time she arrived home until their bedtime, for their first five years, and she often fell asleep in bed with them after the evening's activities. One strategy that worked for her was getting to work very early and, at night, doing smaller tasks, for example, minor administrative tasks or skimming publications.

Be inspired by and proud of what you do. Put aside any parental guilt and remind yourself of some of the positive impacts of your career choice on your family. NH often felt guilty in comparison to some of the neighbourhood moms who worked in the home for their children's first years. Her daughter regularly questioned why some moms (not NH!) always had fresh-baked cookies as an afterschool snack. NH was surprised to learn that some neighbourhood kids were also questioning their moms, but instead asking why their moms weren't leaving the house each day, looking cool with a laptop case. Even when NH, for lack of proper child care, guiltily dragged her daughter into work, her daughter had special times — from sitting in on seminars eating cookies and taking notes like the other attendees, to collecting swag at the Joint Statistical Meetings. Now that her daughter is in her teens, NH sees a bigger payoff — her daughter takes it for granted that excelling in math is not related to gender, that becoming a leader is something within her reach. CD has raised sons who are not surprised to see a woman — and mother and wife — in a strong leadership role. Women leaders give their children the precious gift of unlimited possibilities, of acceptance that achievement is not limited by gender.

Institutional Responsibilities

Much of this discussion has considered how an individual may work to achieve success, and some of the strategies proposed may be useful to particular individuals. However, the formation of effective leaders is also an institutional responsibility. Structural supports by the university to promote success for women leaders are foundational, and include equity guidelines, removal of social barriers, opportunities for networking, mentorship, and providing flexibility related to family care. Institutions should consider how practices and polices affect a nontraditional career path, understand particular challenges and inhibitors to success in the context of a specific university or department, appreciate the difficulties for women working in a male-dominated field, provide transparency in evaluation and promotion processes, and be diligent in providing pay equity and monitoring gender equity carefully through metrics. Such practices enable both men and women to achieve success in their leadership roles.

Institutions must recognize all leadership contributions, even those carried out without a specific title or without recognition. While women can and should be proactive in getting credit for their work, the organization must also be aware that some contributions may not be recognized, and this is potentially harmful to a career.

While many organizations provide junior members with mentorship, the more proactive approach of active sponsorship is needed to encourage and train future leaders. The sponsor has a role as an advocate in promoting junior people, and by working directly with a junior member, helps her to gain experience and learn about leadership. Sponsorship can provide women with the confidence to know that they have what it takes to be an excellent leader. Sponsors can also be alert to suitable leadership opportunities as they arise, and encourage colleagues to consider taking on these roles.

Conclusion

We are fortunate in statistical science to have many examples of outstanding leaders across the profession. One of the most important contributions of leadership is in serving as a role model for the next generation of leaders. And in learning to be a leader, learn from those, male and female, that you respect most highly. Leadership is demanding, time-consuming, and at times, exhausting, but it also broadens your horizons and brings a great deal of satisfaction.

22

The Many Facets of Leadership

Jacqueline M. Hughes-Oliver and Marcia L. Gumpertz

North Carolina State University

Introduction

"Success is a journey, not a destination" (Ben Sweetland). No matter what you pursue, the journey called success will usually include hard work, determination, ability, and timing (being in the right place at the right time with the right people). But there are other qualities and experiences, fine-tuned to your specific interests, that can make your journey more conducive to success, enjoyable, and even inviting. This article addresses a number of such qualities and experiences. Based on a recently completed survey we discuss the need for women as leaders in statistics, where leadership is painted with a broad brush. This discussion is followed by providing a compilation of strategies and skills for success, and then by a description of several effective programs and policies aimed at developing leaders. We present brief autobiographies that demonstrate some of the issues that have been critical junctures in our own leadership and professional development.

Why Do We Need Women in Leadership Roles?

Female leaders and male leaders who promote the careers of women in statistics are key to making the field hospitable to women and to opening up opportunities for women in statistics. For example, statistics and biostatistics departments headed by women have higher proportions of female tenure track assistant professors than other departments, and the ratio of female to male tenure track faculty is almost twice as high for departments with female chairs compared to male chairs (Gumpertz and Hughes-Oliver, 2014). In fall 2013, departments headed by women averaged 39% female tenure track faculty, but departments headed by men averaged only 23%.

By being a female leader in statistics, you can have an enormous impact on who pursues a career in statistics, what kinds of experiences they have, and the level of appreciation of statistics and statistical thinking in society. We find examples of extraordinary impact among the female faculty at North

Carolina State University, where in fall 2013, females accounted for 31.6% of the 38 faculty members. As the 2013 president of the American Statistical Association, Marie Davidian has been an important champion of statistics; she also is executive editor of *Biometrics*, the flagship journal of the International Biometric Society, and she serves as the coordinator of the Personalized Medicine Discovery Faculty Cluster, which develops quantitative methods toward the promise of personalized medicine. Montserrat Fuentes, as head of the Department of Statistics, has implemented family-friendly policies and processes in the department; she is also editor of the *Journal of Agricultural, Biological and Environmental Statistics*. Marcia Gumpertz is assistant vice provost for Faculty Diversity, a role that impacts all units on all campuses of NC State; her work spans the range of mentoring doctoral candidates from across the country regarding faculty expectations, to assisting faculty recruitment efforts, to creating receptive climates in departments on campus to help with retention. Jacqueline Hughes-Oliver was the recipient of the 2014 Blackwell-Tapia Prize, which recognized her work to simultaneously excel in research while informing, empowering, and encouraging undergraduates to pursue graduate education in the mathematical sciences and to expand recruitment, retention, and mentoring of minority graduate students in statistics at NC State. Alison Motsinger-Reif serves as assistant department head and is also director of the Bioinformatics Consulting and Service Core. Kim Weems, as codirector of Graduate Programs, leads the department's efforts to recruit and mentor graduate students. Alyson Wilson is Principal Investigator of the Laboratory for Analytic Sciences, which focuses on advancing the field of big data analytics and intelligence analysis.

In a university department, the senior faculty jointly decide on the curriculum, which kinds of students are admitted, and which faculty are hired. They also serve as role models to the students. In the survey of heads of departments of statistics mentioned earlier, one department head had this to say about the importance of a critical mass of women faculty (Gumpertz and Hughes-Oliver, 2014):

> Critical mass for females seemed to be very important. The single female was not able to convince the faculty to hire more females. When a 2nd female came..., that completely changed the climate and we were able to increase our numbers.

Critical mass is often viewed as either 15% of a group or 30%, and is defined as the point at which a qualitative shift occurs in the climate or environment (Etzkowitz et al., 1994; Nelson and Brammer, 2010). Most departments responding to our survey — 69% — did not have a critical mass of women faculty, according to the 30% definition, and more than a fifth of departments — 21% — did not even reach the 15% mark. **You can see that we still have a great need for women leaders in statistics.**

Leadership comes in a variety of forms. The classical and most often-thought-of variety is the person in charge, the one who makes all final decisions, the person who tells other people what to do, the one who defines the goals and sets the direction. In academia, this role typically comes in the form of department head, college dean, provost, or university president. In industry, this role typically comes in the form of team leader, project manager, department lead, regional director, vice president, or chief executive officer. But successful people in these classical roles often share that although they have extensive decision and delegation authority, their level of effectiveness is highly defined by their ability to engage others to become accountable and committed. For this, the classical leader often relies on assistance from a core inner circle of individuals who have influence.

The core inner circle of individuals who have influence defines the other extreme of leadership: a leader without an official title. This person is present, involved, outspoken, and well respected by peers. Their willingness to say what they think in open, candid, professional environments demonstrates a willingness to work for the common good. Consequently, others are willing to thoughtfully consider their opinions and be influenced by what they have to say. While this type of leader is not often recognized as such, her or his contributions are critical to the success of any organization. In other words, it is not just heads of departments and people in designated positions who play leadership roles. All senior faculty and senior professional statisticians are leaders within their sphere of influence.

All levels of leadership are important, from the totally-in-charge designated leader to the no-official-title-but-influential leader. Finding your fit on this continuum is important. Perhaps your role is to chair a search committee, serve as an officer in a professional statistics organization, or maybe even serve on a panel to assess and make recommendations regarding research grant proposals. To find this fit, you must have a good idea of your interests and strengths and be willing and open to new roles and responsibilities. Perhaps serving as director of graduate programs will reveal an affinity to personnel and budgetary decision making, and this may suggest that a research center director or department head or higher-level administrative position is well suited to you. The point is this: Opportunities avail themselves to people who are willing and open to new experiences. Being available means being flexible on timing, the types of activities you are willing to engage in, and perhaps the location. It is a myth that leadership opportunities are of a certain type or follow a predefined time schedule.

Another myth about leadership is that you have to choose to either be a leader or a researcher, but not both. A long roster of statistics leaders show that it is possible to assume the classical leadership role while continuing their accomplishments as superb researchers. For example, the female statistics faculty leaders at NC State include those who are world-renowned in their respective research areas: Marie Davidian is part of the leadership team of IMPACT (Innovative Methods Program for Advancing Clinical Trials), which is a joint venture between NC State, the University of North Carolina at Chapel Hill, and Duke University; Montserrat Fuentes is well recognized as a top expert in spatial statistics, and she has maintained this footing even after several years of leading one of the largest statistics departments in the country; Alison Motsinger-Reif is one of the most highly cited and prolific researchers in statistical genetics and pharmacogenomics; and Alyson Wilson's expertise in reliability and Bayesian methods have led to integral involvement in defense and national security, with her even providing research support to the Office of the Secretary of Defense. Once again, determining your fit with regard to balancing research and leadership is completely personal and can only be accomplished by you. You have the ability to tap various resources in support of your goals; perhaps additional support staff and greater levels of delegation of duties can facilitate your desire to spend more time on research.

Leadership or research excellence does not have to come at the cost of having a family. In the same way that balance is required to juggle leadership and research activities, those same skills are useful for balancing child care and other family obligations. Many (though not all) institutions have made great advances in the resources offered to help individuals pursue a balanced life. The ultimate decision rests with each individual to position herself in a supportive and progressive environment, one where personal goals are encouraged and respected.

Strategies and Skills for Success

How do you get from where you are to a position of leadership? In our experience, important factors are: being willing, and offering, to take on responsibilities; participating in activities that are particularly meaningful to you; taking advantage of leadership development opportunities; and being open to taking on new and different roles. By actively participating in the realms that are important to you, you will get to know the people involved and the opportunities available, and others will get to know you and know your talents. When you speak up in meetings, when you make presentations at conferences, others with similar interests will get to know your abilities and your work. If your name is already known to decision-makers as a person with a successful track record who is willing to accept challenges, you will be in a position to learn about and to be considered for leadership opportunities.

Once you have found a leadership position, how do you lead successfully? Many people have written on the shared traits of successful people (e.g., Covey, 1989), effective leadership (e.g., Maxwell, 2007), key skills required to lead change (e.g., Kotter, 2012), and daily words of wisdom regarding leadership (e.g., Drucker, 2004). We include here skills and strategies that we have personally found useful in our careers. Our choice of the word skills, and not traits, is deliberate, because some of us have to work to develop these skills; effective leaders are not all born fully equipped to lead. Recognizing this, many workplaces offer training in skills needed for leadership, such as how to delegate responsibilities, how to deal with conflict among employees or factions, how to give and receive feedback, and coaching skills. Abilities such as public speaking, rather than requiring innate talent, involve specific skills that can be broken down into steps to master. We highly recommend taking advantage of the skills training that is available to you. These skills cover a wide array of areas of expertise that are new to most people entering leadership roles.

- **Flexibility.** Planning is critical, but life does not always go as planned. Being able to adapt to new circumstances will allow you to focus on moving forward instead of lamenting about what could have been. There's no need to sell your soul for every new idea, but organizations must evolve to remain relevant, and that may bring changes to your domain. Change can be great.

- **Drive/Focus/Grit/Fortitude/Determination/...** It does not matter what you call IT, the fact is that without IT you will not succeed. IT is the ability to push forward no matter what, to keep trying even though you come across major hurdles, to not quit. You may be gifted, but if you fold at the first obstacle, or the tenth, you will not fully realize your potential. Failure is inevitable; if you never fail, it means you are not trying hard enough. The question is what comes after a failure — how do you react to a fall? Some of us secretly doubt our abilities (despite years of hard work and accomplishments), so when failures occur our minds make mountains out of molehills, and this puts us in a paralyzed state of mind. This reaction zaps our energy, our creativity, and our future success. Yes, it can be hard to do, especially for those people who have had very few failures, but shaking off the dust and getting back at it is the way to move forward. Quoting Winston Churchill, "Success is not final, failure is not fatal: it is the courage to continue that counts."

- **Passion.** If you are not excited about your cause, how can you expect to do an exceptional job at it or for anyone else to be excited about it? Buy-in and commitment from those you lead will be critical to your success as a leader, but why should someone commit to an idea to which you are only lukewarm? Leaders have to work long hours; how will you muster the energy necessary to work on the project if you do not really care about its success? Sometimes, we are handed projects and asked to lead them even though we are not passionate about them; this will typically be an unpleasant experience. To the best of your ability, seek to lead endeavors for which you are very passionate. For those situations where you do not get a choice, this is where drive, focus, and flexibility must take over to get you through it successfully.

- **Confidence in One's Ability.** Everyone talks about self-confidence, but if we are truly honest, we will admit that it is often nowhere in sight! How can you be confident in yourself when you have just failed, your drive is almost gone, and your passion is waning? *This is the time to remind yourself* *of what you have already accomplished.* Your accomplishments came even though you had periods of doubting whether you could keep going. Your *abilities* have taken you this far, and they have prepared you to take the next steps to move forward. Admittedly, it is a very subtle distinction — have confidence in your abilities, and focus less on the feelings associated with "self." And if none of this works in how you project yourself, you can always use the strategy of faking it — talk a good game in public, shake in private. Seriously.

- **Argumentation and Public Speaking.** Develop skill at building comprehensive arguments founded on unassailable logic. Your ability to present a strong case for your ideas will be key to success in influencing approaches, getting programs adopted, gaining funding, and winning awards for your protégé. All leaders are called upon to speak in public arenas. Programs like Toastmasters International provide a supportive, painless way of learning the skills for public speaking and logging the hours of practice needed to hone those skills.

- **Ability to Delegate.** Leaders lead others; if you do everything yourself, what will your team do? More importantly, one person cannot do everything, at least not very well. If you believe in your team, give them control over certain aspects of the project. People tend to be more committed and work harder for success if they have responsibility in some way and have a feeling of ownership. Yes, this means that you cannot micromanage, and you need to let go. Aside from giving your team breathing room, delegation will also free up some of your time to be able to do other things (perhaps research?)

- **Do Not be Afraid of Conflict.** If you lead long enough, you will come across conflict. Even if you do not go looking for it (and you should not go looking for it!), conflict will find you. There will be others who disagree with you, openly or in back rooms. Some battles are best handled by ignoring them. Without intervention, other battles will continue brewing until explosion. You should be prepared to deal with these before they get out of control and take over your team. An ideal conflict may come in the form of a team member who disagrees about the direction you have laid out. Do not get flustered. Instead, make this a way to strengthen the team. Perhaps you can ask that person to lay out his/her vision, discuss strengths and weaknesses of both visions, all while incorporating feedback from other team members. The result will often be a vision that has more buy-in from the team and hence a greater chance of success. In the process, you have shown strength (because only strong leaders allow others to lead) and flexibility — a win–win for everyone. At the other extreme, conflict may arise from an employee who is disgruntled and not truly interested in seeing the team succeed. Such a conclusion should not be arrived at quickly — there should be history before you become willing to conclude you have a source of poison. But after you reach this conclusion, the poison must go; it must not be allowed to bring the team and the mission to ruin.

- **Get Along With People.** People can be difficult, unreasonable, rude, and obnoxious. We are not saying you should necessarily go along with what they say or do, but you should be responsive and tolerant. Hear (that means truly listen and absorb) what they have to say, what is important to them. Address their concerns: if this helps develop the path and can be incorporated into the mission, then great; if their concerns cannot be addressed, then explain why not. Do not participate in or encourage gossip. This sounds obvious, but if a leader engages in gossip it will destroy trust within the unit. Some of the leaders we most admire consciously find a way to make positive remarks and eliminate negative comments from their speech. This positive attitude propagates through the unit, resulting in a sunnier climate for all.

- **Continuous Improvement.** We started the list with flexibility and end with continuous improvement. Indeed, they are related. Flexibility is the willingness to adapt to changing needs. Continuous improvement of one's abilities through training, exposure, and new opportunities presents the path to being flexible. The person who does not continue to learn and improve becomes stagnant, stale, and irrelevant.

Mentoring and Workplace Policies

Having a "sponsor," someone who advocates for you, introduces you to senior colleagues and suggests your name for positions, can give a powerful boost to your career. We have been fortunate to have two or three such advocates in our lifetimes. Luck had a good deal to do with meeting these sponsors, but it was luck engendered by our active participation in activities that we were serious about.

Many workplaces have mentoring programs or leadership development programs. For instance, at NC State, the Provost's Office offers a workshop series titled Leadership for a Diverse Campus that aims to motivate women and minority faculty to consider line leadership positions, such as department head and dean. In this program, which was developed under an National Science Foundation (NSF) ADVANCE grant, participants meet university leaders and learn about their responsibilities and career paths, and the challenges and rewards of the different kinds of positions. Several faculty who have participated in this workshop series have gone on to become faculty senators, department heads, and directors of centers. Participants in this series are also often tapped by senior administrators to chair committees and serve in various leadership capacities around campus.

One particularly important leadership development program for university faculty aspiring to senior administrative positions, such as associate dean,

dean, vice provost, vice chancellor, provost, and chancellor, is the American Council on Education (ACE) Fellows Program. This is a yearlong program in which participants are immersed in the decision-making processes of another institution.

Our Personal Journeys

Marcia Gumpertz began her career in statistics with a master's degree from Oregon State University, then served as a statistician at the Environmental Protection Agency research lab in Corvallis, Oregon, for five years. With this background, she returned to school for a PhD in statistics from North Carolina State University. She started as an assistant professor in statistics at NC State in 1989. Her position, which in addition to teaching and research involved consulting with faculty and students in agricultural and environmental sciences and forestry, drew heavily on her experience at the Environmental Protection Agency. She taught graduate courses in applied statistics, design of experiments, and spatial statistics, and did collaborative research involving statistical applications to environmental and agricultural studies.

In 2002, two years after promotion to full professor, Gumpertz was given the opportunity to serve as director of Undergraduate Programs in Statistics. In this role she: advised undergraduates; served on and chaired departmental, collegewide, and university committees regarding the undergraduate curriculum; coordinated assessment of the undergraduate program; directed scholarships for undergraduates; developed a mentoring program; and served as faculty advisor to student organizations related to the field of statistics. This role provided a wider view of the student experience and gave her an understanding of student recruitment and admissions, course and curriculum review and approval processes, processes for awarding scholarships, college administration, and university program assessment goals and procedures. Gumpertz found working with undergraduates to be refreshing and she found advising the students and the student organizations to be rewarding and very enjoyable.

Gumpertz served as co-chair of NC State's Association of Women Faculty from 2002 to 2004. As a result of this, in 2006 she was invited to apply for a new position at the university, assistant vice provost for Faculty and Staff Diversity. Applying for this position, and ultimately accepting the position, required being open to stepping off the charted path and trying something that she had never before imagined doing. This represented a major change in direction; however, the position involved advocating for women faculty and faculty from underrepresented groups more generally and assessing the university's progress with respect to diversity and climate. These were both areas close to her heart, and Gumpertz had also discovered that she enjoyed university administration. The lessons that Gumpertz took from this are that for her it is important to be open to new and different possibilities, important to know and follow your own interests, and that change is good. The role of assistant vice provost for Faculty Diversity requires working with faculty, department heads and deans across the university, chairing or serving as a resource person for several committees and task forces to try to understand the issues of concern for various constituent groups, then connecting people with the right people and offices to address those concerns, writing proposals, and assessing progress. Gumpertz' years of consulting with faculty in a wide variety of departments and serving on university committees unexpectedly turned out to be an important asset and precursor to her current role. Skill in facilitating meetings and communicating well with people in all disciplines is a must.

Jacqueline Hughes-Oliver joined the Department of Statistics at NC State as assistant professor in August 1992. Following the usual tenure-track academic expectations, Hughes-Oliver progressed by leading her research program (with her doctoral advisees and a few collaborators), teaching, and serving her institution and the statistics profession. With diverse research interests, Hughes-Oliver has been funded as principal investigator by the National Institutes of Health (NIH), the National Science Foundation (NSF), the North Carolina Department of Transportation (NCDOT), and a variety of other organizations. Her methodological research focuses on prediction and classification, variable and model selection with dimension reduction, design of experiments, and spatial modeling. Application areas range from drug discovery, to engineering manufacturing, to environmental modeling, to transportation modeling, and even to genomics and metabolomics.

A major shift occurred in 2005 when Hughes-Oliver was awarded a large federal research grant to create an exploratory research center that, at its height, consisted of 22 people, most of whom had PhD degrees. The group was highly multidisciplinary (including statisticians, mathematicians, chemists, and computer scientists), and international (team members lived and worked in Canada and several states in the US). Hughes-Oliver had to develop new skills to: simultaneously direct many research projects; keep the overall project on schedule; facilitate communications and cross-disciplinary learning; hire and manage personnel (GRAs from statistics and computer science); manage a budget much larger than previously accustomed; and interact with several other similarly funded research groups. Although challenging, Hughes-Oliver was energized by this experience. Research collaborations with talented investigators was stimulating, providing more than enough enjoyment to stave off doldrums that came from the more mundane activities. This was an exhausting but very fun leadership experience.

Another major shift occurred in 2007 when Hughes-Oliver (with hesitation) agreed to direct the statistics graduate programs at NC State. As one of the largest and most well-respected, these graduate programs involve a large body of continuing students, plus a much larger number of students who apply. To manage the programs as they existed required major commitment and management of an extremely large budget. But the department was actually looking not to maintain the status quo but to also increase the diversity of the graduate programs. Being extremely passionate about: (1) the role of education in impacting a person's life; (2) the call academicians have to educating the next generations; and (3) the incredible need within the US to develop talent in people of ALL backgrounds, Hughes-Oliver accepted the challenge. This leadership experience required development of additional skills: counseling through crises that sometimes had nothing to do with academics; assessment of human potential to the point of having to rank-order individuals; and strategies for recruiting students into the discipline of statistics and to pursue graduate studies. Knowing the positive impact she made on many lives, Hughes-Oliver found this position very rewarding. On the other hand, providing extensive counseling through personal challenges and rank ordering individuals with the result of having to turn some people away posed emotional strains that were not to her liking. Hughes-Oliver now follows her passion regarding education and diversity by being available to students on an individual basis; she frequently offers advice, feedback, and motivation in a variety of settings. She also participates in programs and organizations that share her goals.

Final Thoughts

Leadership is both personal and public. Should everyone develop leadership skills to be flexible, driven, passionate, confident, a good speaker, able to delegate, able to deal with conflict, get along with others, and always improving? Absolutely! Does everyone need to lead in the same way? Absolutely not! Growth requires cycling through at least two steps, self-development and finding your path. Embedded in these two steps is the need for flexibility in responding and reacting to what speaks to, and works for, YOU.

Bibliography

Covey, S.R. (1989). *The 7 Habits of Highly Effective People*. Free Press.

Drucker, P.F. (2004). *The Daily Drucker*. HarperCollins.

Etzkowitz, H., Kemelgor, C., Neuschatz, M., Uzzi, B., and Joseph A. (1994). The paradox of critical mass for women in science. *Science, New Series*. 266(5182):51–54.

Gumpertz, M. and Hughes-Oliver, J. (2014). Women faculty in US stat/biostat departments. *IMS Bulletin*. 43(6): 14–15.

Kotter, J.P. (2012). *Leading Change*. Harvard Business Review Press.

Maxwell, J.C. (2007). *The 21 Irrefutable Laws of Leadership: Follow Them and People Will Follow You (10th Anniversary Edition)*. Thomas Nelson, Inc.

Nelson, D.J. and Christopher N.B. (2010). *A National Analysis of Minorities in Science and Engineering Faculties at Research Universities*, 2nd Ed. Jan 4, 2010. [http://faculty-staff.ou.edu/N/ Donna.J.Nelson-1/diversity/Faculty_Tables_FY07/07Report.pdf].

23

Leadership and Scholarship: Conflict or Synergy?

Roy E. Welsch

Massachusetts Institute of Technology

Introduction

There is a saying that as a professor, you work 16 hours a day, but you get to choose which 16 hours. This flexibility sounded good to me, and I became a professor and recommended my profession to many PhD students, both men and women, but especially women because of the flexibility.

However, now many years later, I see that my guidance has not worked out as I might have hoped.

In many STEM areas, we see some of the best women at MIT opt for positions in second tier universities (or outside academia) where the tenure process is somewhat less demanding and schedules more flexible. This may effectively remove them from national leadership positions over the long term. This is a serious problem, but I am more concerned about what happens to women in top tier universities.

To get through the tenure process, many women delay having children. This means they are having children when they are 33–40 and facing all the demands that go with raising children in the critical years when they could be moving into leadership positions. The male-oriented schedule of late afternoon meetings and seminars makes taking on leadership roles extremely difficult. In addition, it is hard to get extra support and child care to travel to attend meetings and participate in the activities of professional organizations.

I remember a conversation with a faculty member several years ago who said that he had completely missed seeing his children grow and change from birth to age three because he had been so busy running his lab, writing papers, and traveling. He had delegated child care to his wife. My experience with my own children and with my grandchildren leads me to believe that this is a time in a child's life that is not to be missed. I suspect mothers would feel this even more strongly. I should also add that my only children are women, and my only grandchildren are girls. No career accomplishments can replace the experience of seeing one's children grow and develop. If this is a female instinct, then it is time it became a male instinct as well, and we design our organizations to make this possible.

I recently proposed that we increase support and child care so our female faculty with children could attend more meetings and conferences. As we went down the list of female faculty in one part of the Sloan School, we had no women who needed such services. Why? Because they either did not have children or had a stay-at-home partner. This says to me that we may not be able to attract women who have children and no stay-at-home partner. That must mean we are losing some great talent that we should be attracting. We need leaders who will get the changes we need done. My belief is that those leaders will mainly need to be women, since men have so far defined the rules of the game, including expectations and qualifications, and operate almost on instinct. A gender-neutral world does not seem to be a natural male instinct.

As a society, we should think carefully about the benefits and costs of gender-neutrality. I cannot have children and women can. That is hardly gender-neutral. Clearly, men and women have important roles to play in all aspects of society, especially the development of society's future — the nurturing and education of our children. If some specialization is necessary, then we need to allow that specialization and build organizations that foster and reward it.

Barriers

The are some barriers that contribute to the lack of women in leadership roles, and I will comment on a number of them later. Barriers often mentioned are: lack of self-confidence, not realizing the need to provide greater encouragement to younger women and peers, less awareness among educators to think of women as potential leaders, less training in speaking up (taking the floor), not being aware of leadership opportunities when they arise, different expectations for men vs. women, and, perhaps, some distrust by both men and women of women in positions of leadership.

Gaining Experience

I have encouraged our administration to arrange for younger women faculty to get involved with leadership by being group heads and associate deans (some administration and reduced teaching along with some released time for work with professional organizations). This allows for scheduling flexibility and experience. It also allows women to be role models in leadership for younger women faculty.

Experience in leadership and being near leaders is important. First, you get to see if you like taking on a leadership role. If not, you can return to what may be the best job of all, being a professor. However, you may find that, often very slowly, I admit, you can make changes in your group, department, and university. This can be very rewarding over the long term and, looking back, as important as the paper you did not get to write because you took the risk of taking on an administrative role. Perhaps you will learn more about taking risks and proposing your own solutions to problems and the risk of seeing them turned down. However, some of your ideas and solutions will succeed.

You will also interact with more people in your organization. The tenure process encourages isolation and singularity — did you, not your thesis advisor or coauthors, make important contributions to science? In fact, coauthored papers may be discouraged, which means working and interacting with other faculty may appear to be discouraged. You may find that you like working with others in your leadership role and finding ways to allow them to succeed to the maximum of their ability.

On the other hand, you will need to deal with people problems rather than just statistics. People have real personal, family, and professional problems. Addressing them and helping to work them out is a part of leadership positions.

There is another important aspect of getting some experience in leadership positions. Generally, you will have to learn more about delegation and choosing colleagues to work with and for you. As professors, especially ones comfortable with technology, we tend to try to do everything ourselves. While this may work while just being a professor, it most likely will not work when taking on a leadership role and trying to move the organization you are leading forward and making it a better place for women (and men) to succeed in all aspects of their lives.

Entrepreneurship and Innovation

In a school of management, we talk and teach about entrepreneurship and innovation. As a professor, there is not much opportunity to actually do these things in a corporate way. However, the academic organization you are a part of (or the professional associations that you participate in) can offer very interesting opportunities for entrepreneurship, innovation, and creativity. For example, suggesting new courses, new or revised academic programs, distance learning opportunities, schedule and course timing changes, summer programs, return to campus to learn short courses, etc. If you have ideas and are frustrated that they are not getting the attention, experimentation, or funding they deserve, then it is time to consider investing energy in a leadership role and becoming a change agent. To get ideas, think about what the world would look like if the world were dominated by women leaders and male leaders were sparse.

Speak up

At meetings with both men and women present, especially where the acoustics are not perfect, it is very important that women make themselves heard. Sit in the middle of the table, not at the end. Take some public speaking classes to learn how to project your voice or even acting classes (good for teaching as well). For whatever reason (designed by men?), most rooms and audio-visual systems seem to be better tuned for male rather than female voices.

Try to sit down to address men at eye level and operate on an equal footing, so to speak. As someone who is 6'5", I have had to learn to do this with men as well.

In more technical meetings (and maybe in all meetings) be careful not to focus too much on details without realizing that others are not following what you are saying. Recognize and address the big picture issues before diving into technical details. Men seem to be less patient than women when too many details are presented. When talking about statistics, both men and women need to be aware that others in the room may not know what you are talking about and, in fact, may have a negative view of statistics and statisticians.

Use Your Power

As tenured professors, women should have just as much influence as men in departmental and university decisions. Often, I see a certain shyness or reluctance to take forceful positions or voice opposition to the status quo. This can delay implementation of or experimentation with the great ideas women have that reflect their differences from men in experience and personal value formation. In academia and, perhaps, more often in industry, women do seem to hit some sort of ceiling. There is evidence that women do not ask for promotions (or salary adjustments, etc.) as often as men do. This may be a self-confidence or fear of rejection issue. Due to family concerns and male-dominated families, women may be perceived as less likely to change location or threaten to leave and, therefore, less likely to accept higher offers or better positions.

Informal Interaction

With the advent of in-office communication (email, etc.), there seems to be less personal interaction. Many of us grab lunch and eat in our offices. Some days, almost all of my interaction occurs in the men's room, where some important business is transacted in sound byte fashion. I have often wondered if there should be a common washstand area but a separate stall area, so both men and women would see each other in the washstand area.

Good building design leads to choke points like hallways, stairways near elevators, and areas outside elevators where people can engage in quick conversations. Small kitchens can also play such a role. The days of everyone coming together at 3:30 or 4 to have coffee and play Go or chess seems to have given way to the constant pressure of email, etc.

I still see groups at lunch that are all men or all women. I am not sure this is good or bad, but informal interaction is an important part of preparing for and of leadership itself. I am often reluctant to sit down at a table that is mostly women, and women may feel the same way about a table that is all men. Perhaps a good idea for both men and women is to make sure there is a gender mix at a table, making it easier for either gender to join and participate. Make lunch count.

There probably is still an old boys network in some cases. A woman commented to me recently that she was at a diversity conference internal to her company, and afterwards four male colleagues got together and went to dinner without her. She interpreted this as networking that appeared to be male only even after a conference that was focused primarily on female inclusion.

Mentors are important. Most mentors for women are women, and that probably is a reasonable solution, when possible. However, there probably should also be a male mentor in situations where most of the leadership positions are held by men.

Closing Note

As a PhD student in mathematics more than 40 years ago, there was one woman and, I believe, she did not finish her degree. There were few role models for her in mathematics, and certainly very few women in any leadership roles in the STEM areas. Progress has certainly been made, but not at the pace we should be seeing now that it has been more than four decades (two generations) since Title IX. As in the past, women will need to learn more about leadership, how to advocate for and obtain leadership positions, and how to transform their organizations to remove some of the impediments that prevent women from taking on leadership positions. I am optimistic that women and men working and leading together will bring full leadership equality before another generation passes.

Part VII

Institutional and Network Strategies

24

The Value of Professional Champions and Mentors in an Academic Environment

Katherine Bennett Ensor

Rice University

As a senior full professor and department chair for fourteen years, I have grown to truly appreciate the value of professional champions and mentors. Mentors and champions are individuals who know you and understand your professional profile and interests. Mentors serve as a source of advice to you, whereas professional champions are individuals who keep up with your professional growth and recommend you for opportunities or bring opportunities to your attention. Of course, the same person can, and often does, serve both roles. I have mentored many people throughout my career. But I also find that I myself am a "champion" for those whose research and accomplishments and professional profile "grab me." The people I champion may not even know that I am one of their distant fans and often a "champion in the room" on their behalf. I expect I myself have benefitted from many champions over the years, some I know and some I infer. In this article, I will speak further to these two basic ideas as well as a bit about leadership in general.

The Importance of Mentors

As educators and researchers, we understand the importance of mentoring. We guide our students not only intellectually, but also through the early stages of their professional life choices. Once you have your desired degree and you enter the workforce, the need for mentorship changes form, but does not disappear. As a profession, the statistics community learned that mentoring faculty in the early stages of their careers greatly improved the outcomes mid-career and set the stage for strong leadership as the individual grows in the profession. This is especially true for underrepresented groups. In other words, we are not a profession of individuals pursuing unconnected goals, but we are a community and the community thrives as the "young guns," using the words of Ingram Olkin, continue bringing ideas, strength, passion, and enthusiasm to the community.

While an assistant professor, I was the recipient of some of this early mentoring initiated by leaders of the statistics profession. I attended one of the early faculty Pathways to the Future workshops, organized by Lynne Billard and Nancy Flournoy, and supported by Mary Ellen Bock. These workshops are now a standing entity at the Joint Statistical Meetings and have helped our profession tremendously over the years. Many of the leaders of today attended these early workshops. For me, attending the pathways workshop also served as a mechanism for me to attend JSM that year. I began my career before the time of generous start-ups for assistant professors (also before salary parity improved with respect to gender) and funding for travel was difficult to obtain. The year I attended the pathways workshop was the year I was coming up for tenure at Rice. I learned about many aspects of the profession that I had not yet understood. In this workshop, I learned that I was one of just a few female faculty members in tenure track positions in departments of statistics across the nation. I certainly understood that there were never very many female students in any of my classes, but had not considered the translation of this reality to the national scale. The talks by Lynne, Nancy, Mary Ellen and others, on the scarcity of women and the bias against women in the publication process, were illuminating. Again, I had never really considered such issues. The workshop encouraged the new professors attending to focus on their research and delay getting involved in university service, and if you did, to do so at the highest levels. Further, we learned about the overall process of tenure in American Tier I research universities, especially the critical role of external letters. The advice and knowledge I gleaned from this workshop was tremendous, and mirrors advice I now give those that I mentor. As I previously noted, I attended this workshop the year that I was up for tenure at Rice, and so the stage for my tenure evaluation was pretty much set. I recall sitting in that workshop listening to what one should do for success and thinking, *wow, there is no way I will succeed, because I have done everything wrong.*

I would like now to give some cautioned counter advice to the strict focus on research during those early years as a professor. While my path to full professor and a leader in statistics has not been ideal, it has been strong. I was in a unique position as an assistant professor in that I was the first hire of the new department of statistics, and the success of the department came before the individual success of all the founding members. It was a goal the faculty collectively agreed to, and all contributed significantly in those early years to bring statistics at Rice to the international prominence it is today. One aspect of that visibility was to become visible at Rice. It was both natural for me, and relatively easy for me, to help in that regard. At one point as an assistant professor, I actually served as speaker of the Faculty Senate. I learned a great deal about how universities function and met people from all over campus, all the time championing the strengths and contributions of our department. In a sense, I was the *champion in the room* for statistics, and one of the first at Rice. Yes, my research took a big hit due to this early service role, but there was a significant upside. I was afforded the opportunity to become an instrument of change at Rice, together with my colleagues. The change surpassed just statistics, as I helped with respect to the treatment and climate for women faculty, students, and staff. Further, my ability to connect across the university directly benefitted our department in those early years, and was definitely beneficial when I became chair, a position I held for fourteen years.

I am proud of what we have been able to accomplish in statistics at Rice, as well as the significant and measureable positive impact our department has had on the entire university. We are a leadership-driven department, and the attitude of the faculty filters to students at all academic levels. This type of impact and contribution does not happen by faculty keeping their doors continuously shut and not interacting with the world around them. In addition to myself, the early crew consisted of Jim Thompson, David Scott, Marek Kimmel, Dennis Cox, and Peter Olofsson (now chair of mathematics at Trinity University, San Antonio); each a leader in research as well as exceptional educators, mentors, and active university and professional citizens. The faculty that joined after these early years also recognized the importance of contributions across the spectrum of activities of a research university. Today, as I watch the department support our new chair, Dr. Marina Vannucci, I am reminded of the power of collaboration and cooperation toward common and laudable goals in research, education, and overall societal impact.

The value of mentorship is well understood by most scientific professions and certainly by the National Science Foundation (NSF). In the mathematical sciences, funding for mentoring programs at all stages of academia have contributed greatly to the overall diversity of faculty, and faculty leaders across our nation. Although there is still much to accomplish, there is improvement over the last two decades. At Rice, we benefitted from the long-standing NSF Alliance for Graduate Education and the Professoriate (AGEP) (Tapia, 2014) program and ten year Division of Mathematical Sciences (DMS) Vertical Integration of Graduate Research and Education (VIGRE) (Forman et al., 2003–2008, Cox, Ensor, Wolf, 2008–2013) program.

The AGEP program provided us an extended opportunity to bring exceptionally strong, underrepresented students to our doctoral program. AGEP provided additional mentorship, as well as funding for many of these students. Faculty in the Department of Statistics at Rice actively seek a diverse graduate student body, and welcome and encourage all undergraduates interested in the discipline[1]. These alumni have gone on to excellent careers in academia and industry. Although the AGEP program made this possible with additional funding and support, the faculty saw the program as an additional

---

[1] At the Caucus for Academic Representatives Workshop, JSM 2014, Kathryn Chaloner educated the chairs on how to increase the number of strong domestic students, especially diversity candidates, in their programs. She highlighted Rice's success in this regard. Rice was recently recognized as national leaders in the number, *not percentage*, of students from underrepresented groups awarded doctoral degrees.

opportunity to work with the exceptionally strong students, while many of the participating students appreciated the additional social support provided through AGEP.

The VIGRE program provided direct mentoring to Rice mathematical science students, from undergraduate to graduate to postdoctoral faculty. Through the VIGRE program, we were able to encourage more of our own students at all levels to pursue academic careers. For example, one of our undergraduate alumni, namely Genevera Allen, participated in our VIGRE program, completed a PhD in statistics at Stanford, received funding as a PhD student through Stanford's VIGRE program, and then returned to our department as the Dobelman Family Junior Chair in Statistics. Dr. Allen has quickly risen to national leadership and prominence through her strong research, coupled with her outreach and mentoring mentality. The mathematical sciences, in general, benefitted greatly from the VIGRE programs, and I expect the same will be true for the Research Training Group (RTG) opportunities provided to the mathematical sciences community. I personally believe that programs of this type are very important for our nation and worthy of strong support from NSF.

So for all of us senior faculty and successful students who have received encouragement and mentoring in their careers, I encourage you to keep the tradition going by reaching back and mentoring others. Even though so much information now exists at everyone's fingertips, the personal connection is key. It is important to acknowledge also that many individuals do not need professional mentors, as they have a clear understanding of the profession and how to succeed. This point brings us to the topic of professional champions.

Professional Champions

Professional champions guide others toward your work and contributions. They help you advance in your career. Many of us do not have one single professional champion, but several.

The importance of professional champions in academia cannot be overstated. It is easy and natural to promote and speak about people you know, and this knowledge benefits from more than just the publicly available information. At universities, it is important to have a large number of individuals willing to write letters in review of your tenure and promotions as well as various awards for which you might be deserving as you progress through your career. Although for promotion, at least half of your letter writers will most likely be suggested by your department, most faculty are given the opportunity to make suggestions as well. Ideally, your suggestions will be leaders in the field who can speak strongly about your work and your contributions to the scientific community. This is a difficult list to establish if you have not built up a professional network cognizant of your many contributions. Within this professional network, strong champions, who are themselves leaders, can make a clear-positive impact on your own career. In addition to providing letters when required, this group will bring connections to career opportunities for which you might be a natural fit.

In *The New York Times* article "Mentors Are Good. Sponsors Are Better," Sylvia Ann Hewlett (2013) highlights the fact that to break through the leadership network requires a sponsor. Quoting Hewlett,

> Mentorship, let's be clear, is a relatively loose relationship. Mentors act as a sounding board or a shoulder to cry on, offering advice as needed and support and guidance as requested; they expect very little in return. Sponsors, in contrast, are much more vested in their protégé, offering guidance and critical feedback because they believe in them.
>
> Sponsors advocate on their protégé' behalf, connecting them to important players and assignments. In doing so, they make themselves look good. And precisely because sponsors go out on a limb, they expect stellar performance and loyalty.

Hewlett's article highlights the research by the Center for Talent Innovation, a center founded by Hewlett and focused on advancing women and minorities. The center's research based off of a sample of 12,000 men and women demonstrates measurable advantages of professional champions. As Hewlett states, "To get ahead, women need to acquire a sponsor — a powerfully positioned champion — to help them escape the "marzipan layer," that sticky middle slice of management where so many driven and talented women languish." Hewlett is speaking directly to the book *Lean In*, by Facebook CEO Sheryl Sandberg (2013). Hewlett notes, "No matter how fiercely you lean in, you still need someone with power to lean in with you."

Now, both Hewlett and Sandberg bring a corporate perspective to career development. In academia, a single champion will not suffice to help advance a scholar and educator. Decisions in academia are often made by groups of individuals, each bringing their own knowledge and experiences. For example, as I previously noted tenure and promotion requires letters from multiple leaders in your respective field of study. Further, your department will make a collective decision on promotion. If you are nominated for an award, you will most likely have three letter writers, and then a committee of scholars will be reviewing your professional portfolio as well as everyone else under consideration. When academics apply for federal funding, decisions are informed by a panel of scholars selected for their general areas of expertise. For individuals interested in being engaged at a broader level, you will benefit greatly from a champion in the room during any such discussions.

Finding professional sponsors or champions is a relatively natural process, but is not a passive process. Academic leaders naturally seek out excellence and inventive scholarship. For example, sometimes when I read a piece of research that I believe is truly inventive, I will try to learn more about that particular scholar, first by starting with a search of their other papers and their website[2]. If we are both at the same conference, I will try to be in the audience during her or his presentation. As I learn more about this scientist, then I can become her or his unknown champion in the room. When there is an opportunity that fits her or his unique gifts and directions, I speak up, highlighting her or his potential for contributions. I believe this behavior is true in general in academia, and is how the scientific community grows. Scientific leaders try to understand how the pieces of the puzzle fit together, in other words, what is the larger landscape to which this proposed or completed research contributes. Associated with the research is an individual(s), and so it is natural to learn more about that scholar.

Another distinct difference in academia from the single sponsorship ideal put forward by Hewlett is that there is not the "quid pro quo" between champions and recipients, although the expectation of excellence and commitment does persist (as it should). As academicians, we are dedicated to the advancement of science and human inquiry. When we see exceptional talent, then we should and do champion that talent. The issue for women and underrepresented groups is that champions may need to be actively engaged rather than passively acquired. Talent is identified through networks, and women are often excluded from the top research networks. Today, this exclusion is generally not intentional, but rather a biproduct of human relationships. As I advanced in leadership roles, I recognized that once again I was one of few, or the only, female in the room, especially if I move out of the discipline of statistics. So the common issues, of making sure one's voice is heard and opinions respected, return. Over the years, I have honed this skill, and now can step back and reflect on that fact that it is a skill and an important one.

I would like to take this opportunity to identify key sponsorship of my own career. In the early years of my career, I had several professional champions. One in particular was Ingram Olkin. While an assistant professor, I was invited by Ingram Olkin to attend a multivariate conference in Hong Kong. He pulled together a group of "young guns," as he called us, or assistant and associate professors of statistics from around the nation to attend a conference on multivariate methods. It was simply an exceptional conference and experience. Today, such opportunities are much more commonplace, and many assistant professors get invitations of this type. Twenty plus years ago, this was not the case, especially for female faculty. It takes leaders, like Ingram, to step up and secure the necessary funds to invite the younger faculty. I put this opportunity into the "professional champion" category, rather than mentoring. Ingram provided a critical opportunity for professional growth and outreach. He has helped so many in this regard throughout his career, and I am honored to be among this group. It is interesting that he served as a professional champion for Monnie McGee, a very successful professor at Southern Methodist University, and my first doctoral student.

---

[2] As I write about looking to the websites of others for my information, I cringe, as my own website is rarely up to date. Those that are able to properly manage their online presentation reap the benefits.

A year later, I was invited by Ingram to visit Stanford University over the summer. Also visiting that summer were Lori Thombs and Sallie Keller, both of whom remain my closest friends. I followed up the summer visit by spending my first sabbatical at Stanford, expanding my research and developing collaboration on simulation based estimation with Peter Glynn, an international leader on simulation and now an elected member of the National Academy of Engineering. This collaborative research led to a successful NSF proposal, multiple papers and doctoral dissertations, and I continue to pursue ideas along these general lines. My colleagues at Rice University, as well as my family, understood the opportunity for broadening my perspective that visits to Stanford provided, and encouraged me in this regard. It was not easy from a personal or financial perspective to pick up and move, even for a short period, but my family and I knew at the time such an effort would be highly beneficial, and it was. In addition to the mentoring and simply learning more about statistics and the academic profession, as noted I established lifelong friendships and collaborations. By providing me these opportunities for my professional growth, Ingram was a champion in my corner. I expect he has also served as a champion in the room over the years on my behalf, but again that is an inference not a known fact.

A professional group to which I have committed significant time since an assistant professor is the Southern Regional Council on Statistics (SRCOS) and its annual Summer Research Conference (SRC). In June of 2014, Rice hosted the 50th anniversary SRC at Hotel Galvez in Galveston, Texas (Ensor, 2014). Attending this conference were more than one hundred faculty and sixty students. This conference and organization has served as a champion incubator. The goal is not necessarily on mentorship, but rather providing high-level professional opportunities for new scholars to interact with leaders in statistics. Benefitting greatly from these conferences early in my career, I am now pleased to pass along this opportunity to the young scholars behind me. Key participants in the organization consist of ASA presidents, numerous fellows of all leading statistical organizations, National Academy members, deans, and leaders in the National Science Foundation and National Institutes of Health, and the list goes on. How else would I know Michael Kutner, a significant voice in statistics throughout his career and a consistent and strong supporter of SRCOS? At the June conference, I was honored to receive the Paul Minton Service Award, given annually by SRCOS, presented to me by Michael Kutner; clearly, Mike was a champion for me and my contributions to the profession and to SRCOS. The SRCOS SRC is styled after the successful Gordon Conferences. These smaller, focused conferences bring unique opportunities as incubators for professional champions at all levels of the professional spectrum. SRCOS is unique to the southern region of the US, but the model for SRCOS is replicable and valuable.

As you move from the early stages of your career to the mid-stages, champions are equally or even more important. Take my own case, for example. I have been acknowledged by my colleagues and others as an excellent leader;

someone with innate leadership qualities. I certainly understand what I perceive as ideal leadership and what it takes to be this person. However, without others speaking up for me with respect to these skills and talents, any further opportunities for me to lead larger groups will be limited. For example, if I wanted to pursue further academic leadership, such as a position as dean or provost, I would need strong professional champions to help me achieve that goal. I do know who these champions would be, and could initiate such an effort. In other words, once you identify your goals, it is equally important to identify and engage the champions who can help you toward those goals.

Today, I serve as a champion for many, and as previously stated, I have benefitted from multiple champions over the years. For every opportunity realized, there is a champion in the room. For every award received, there are a group of letter writers informing the champion(s) in the room. Key known statistics champions for me over the years include my current and former colleagues, especially Jim Thompson and David Scott, as well as friends and professional champions such as my advisor, Joe Newton, collaborators Peter Glynn, Mary Ellen Bock, Linda Young, Lori Thombs, and former dean, Sallie Keller. As a profession, statistics is very open, and there is good opportunity to engage with current and future leaders. As good leaders get to know you, they will naturally serve as professional champions on your behalf. I encourage current leaders to proactively seek out future leaders, or individuals they can themselves champion. Further, for those starting their career, seek out your champions and help them to understand your strengths and potential for future contributions. For women, the proactive nature of this process is again very important.

Leadership

Although the focus of this paper is primarily on the role of mentors and champions, I would like to spend a moment discussing leadership generally.

Leadership consists of vision, focus, guidance, and management. Great leaders bring vision and focus to groups they lead. In academia, the vision should not be top down but rather a collective vision of the faculty. Strong leaders are positive, encouraging, and supportive of initiatives and activity of others. They do not view their colleagues as competitors, but rather as colleagues. The management side of leadership is also key; individuals within organizations produce their best work if that organization functions well. While bad managers suppress the creativity of those they manage; great leaders do the opposite. I have closely studied many academic leaders throughout my career. Leaders come with many different styles and talents, as do all people. I have also participated in leadership training programs to better understand this important component of academia, and regularly read leadership-oriented articles and books. The leadership programs targeting MBAs and the corporate world do not automatically scale to academia. In academia, with tenured faculty, the dynamics are different and must be managed differently. It is important to engage faculty in the university collectively throughout the duration of their careers. Further, the structure of academia is changing exponentially fast. Leading and mentoring a new set of faculty through this landscape means that leaders must expand the traditional scope of academia.

In general, I am a firm believer that if you work with smart, creative, and engaged people, supporting them as much as possible, they will find new solutions and innovations that appropriately bring strong science and educational opportunities to the forefront for a better world. This belief has proven a reality today. Much of the innovation bringing positive change to the global society comes from university research and education. It is important that tomorrow's educators and researchers continue this positive direction.

In conclusion, I encourage us all to continue our mentoring, whether mentee or mentor, for each of us to identify and engage the professional champions that can help us achieve our professional goals, and to serve as professional champions for others we respect. Champions are a key component to one's success, and may change through one's career. With the dramatically changing landscape of academia, leaders who can empower, engage, and guide academicians and students are a needed international commodity.

Acknowledgments

I acknowledge and thank Allyne M. Ensor for her guidance and help with this manuscript. The conversations and editorial assistance greatly improved the message and the presentation.

Bibliography

Cox, S., Ensor, K.B., and Wolf, M. (2013). Vertical Integration of Research and Education in the Mathematical Sciences. NSF DMS #0739420. url: http://www.nsf.gov/awardsearch/showAward?AWD_ID=0739420. July 2008 through September.

Ensor, K. (2014). *Local Host 2014 SRCOS 50th Anniversary Summer Research Conference. Southern Regional Council on Statistics.* urls: http://srcos2014.rice.edu and http://www.sph.emory.edu/srcos/, Galveston, Texas. June.

Forman, R., Ensor, K.B., Symes, W., Wolf, M., and Cox, S. (2008). Vertical Integration of Research and Education in the Mathematical Sciences. NSF DMS #0240058. url: http://www.nsf.gov/awardsearch/showAward?AWD_ID=0240058. July 2003 through June.

Hewlett, S.A. (2013). Mentors Are Good. Sponsors Are Better. *The New York Times*, April 13. url: http://www.nytimes.com/2013/04/14/jobs/sponsors-seen-as-crucial-for-womens-career-advancement.html?_r=0.

Sandberg, S. with Scovell, N. (2013). *Lean In: Women, Work, and the Will to Lead.* Alfred A. Knopf, New York.

Tapia, R. AGEP Graduate STEM Rice program. *Richard Tapia Center of Excellence.* Rice University. url: http://tapiacenter.rice.edu/programs/agep-graduate-stem-rice-program/.

25

Mentoring: It Takes a Village. Personal Story

Sastry G. Pantula

Oregon State University

As an African proverb says, "It takes a village to raise a child." I strongly believe that many of us benefit from having many mentors in our lives, formal or informal. Not a single mentor, but a collection of them. It is important that we take advantage of all of them. They are all around us, and there is much to learn. Mentoring can benefit both the mentee as well as the mentor. There are so many opportunities to mentor or to be mentored with extraordinary benefits. It is a win–win relationship! No one needs to feel isolated or miss the benefits of mentoring.

I was a bit surprised that I was asked to write an article about mentoring. It is ironic, since I have no expertise being a mentor. Perhaps it is the passion I have for recognizing the important role mentoring has played in my own life. Or, perhaps it is the passion and commitment I have to see women in leadership roles and believe that it is everyone's responsibility to enhance diversity through mentoring. Because my mentors have influenced me and shaped my life in such important and positive ways, I am motivated to put my gratitude on paper. In this article, I will focus on many mentors in my life. They are my parents, my friends, my supervisor, my colleagues, my department head, my dean, and my provost, among others. Some of the mentoring happened informally, and the others more formally. I will also emphasize the importance of recognizing your mentors, though many of my mentors do not like the word mentor, let alone being recognized publicly as being an outstanding mentor.

But it is important to do. My mentors are the wind behind my wings who helped me to reach new heights. I sincerely hope this article motivates others, men and women, to take the time to mentor others, especially women, to reach their full potential. Each one of us should help break any glass ceilings for women and underrepresented minorities, and be the wind behind the wings for many others. Our profession is fortunate to have many wonderful role models, especially a number of female presidents and vice presidents of the American Statistical Association and the Institute of Mathematical Statistics, and a number of department heads throughout the country. I am proud to have had the opportunity to start my academic career in a department founded by a great visionary and a mentor for many, Gertrude Cox.

Before I delve into my story, I want to point out an excellent article on mentoring that had a significant impact on me. The article is "Mentoring: A skill professional statisticians can develop" by Rich Allen (Allen, 2007), which appeared on the American Statistical Association's website in 2007. Allen begins by defining who a mentor is not. A mentor is not a teacher, not a supervisor, not a parent, not a role model, not an older person. You are probably wondering, why, then, have I just stated that my parents, teachers, supervisors, and friends were my mentors. However, Allen (2007) goes on to say:

> Of course, few things are absolute. Teachers can be successful mentors. Many supervisors may be imparting important lessons for solving future problems (and therefore mentoring) while directing current work assignments. Most parents and role models hopefully provide valuable mentoring as they demonstrate how to handle life and career challenges. Thus, we might define a mentor as one who helps others develop the skills to reach professional and personal goals. The mentor acts as an informal sounding board to lead the one being mentored into discovering the best courses of action.

In this sense, I also very much appreciate this quote by Steven Spielberg, "The delicate balance of mentoring someone is not creating them in your own image, but giving them the opportunity to create themselves." I want to point out that I have used mentor as a verb and as a noun. It is important to understand both. Gina Luna (Luna, 2014), a senior executive at JPMorgan Chase, favors mentor as a verb.

> We all develop relationships, and mentoring — advice, coaching, perspective and help — is a component of almost every relationship. I mentor someone when I see in that person the potential to do more, make a change and become a high performer. When I see someone I know has what it takes, I want to maximize the potential.

Others lean toward mentor as a noun. Walter C. Wright, a senior fellow at The Max De Pree Center for Leadership, sees it this way (Wright, 2011):

> Because the verb "to mentor" places the initiative, and perhaps the responsibility, in the hands of the mentor, and from my perspective that undercuts the power of the mentoring process.... Mentors are important resources for our learning, they are guides for our development, they are models for our choices. But they are not responsible for our growth. The power of

mentoring rests in the decision to select mentors, the choice to learn from them, and the responsibility to act on our learning.

Mentoring helps us navigate the world, and as we shall see, mentoring can take many different forms. And because I am a statistician and love data, I want to share some interesting numbers. A Google Ngram shows a sixfold rise in the use of the word mentor from 1960 to 2008. A recent book search using the word "mentoring" on Amazon shows 8,821 books available! I believe mentoring will continue to become more necessary and more widespread as life becomes more complex and harder to navigate. It is not a weakness to lean on others, in fact, it is a strength to lean in.

My Mentors

My parents: My mother continues to play an important role in my life, with the values she has been teaching me since I was a small child. Of course, in her eyes, I am still a small child, and she continues to offer unsolicited advice on various topics, weekly. One of the important lessons she taught me is that a building needs a strong foundation. Without a strong foundation, a building will not last long, and it will collapse. The early years of childhood are so important. She insisted that we take pride in everything we do, whether it was doing our homework, how neatly we kept our textbooks, or how we dressed. We communicate our values through the quality of work we do and how we present ourselves. She is no tiger mom, but she certainly helped us by laying a strong foundation for us to build on — build our own buildings. She is a good listener and a good guide in helping us find our own passion — not molding all of her children to be the same. The diverse paths that my four siblings and I took are a testimony to her strong mentoring skills in helping us create ourselves.

My dad was a mathematics teacher, and an outstanding teacher, I must say. He used to wake me up at 4 a.m. in the morning to teach calculus to me and to several of my classmates who used to come on their bikes from a great distance. He often taught kids who could not afford to pay for private tutoring or who came from families where no parent attended high school. He instilled in me the importance of education and sharing, and the value of diversity. I appreciated his teaching style, where he would not give out the solution, but cleverly draw the solution out of us. While deriving a result, he would ask us to tell what the next steps were. Sometimes he would even deliberately let us follow the wrong direction of a proof, only to teach us that we learn through our missteps. There are valuable lessons in failure.

I always enjoyed going for evening walks with my dad through crowded markets. It was interesting how many people he knew in the town, and how many would stop him and seek his advice. As a kid, I was so impressed that he had a solution for everyone's problem, whether it was about a child getting into a college or finding a match for a happy married life, or balancing a family checkbook. He was the go-to guy in town! Why? Because he was a good listener. He listened patiently to understand the underlying problem rather than the words someone used. He accumulated his knowledge over the years to understand what works and what does not. People fondly remember him for how he made them feel. I must have soaked that up like a sponge, watching him in action. Whether he ever intended to be a mentor to me or not, I feel that he was one of my great mentors.

My father used to tell us that he would help us climb the tree by gently pushing us up the tree, but only up to a certain height. Then, we were on our own to climb the rest of the tree and find our own fruits or a northern star. Of course, we had the confidence that he was there to catch us if we would fall. After high school, I had an option to go to an engineering college or to the Indian Statistical Institute (ISI). It was not my father's passion for mathematics that led me to statistics, but rather my family's finances that helped me choose statistics over engineering. We were neither rich nor poor, but with five children in the family, we did have to make some choices. ISI provided free education and a stipend to live on. I credit much of my mathematical foundation to my father, and his subtle mentoring and advice were priceless throughout my college days and my career.

My father was not only a professor, but also an assistant department head and an assistant principal of a college. He had to retire due to his age before he could ever become a department head or a principal. He encouraged me to be a department head or a principal to have an impact on students and faculty, to build future leaders. He talked about the gratification in service, and the secret pleasure of seeing others around you thrive. It is the many conversations I had with him, and his incredible modesty that have left the strongest impression on my mind. Also, it is a great example to me the way he helped my older sister and my mother's younger sister, through subtle persistence, to be leaders in their own fields and to thrive. I would consider myself successful if I could have a positive impact on even half the folks that he positively impacted.

My peers: I believe strongly in peer mentors. As a student at ISI, I was fortunate to have many peer mentors who were a year or two my senior. They are still some of my best friends, and we are connected forever in an *ISICAL73* Yahoo Group! They helped me navigate how to thrive at ISI by sharing their own experiences. I vividly remember sitting on wooden benches in our hostels near a pond, and talking about the periodicals (monthly quizzes) and the many unwritten rules. We came to ISI from various states of India with no common language other than broken English. Those therapy sessions on the wooden bench, fondly referred to as *adda* sessions, were actually the beginning of a lifetime of bonding, as well as mentoring through what is known as *vertical integration*, or peers mentoring each other informally.

I noticed similar mentoring when our daughter, Asha, joined the *Follow the Child Montessori School* in Raleigh, North Carolina. Her classes had children of three different age groups, e.g., children in first, second, and third grades all sitting in the same classroom, helping each other, and learning from their own curiosities, with the teacher guiding them through the learning process. We used a similar model in our statistics department at North Carolina State University, where we had undergraduate students, graduate students, and postdoctoral fellows in vertically integrated teams, mentored by each other. This was a part of a successful National Science Foundation (NSF) — Vertical Integration of Research and Education in the Mathematical Sciences (VIGRE) grant for 10 years that helped develop a mentoring culture in the department. Graduate students started a peer mentoring program in the department, which was extremely beneficial to students arriving in the department from around the globe.

A sense of belonging is important and happens as a result of a social process of physical congregation, a clear focus of attention and regular, ceremonial-type activities, according to Chamblis and Takacs (2014). The authors studied a small group of 100 students over an eight-year span to find that personal relationships play a clearly decisive role in determining a student's success in college. They noted that there were several layers to students' social worlds and found that having learning communities, undergraduate research opportunities, and activities that create a connection and a support network matter.

Let me digress and share an experience I had as the director of Graduate Programs at NC State, which helped me deepen my commitment to peer mentoring. During my first year as the director of Graduate Programs, my department head, Tom Gerig, encouraged me to attend a diversity workshop. At the workshop, we had about 50 faculty. We were asked to sit around round tables with six of us per table. At the beginning of the workshop, the leader asked us to read the instructions for a card game that were left on the table. Once we finished reading the instructions, he collected them and asked us to play the game. Once the game started, he said no talking at the tables. After playing for about five minutes, he asked us to stop and asked the winner to move to the table on his/her right, and the loser to move to the table on the left. I won at the first table and moved to the table on my right. I began playing as competitively as I had at the first table. However, when I went to pick up a trick I thought I won, one of my colleagues grabbed the trick. Since we could not talk, I thought it was OK, that she did not understand the game. I lost this time, and moved back to my table. I was relieved to move back with the people who know how to play the game, and concluded that the folks at the right table are dumb and did not know how to play the game. Unfortunately, this time my luck with the cards turned and I was the loser, so had to move to the left table. I got the best card possible and was sure I was going to win. But, as I was playing, I started losing my tricks. My blood pressure was going up, and the only thing I could think of was how dumb some of my colleagues were. They cannot read the rules or understand them very well. I was so glad when the moderator stopped us from playing. He asked us what the experience was like. I was the first one to open my mouth and say that my colleagues did not know how to read the rules or play the game properly. I felt so dumb when I realized that each table had a different set of rules and they were all playing by their rules.

Even after all these years, the part that was striking to me was that my first impression was that my colleagues did not know the rules, and hence they must be thick skulled. The important lesson of the game was that students come from all different types of schools, small and large, from diverse cultures where the rules may be different. To apply that lesson learned to teaching, if we do not communicate, we will end up with a wrong first impression — that students are dumb. Worse, students may believe that they are dumb, and spiral down even before they have had a chance to show their potential. This is particularly relevant to students coming from institutions serving minorities and from smaller liberal arts colleges where faculty are able to provide much more one-on-one mentoring and advising to their students.

This experience helped me to share the rules and expectations with our students and to set up a peer mentoring program with the help of the Graduate Student Association. Peer mentors are there to help new students learn, navigate, and integrate into the departmental culture.

As we were developing a peer mentoring program for graduate students, our undergraduate director at the time, Marcia Gumpertz, also created a mentoring program for undergraduate students that included both graduate students and alumni working at local companies as mentors. These mentors met regularly with our undergraduate students for coffee or a meal, and some have taken their mentees to shadow them at their work. It was a rewarding experience both for our undergraduate students as well as to our alumni who are eager to find ways to give back to the university.

Peer mentors and advisers are essential for success. One good example is the North Idaho College's It Takes a Village Project (Thompson, 2014).

> The university recruited high school dropouts from the streets, tattoo parlors and the Skate Park and put them into "villages" of students. Statistically, only 1 in 20 should have succeeded, but almost all did. Interestingly, the students did not just perform better than expected, they performed better than average, achieving higher GPAs and completing more credits than a typical North Idaho College student. The reason the students were able to overcome adversity and succeed was because of each other. Villagers went to class together, learned study and life skills and had peer mentors and advisers providing encouragement. They could lean on each other. They served as their own support system.

Advisors also play a crucial role by looking out for you, helping you make sense of things, goal planning, and coming up with a strategy for you to reach your goals.

My advisor: I am so fortunate to have had an opportunity to work with Wayne Fuller at Iowa State University as my PhD advisor. Even today, he would refuse for me to call him a mentor. However, he is a mentor to me in so many ways, professionally and personally. I have learned a great deal from him and continue to benefit from conversations with him. We make it a point to have breakfast at the Joint Statistical Meetings every year. He has never given me a solution directly, but has always helped me figure out what might be the best solution for me. He worked with so many students at one point that we actually had to take numbers just to see him. These were the days of punch cards before Microsoft Outlook Calendar. A few times, I figured out the answer myself while waiting in line. Maybe it was a part of his strategy to keep us waiting!

Wayne never sugar coats anything, nor does he speak ill of others behind their backs. He has tremendous integrity, which is why so many respect his guidance. He has a great way of helping us analyze a situation and decompose the problem into solvable bites. Just like walking to the market with my father and watching him help people asking his advice along the way, I enjoy standing next to Wayne at the opening mixer of any Joint Statistical Meetings. He is a magnet! Many statisticians come and talk to him. He made sure to introduce his students, and now he is vigilant about introducing us to others as colleagues. I continue to learn so much from these conversations and the connections made from networking. I call on him every time that I am considering a life-changing decision. He is a good listener and helps me choose my own fork in the road without being judgmental. I know he is a great cheerleader for me. I recall how he helped me get through difficult times when my papers were rejected by journals by sharing his own personal experiences.

Wayne was also enthusiastic about getting us involved with the refereeing process early on as graduate students. One thing I also realized is that he is not flowery or effusive in his recommendation letters. If he says the person is good, people know that he/she is good. He does not take two pages to say what he can say in two sentences. It is his way of conserving paper and being green.

While advisers are so important to student success, they are not as pervasive on college campuses. Recently, Busteed (2014) reported on findings from the largest representative study of US college graduates. The Gallup-Purdue Index surveyed more than 30,000 graduates to find out whether they are engaged in their work and thriving in their overall well-being. In simple terms, did they end up with great jobs and great lives? Surprisingly, only about two in 10 strongly agree they had a mentor who encouraged their goals and dreams, or 22%, which means that eight in 10 college graduates lacked a mentor in college.

Support can come from those in the university family-alumni. We can certainly tap one of the greatest assets in terms of human capital. Recruiting alumni to mentor a current student can be a simple solution. A mentoring relationship with alumni does not have to be complicated. It would just take two to three calls, Skype meetings, or Google Hangouts between an alumnus/ae and a student each year for coaching, career guidance, or professional advice and networking, according to Busteed (2014).

Universities have been studying and supporting mentoring for about 35 years. Specifically, minorities and underrepresented groups need mentors the most. Lopez (2014) wrote in a new blog on MLA Commons by Members of the Committee on the Literatures of People of Color: Since the 1980s, higher education's increased complexity illuminated the need to offer young faculty members better and more-explicit guidance, particularly the rising number of women and minority scholars entering academia.

Young scholars on the tenure track need mentors they can trust. It is important for them to have help navigating the demanding career path of academia and to understand the culture and standards of a university. Academics are given a lot of latitude regarding their research, teaching methods, and campus service opportunities, but as Lopez (2014) points out, there are many potential land mines on the path to success. For example, faculty need guidance about which meetings a department, college, or institution deems important to attend, which relationships are important to cultivate, etc. That is why having colleagues you can trust is invaluable.

Back to my opening statement, it takes a village. As good as it sounds, there is no "super mentor," the one wise adviser or guru, who guides you from elementary or high school and throughout your life. The reality is, you can and will have many mentors over the course of your career.

Colleagues who are good often have an open-door policy. They invite younger faculty members, know that they are available to talk to them, read drafts of papers or grant proposals, or share contacts in their professional networks. It is important to be proactive with new faculty. They may not wish to show their ignorance, so be sure to reach out to them and invite them to do things with you. It is about building their trust and confidence in you as a colleague. Sometimes, it is just asking them to go for a cup of coffee and asking how things are going. It can be that simple. Just be brave enough to take the first step.

Establishing a solid, thoughtful mentoring relationship does not happen overnight. It takes lots of little moments. But it can have a tremendous impact on the culture of a department, not to mention on the younger faculty's career and life. The chemistry department at the University of Michigan found that deliberate, structured mentoring paid off for them after 10 years. It changed from a "dog eat dog, sink or swim" mentality to a "we are all in this together, we're trying to build a great place" department (Lopez, 2014).

I have had much good fortune in my own career with excellent colleagues.

My colleagues: Colleagues play a very important role in a junior faculty member's life. I was fortunate to have many wonderful colleagues at NC State. They were not assigned to be my formal mentors, but I have found several of my colleagues to be outstanding mentors. Francis Giesbrecht was a great mentor who helped me set high expectations for students in my classes. He helped me get over a bad lecture day, and helped me figure out how to correct it for my following lecture. He was always there as a mentor when I needed him, especially during my early assistant professor days. Similarly, colleagues like Dave Dickey, John Monahan, and Peter Bloomfield helped me navigate the research publication process and grant proposal writing by reading drafts, being coauthors and coprincipal investigators. The help of folks like John Rawlings and Dennis Boos, who helped me become an advisor to graduate students by serving as cochairs on doctoral committees, is invaluable.

As the department grew, it became more important that we had a more formalized mentoring program for our junior faculty. Otherwise, only a select few were finding their own mentors while others felt isolated. When I became the head of the Department of Statistics in 2002, I worked with our faculty to develop a formal mentoring program. Junior faculty began to have a teaching mentor and a research mentor who met regularly throughout the year. Faculty were also reporting their mentoring activities in their annual reports. We began to recognize, assess, and reward mentoring of our colleagues, and established a mentoring award for faculty. Most mentors found their experience rewarding without being recognized with formal awards, but the awards show that mentoring is valued in the department. It is a simple pat on the back that goes a long way. I am very proud of our faculty mentors at NC State who took on the responsibility to see all our faculty succeeded as a team.

My administrators: I am blessed with having outstanding bosses. Dan Solomon was the department head at NC State when I joined the department in 1982. Dan always looked out for the junior faculty in the department and encouraged to strive for excellence in research, teaching, and service. I recall when I was thinking of going on a sabbatical leave to a university, Dan suggested that I consider going to industry instead. He understood that such a leave would not necessarily enhance my publication records, but would provide me an outstanding experience of working in a team environment and offer practical problems to solve. It was a fantastic experience to work with engineers at SEMATECH for a year during my sabbatical leave.

When Dan moved to become an associate dean of the college, Tom Gerig, who was then a director of Graduate Programs, became the department head. I flew back from the Joint Statistical Meetings in Canada to request that Tom select me as the director of Graduate Programs. Tom not only gave me the opportunity, but mentored me for eight years in that role, the job that he held for more than a decade. His approach to mentoring was wonderful and very supportive. He allowed me to create the job the way I wanted it to be and allowed me to take risks, while he provided me with an excellent sounding board. When he decided to step down as the department head after eight years, I was fortunate to be appointed head of the department by Dan, who by then was the dean of the college. A phenomenal leader, Dan was an excellent role model and mentor to me while I was the department head. I have learned so much by observing him, engaging in many conversations with him, and experimenting with new ideas with his support. Like my dad, Dan helped me strengthen my commitment to strive for excellence, to enhance diversity, and to foster harmony. Dan is a champion for finding ways to advance women and minorities to leadership roles, and I share his values and commitment. I have learned many skills by watching the master. His mentoring throughout my three decades at NC State is priceless.

When headhunters approached me about positions in upper administration (dean or vice provost/president for research) at other universities, I asked them how they got my name. I was surprised and shocked to find out that Dan was behind it. I had to ask Dan if he thinks that I am doing a poor job as the department head that he was suggesting my name to others as a way of letting me go. He said that on the contrary, he thought that I should look for higher positions given my experience, and that he is genuinely proud of me. I was elated to hear his confidence in me and thought maybe it was time to cut the umbilical cord.

As the dean of the college of science now at Oregon State University, I am fortunate to have a great provost, Sabah Randhawa, who proactively looks out for the success of all the deans at the university. His calm personality, good listening skills, and proactive approach in looking for mentoring opportunities are much appreciated. It is great to be a part of the leadership team with a diverse set of deans and vice provosts who are committed to mentoring future leaders.

My turn: Striving for excellence, enhancing diversity, and fostering harmony are ingrained in me now. I have been trying to put many of the lessons I learned from mentors into practice. I can only hope to follow the footsteps of many of my mentors.

Mentoring became a part of the culture of the department of statistics at NC State. I have enjoyed working with a number of females, like Marcia Gumpertz, Pam Arroway, and Jackie Hughes-Oliver, on my administrative team in leadership roles, and I continue to build a diverse leadership team now at Oregon State. We have recruited Julie Greenwood as my associate dean for Undergraduate and Academic Affairs, and Debbie Farris as the assistant director for Marketing. We are fortunate to have recruited the first female head of the department for our physics department, and have the pleasure of working with two other female department heads of the Integrative Biology and Statistics Departments at Oregon State. What a wonderful diverse team I have to work with. Currently, we are working on recruiting two other department heads for the Mathematics and Microbiology departments. Both these search committees are chaired by a couple of our female leaders at Oregon State. We are currently recruiting five new faculty members, and we are focusing on enhancing diversity and recruiting faculty who are committed to diversity and student success. I share my expectations of having a diverse pool of candidates as I give the charge to the search committees. I am also looking forward to working with my colleagues on the recently funded NSF-ADVANCE grant that increases the participation and advancement of women in academic science and engineering careers. I have been making use of some of the things I learned from my mentors. I recall how my former dean,

Dan Solomon, doubled the number of females in our college, proactively. He used to share his expectations with the search committees, and in one case, even providing the list of ASA fellows who are women, as an example to indicate that there are well-qualified females in our profession. I have used similar approaches to enhance diversity in my roles at NSF as the director of the Division of Mathematical Sciences, at NC State as the department head, and now at Oregon State as the dean of the College of Science. We were successful in recruiting ten female faculty at NC State while I was the department head, and also recruited ten female program officers while at NSF. NSF has been supportive of conferences like the Infinite Possibilities Conference, which celebrates the successes of women of color in mathematical and statistical sciences, and provides many mentoring and networking opportunities for women. We hosted the Infinite Possibilities Conference on March 1–3, 2015, at Oregon State. I am proud of the Women in Statistics Conference in May 2014 which also celebrated our female leaders and provided opportunities to learn and grow. I look forward to the next one in 2016. We brought Laura Liswood, secretary general of the Council of Women World Leaders and author of the book *The Loudest Duck* to conduct a workshop on moving beyond diversity for the Division of Mathematical Sciences at NSF. She interviewed 19 women heads of state and heads of government, summarized in a book, *Women World Leaders Book*. I was so impressed with the workshop and the impact it had at NSF, I brought her recently to Oregon State to conduct a couple of workshops on our campus. The workshops were a great hit and will have a long-term, positive impact.

I have made mentoring a part of my theme while I was the president of the American Statistical Association in 2010, and also joined the MentorNet to help young scientists around the globe for several years. I had the pleasure to work with the ASA presidents Sally Morton and Nancy Geller, in my roles as the president-elect and the past president of ASA, respectively. I had the opportunity to appoint and work with a number of female members on various ASA committees, and work with female leaders like the IISA president Nandini Kannan, IMS President Ruth Williams, and the ICSA President Naisyin Wang. Our profession is fortunate to have many role models for our future female leaders in statistics.

I never pass up an opportunity to talk to students and junior faculty about the importance of finding mentors who are all around us, eager to help. I always let them know that it is not necessarily a single mentor who will help you climb that tree, it is a group of villagers. Seek them out!

Mentors are important, and may be the wind beneath our wings. The importance and value of mentoring relationships cannot be overestimated. Indeed, it takes a village. I will conclude with this quote from Sheryl Sandberg's book (Sandberg, 2013) *Lean In: Women, Work, and the Will to Lead* as a caution that mentoring is not a magic bullet for everyone and for everything.

I realized that searching for a mentor has become the professional equivalent of waiting for Prince Charming. We all grew up on the fairy tale "Sleeping Beauty," which instructs young women that if they just wait for their prince to arrive, they will be kissed and whisked away on a white horse to live happily ever after. Now young women are told that if they can just find the right mentor, they will be pushed up the ladder and whisked away to the corner office to live happily ever after. Once again, we are teaching women to be too dependent on others. We need to stop telling [women], "Get a mentor and you will excel." Instead, we need to tell them "Excel and you will get a mentor."

Acknowledgments

I want to thank Debbie Farris for excellent input and help writing this article.

Bibliography

Allen, R. (2007). Mentoring: A skill professional statisticians can develop. American Statistical Association's website. http://www.amstat.org/sections/sgovt/JEGAllen.htm.

Busteed, B. (2014). The Blown Opportunity. Inside Higher Education. https://www.insidehighered.com/views/2014/09/25/essay-about-importance-mentors-college-students.

Chamblis, D.F. and Takacs, C.G. (2014). *How College Works*. Harvard University Press, Boston, USA.

Lopez, M. (2014). On Mentoring First Generation and Graduate Students of Color. Committee on the Literatures of People of Color, MLA Commons. http://clpc.commons.mla.org/on-mentoring-first-generation-and-graduate-students-of-color/.

Luna, G. (2014). Mentor is a verb, not a noun. *Houston Business Journal*. http://www.bizjournals.com/houston/print-edition/2014/03/07/gina-luna-mentor-is-a-verb-not-a-noun.html?page=all.

Sandberg, S. (2013). *Lean In: Women, Work and the will to Lead*. Alfred A. Knopf, New York.

Thompson, T. (2014). Learning at NIC? It takes a village project. CDAPress.com. http://www.cdapress.com/news/local_news/article_64e5cf14-fd37-53f4-bc56-740e06dd5a72.html.

Wright, W. (2011). Mentor was not a verb. Max De Pree Center for Leadership website. http://depree.org/mentor-was-not-a-verb/.

26

"If You Would Consider a Woman ... "

Daniel L. Solomon

North Carolina State University

Preamble

You might say that I started working on this chapter on a Wednesday. It was July 1, 1981, and I found myself in the terrifying situation of being a rather young statistician who had agreed to take on the position founded some forty years earlier by Gertrude Cox. That was the day on which I became head of the large Department of Statistics at North Carolina State University. At my previous institution, I had been put in charge of a group of six biometricians (probably because I was the only one on speaking terms with each of the other five at the time). Perhaps it was that minor administrative experience, but somehow I had tricked the people at NC State into thinking that I could succeed in their outsized endeavor. More likely, they had tricked me to this view, and surely I would quickly be found out.

Many know the story of George Snedecor's letter dated September 7, 1940, suggesting candidates to lead the new program in North Carolina in which he "recommended five young men, half-heartedly adding, 'If you would consider a woman, I know of no one better qualified than Gertrude M. Cox'" (Greenberg et al., 1978). Less well known is what happened next. Her diary has the following verbatim entry for Saturday, September 14, 1940. The final sentence is especially telling.

> Proof reading most of morning. A telegram "Would you consider appointment head Stat. Lab. Here (sic) Write Experience Training Salary if interested." I wrote and mailed it off before I got cooled down. Not expecting anything.

"Miss Cox" would become the first female professor at (then) North Carolina State College and, on January 22, 1941, head of a new Department of Experimental Statistics.

What did Dr. Cox's department look like at age forty? It had grown large for the times, including 33 faculty members. One — an assistant professor — was a woman.

But perhaps my work on this chapter began even earlier. That notion surfaced when I found myself wondering how it happened that my life path was about to have me, a statistician, writing about diversity[1]. Reflecting on my early years has helped me understand, so perhaps it will give some context for what will follow.

I was born in New York City, the only child of working-class parents. My father came with his family from Russia as a sixteen-year-old and ended his schooling at grade 8. My mother was born in East Harlem, but both of her parents emigrated from Eastern Europe. She completed high school. They (especially she) recognized the opportunity afforded by an education and established in me an early expectation that I would get one.

Indeed there was a no-need-to-even-be-stated expectation that I would go to medical school. At some personal sacrifice, they made my going to college financially possible. As I look back from the perspective of my current circumstance, I marvel at what can happen in one generation, and I suspect that my eventual straying from medicine to higher education was in no small way related to my appreciation for its power.

While my family moved to Florida when I was nine, those formative years in the richly multiethnic neighborhoods of the Bronx were influential. Set against the experiences I would have in the then still deeply segregated South, those early years were important in shaping my values. There were segregated water fountains in my junior high school. It also happened that my high school physics teacher was female.

Diversity Leadership

The editors have asked for a chapter about "strategies that organizational representatives can use to help women overcome barriers to leadership." Much of what follows recounts strategies that I have found useful as an academic administrator. While the title of this volume is *Leadership and Women in Statistics*, there are commonalities with dimensions of diversity other than gender and with disciplines other than statistics. Furthermore, facilitating diversity in our workplaces and in our leadership ranks requires what one might term diversity leadership; that is, enhancing diversity in our leadership requires leadership in diversity. With these matters in mind, I will broaden and tease apart the pieces in a somewhat different way than the book title might suggest, addressing:

- What is diversity?

- Why is it important?

- How can leadership help achieve it?

---

[1] Actually I did briefly engage in diversity research, but it was in mathematical measures of species diversity in community ecology. Coincidentally, it depended heavily on the work of one of the editors of this volume.

In working toward a characterization of "diversity leadership," I do not mean to equate leadership with administrative position. While such positions do provide a pulpit, visibility and, in some cases access, authority or resources to facilitate change, there are many opportunities for what are sometimes called "inside out leadership" and "leadership in place" as agents of change. For example, I have a colleague who has become a high-impact, while informal, resource for many women on campus — serving as confidante, trusted advisor and role model. Some readings on these notions as they play out in an academic environment are Braskamp and Wergin (2008) and Wergin (2004), respectively. In addition, not all administrative positions are able to exercise the same influence in advancing diversity. When we assess diversity in the administrative ranks of our own institutions, it is informative to look carefully at "line" positions — those with significant authority over budgets and hiring.

I write from the perspective of an academic dean, moreover as dean of a particular college (science) at a particular American university at a particular time in American history. Although I do read, somewhat unsystematically, on the subjects of diversity and inclusion, and I have learned much from colleagues[2] — both formally and by example — I do not pretend to know the immense scholarly literature. So I am certainly not an expert, but rather what one might call a "practitioner." While that practice has been higher education, I conjecture that most of what I describe applies with appropriate modifications more broadly.

I will begin by describing what I mean by diversity in higher education and how I think about it from an academic perspective. Then I will try to make a case for it; i.e., address the "why is it important" item, and finally offer some examples of ways in which we have been working to enhance it in our institution — the "how" part.

What Is Diversity?

So what is human diversity? Certainly, the most common meaning is the demographic one ... the numerical mix of people within a group with respect to some characteristic. Gender, race, and ethnicity are probably the most commonly discussed, but other examples include age, religion, disability, sexual orientation, gender identity, national origin, urban/rural upbringing, and so on.

The everyday meaning of demographic diversity in higher education, and in the workplace, will vary by location and context. For example, in the Southwestern US, diversity conversations tend to focus on Hispanic and Native American issues. In the Southeast, it is more commonly about African-

---

[2]I am privileged to have had multiple master teachers for my gender education. I single out here for special thanks my colleagues and friends Jo-Ann Cohen, Justine Hollingshead, Ingram Olkin, Laura Severin, and Mary Wyer.

American, but increasingly Hispanic, issues. In physical sciences and engineering contexts, add gender.

So demographic differences are a dimension of diversity, and people of good will would agree that discrimination, harassment, or disrespect on the basis of one's membership in some demographic group is unconscionable (and, in some cases, illegal) and should not be tolerated.

However, that does not make a case for enhancing diversity in the university or any other workplace. Neither does a call for fairness. Even the best intentioned of us have to get beyond arguing for diversity because "it is the right thing to do." That's not enough. It really does not sell. It may even demean.

Changing the Conversation: What Are the Diversity Imperatives?

If there is any takeaway message in this chapter, it is that leaders must change the conversation from increasing diversity as a moral imperative to increasing diversity in the university as an economic and academic imperative — or, more broadly, to increasing diversity in the workplace as a business imperative.

One case for diversity in the university is national need. Respected economists such as Alan Greenspan "... have identified scientific and technological progress as the single most important determining factor in U.S. economic growth, accounting for as much as half of the Nation's long term growth over the [preceding] 50 years" (Solow, 1987; Greenspan, 2000). But there are well-publicized shortages of American talent in science and technology fields, what we now call the STEM disciplines: science, technology, engineering, and mathematics. Congressional commissions, academy, and other studies have cited the shortages. We also know about the poor performance of our children in science and mathematics and the low proportion who seek undergraduate degrees in these fields by international standards.

Furthermore, the American science and engineering workforce is aging. More than half of its members are over 40 years old (Age and Retirement, NSF Science and Engineering Indicators, 2012), and they are predominantly white (Racial/Ethnic Differences, NSF Science and Engineering Indicators, 2012) and male (Sex Differences in the S&E Workforce, NSF Science and Engineering Indicators, 2012). It is evident that women, blacks, Hispanics, Native Americans, and other groups are significantly underrepresented in STEM careers. Indeed, one congressional report concludes that "if women, underrepresented minorities, and persons with disabilities were represented in the US science, engineering, and technology (SET) workforce in parity with their percentages in the total workforce population, this shortage [of skilled American workers] could largely be ameliorated" (CAWMSET, 2000).

This economic imperative to include more diversity in our STEM workforce (and therefore in our universities that prepare that workforce) becomes even more urgent now that the demographic of this country has moved to one in which such underrepresented groups have achieved majority status. Including women, groups underrepresented in STEM disciplines comprise more than two-thirds of the total U.S. population (State and Country Quick Facts, US Census Bureau, 2014).

Finally, although we have outsourced manufacturing and certain services, there are technical areas, including defense and homeland security, where we cannot outsource our needs to foreign countries. If we do not incorporate currently underrepresented groups into our STEM workforce, we may risk our national security *as well as* our economy (Sega, 2004).

So, although this may make a national, economic case for enhancing the diversity of our universities, at least in STEM fields, it still does not make an *academic* case for it.

But the academic case can be made; it is compelling, it is research-based, and it applies to all fields, not just science and technology. It derives from the fact that a demographically diverse population is diverse in other ways — what you might call a cultural or experiential diversity.

Not only does a black student likely bring a different worldview to the classroom than a white student, but a white student who grew up on a farm in eastern North Carolina likely brings a different worldview than one who went to a prep school in Charlotte, and they both bring experiences different from that of a Hispanic male student who went to high school in a Latino neighborhood of New York, or a female student whose parents immigrated to the US from Baghdad or Beijing.

Why does a multiplicity of world-views provide an academic imperative for ensuring a diverse campus community? Our jobs at the university both as educators and researchers are to see the world in new ways and to help our students do the same.

Marcel Proust is often credited with the notion, "The real voyage of discovery consists not in seeking new landscapes but in having new eyes." A more literal translation of his words would read, "The only true voyage of discovery, the only fountain of Eternal Youth, would be not to visit strange lands but to possess other eyes, to behold the universe through the eyes of another, of a hundred others, to behold the hundred universes that each of them beholds, that each of them is" (Proust, 1929). That is what a diverse community helps us all do, as I will explain.

As faculty, we seek to develop critical thinking skills in our students. For example, in connection with the current reaffirmation of its accreditation, my own university has undertaken a quality enhancement plan that is called "critical and creative thinking." Formal models for critical thinking call for thinking from multiple points of view (intentionally putting yourself in someone else's shoes). Philosopher Richard Paul, who is a proponent of one such model, writes:

Whenever we reason, we must reason within some point of view or frame of reference. Any "defect" in that point of view or frame of reference is a possible source of problems in the reasoning. A point of view may be too narrow, too parochial, may be based on false or misleading analogies or metaphors, may contain contradictions, and so forth (Using Intellectual Standards to Assess Student Reasoning, Foundation for Critical Thinking, 2014).

So to think critically, we must intentionally think about the problem from multiple perspectives.

Mathematician William Byers uses binocular vision as an apt metaphor for ambiguity, but it seems to apply equally to critical thinking:

When you look at things out of one eye, the world seems flat and two-dimensional. However, when you use both eyes, the **inconsistent** viewpoints registered by each eye combine in the brain to produce a unified view that includes something entirely new: depth perception. In the same way, the **conflicting** points of view in an ambiguous situation may give birth to a new, higher-order understanding (Byers, 2014). (Emphasis mine)

Now here is the point: Research shows that development of this critical thinking skill requires serious engagement with others in a diverse campus community, and this provides a fundamental academic imperative for ensuring that we have such a community (Pascarella et al., 2001).

And the impact extends beyond critical thinking; for example there is a decades-long history of research that demonstrates that "racially diverse educational environments are associated with positive intellectual and social outcomes for college students" (Antonio et al., 2004).

As one example of such research, a report in *Psychological Science* (Antonio et al., 2004) titled "Effects of Racial Diversity on Complex Thinking in College Students," describes a randomized study of the effect of varying racial composition in small-group discussions on a cognitive measure called integrative complexity. Higher levels of integrative complexity had earlier been shown to be associated with higher grades. The authors also looked at integrative complexity as a function of the individuals' self-reported diversity of their friend groups. Among other things, the results show that racially diverse friend groups are indeed significantly and positively related to integrative complexity.

Another finding comes from research in stereotype inoculation that demonstrates that contact with successful female scientists, including faculty, benefits female students' self-perception, and motivation to pursue science careers (Stout et al., 2011).

Finally, work in group dynamics shows that heterogeneity of group members typically yields better problem solving than does homogeneity (Nemeth and Wachtler, 1983). This applies across workplace settings, and is particularly important in the *research* mission of the university, especially as it gets increasingly interdisciplinary.

How Can Leaders Enhance Diversity?

Since there are academic, economic, and broader imperatives for enhancing diversity in the workplace and, by implication, in its leadership ranks, how do we move such agendas forward? How do we lead change?

First, as emphasized in the previous section, we need to change the language — change the conversation from a moral one to an academic one or, more broadly, to a business one. That depends on all of us; administrators and "leaders in place."

Second — trite but true — there needs to be a clear and frequently articulated commitment to diversity and inclusion at the highest levels of the organization. But commitment is empty without investment. Examples can be seen in some of the investments being made at my institution; and there is a leadership story behind each one. No doubt most institutions have similar stories.

- The provost invested in a consolidated Office for Institutional Equity and Diversity with a rich set of programs to ensure equity, enhance diversity, and promote an inclusive culture. Note that the entire enterprise resides in the provost's office and that the provost is the chief academic officer. That is, it is characterized as an academic function.

- The provost's office took the lead in a National Science Foundation ADVANCE grant, an National Science Foundation (NSF)-wide program for "Increasing the Participation and Advancement of Women in Academic Science and Engineering Careers" (ADVANCE: Increasing the Participation, 2014). That office then supported institutionalization of some of the leadership programs, climate workshops and recruitment tools developed under the grant so that they continue although the grant has ended (Developing Diverse Departments (D3), 2014). Indeed, some of the material in this chapter was developed in connection with that grant and its continuing programs. Many of the participants in those programs have already moved into leadership positions, including the head of our statistics department.

- We have a Building Future Faculty program that brings in graduate students and postdocs from across the country who are nearing the start of academic job searching. Its explicit goal is to diversify the professoriate. We have an African-American Cultural Center, a Women's Center, a GLBT Center, and study-abroad programs that educate and help give us those "new eyes."

- We regularly undertake campus wide climate studies of faculty, staff, and student constituencies to see how we are doing.

Some of these programs reside in the university's diversity offices, but we also need leadership from others of us who are not primarily in the "diversity business." We cannot leave the championing and the change agency solely to them. So I will offer some examples of the things that have been initiated in our college, some of which have had impacts beyond the college.

Until recently, my college included only the physical and mathematical sciences, and there is no denying that enhancing gender and racial diversity is an especially daunting challenge in many of those disciplines. For example, in the US in 2012, only 127 doctorates were awarded to black students across the physical and mathematical sciences (excluding computer and information sciences), which is about 3.2% of the doctorates to US students and permanent residents in these fields (U.S. Citizen and Permanent Resident Doctorate Recipients, by Race, Ethnicity, and Major Field of Study, 2012). Only a fraction of those black PhDs go into academia, and, of course, such a shortage of role models is self-perpetuating. At the individual department and discipline level the numbers can be minuscule.

For women, the PhD production is also low in the physical and mathematical sciences but at or above 50% in most of the biological sciences. Within the biological sciences, the percentages of women are lower in the more quantitative fields; e.g., bioinformatics (28.1%), biophysics (37.1%), and computational biology (24.5%), with the notable exception of biometrics and biostatistics (52.3%) (Doctorate Recipients, by Sex and Subfield of Study, 2012).

The story is actually more complicated. While only about 19% of physics PhDs go to women, it is 37% in chemistry — but they are not coming to higher education at that rate. No doubt there are work–life issues in academe versus industry at play here, notably the coincidence of the traditional tenure clock and the family building biological clock. But more about that later.

So while the numbers are daunting, what we have tried to do is move the conversation from hand wringing about how hard it is to diversify the faculty because the pools are so small to "how can we do it?" For example, early meetings of search committees for faculty and administrative positions are expected to have a presentation from a member of the institutional equity and diversity office to learn about the compliance rules for conducting searches, including the cautions about permissible questions and equitable treatment of applicants. However, beyond exhortations to diversify the candidate pool, there is little practical advice offered on how to do it.

So for many years, I have joined those early search committee meetings to offer a list of specific, actionable ideas for enriching the search pools by identifying and contacting women, African-Americans, and other prospects from populations underrepresented in our fields and encouraging them to apply. To demonstrate that such information is available, I actually provide names of some prospective candidates. There is a variety of discipline-specific websites, membership lists, and subscription minority database products on which to draw. Some lists of such sources are available at ADVANCE program websites (Faculty Search Toolbox, 2014). For example, in statistics, I have used the Southern Regional Education Board Doctoral Scholars Program Directory, the Ford Foundation Fellowship Program directory, and the membership list of the Caucus for Women in Statistics. For more senior hires, I point the committee to journal editorial staff lists and the American Statistical Association (ASA) fellows directory. Of course, we also insist on advertising in targeted venues that are likely to be seen by members of underrepresented groups. Such advertising placements also announce your institution's commitment to diversity.

In more recent years, I added an introduction on unconscious or implicit bias in evaluating applications and how to interrupt it. There is much helpful and powerful material available, including handouts, videos and PowerPoint presentations. I find especially useful the film and facilitation guide at the University of Washington ADVANCE website (INTERRUPTING BIAS, 2014). There are also resources developed in connection with other ADVANCE grants (Inclusive Excellence, 2014). My own presentation brings its examples from the research literature, which makes it more compelling for its academic audience. These examples provide undeniable evidence that we all suffer from implicit bias, and that awareness is a first step to interrupting that bias.

One such example is a study of real CVs in which some evaluators see a male name and others a female name for the same CV. Both male and female evaluators are more likely to recommend hiring the male-named candidate than the female (Steinpreis, Anders, Ritzke, 1999). Another dimension of bias is seen in the letters of recommendation for candidates. One study analyzed "all the letters of recommendation for successful applicants for faculty positions at a large American medical school over a three-year period from 1992 to 1995" (Trix and Penska, 2003). Letters for women and men differed systematically in several ways; e.g., there were more very long letters for men and more very short letters for women. Letters for women were more likely to include "doubt raisers" as well as stereotypic adjectives, notably "grindstone" adjectives such as "hardworking." In this gender schema, the woman's success is attributable to effort rather than ability (Valian, 1998). On the other hand, the percentage of letters including standout adjectives such as "excellent" was similar for female and male candidates, although the average number of such terms in letters with at least one was higher for males.

There is also evidence that we evaluate female candidates more objectively — and we are disproportionately more likely to hire them — when there is more than one woman in the interview pool. When there is only one woman on the shortlist, she is less likely to succeed than women who are compared to a diverse pool of candidates. There are similar findings for underrepresented minority candidates.

We also commend to search committee members an online Implicit Association Test (https://implicit.harvard.edu/implicit/takeatest.html) offered by Project Implicit (https://implicit.harvard.edu/implicit/) as a way to assess their own association between male and female with science and liberal arts.

Again, making search committee members aware of the literature on implicit bias as well as their own susceptibility makes them better able to interrupt it.

Such search committee orientations are getting more widely adopted across the university. For example, they are now required for search committees associated with a large, interdisciplinary cluster hiring process. If it proves successful, we will press the case to establish these as expected practices in all searches.

Back to our college, we have tried to create incentives for departments to diversify themselves, notably their faculties. Examples include cases in which we figured out how to make multiple hires in a search that set out to hire one but produced a high quality and diverse pool. I sometimes refer to this as letting the department heads know what my "buttons" are. Good heads quickly learn how to push them and everyone wins. The university has used this practice in the cluster hires recently.

There have been occasions in which we have had opportunities to add proven or promising scholar/educators to our faculty and also serve the academic imperative of diversity. In one such case in statistics, we hired one of our own PhD graduates, but arranged for her to spend her first year on leave at a top department in another university. When she returned, it was as colleague, no longer as student. In a second example, we identified a seasoned statistician from an underrepresented group whom we very much wanted to attract but guessed that it would be a hard sell. I was already dean at the time (having spiraled downwards from department head to associate dean to dean), and the department head and I flew up to DC to spend an afternoon with him at Reagan airport making the case. It worked. I must credit a (female) dean of engineering at a neighboring institution with the idea to travel to a top candidate's location as a demonstration of our enthusiasm.

Another example is in physics — the science discipline with the least gender diversity. The strategy here was to hire a proven researcher. The department was able to attract a highly published and well-funded candidate in an area of substantial interest who came as a tenured full professor and quickly accelerated her program here. The notion of having to trade quality for diversity was quickly dealt a lethal blow — and the conversation around the decision table in the department changed.

An important part of the strategy is to keep diversity on the table continuously — and not as a perfunctory afterthought. If you look at the table of contents of our recent college strategic plan, you will find a chapter on "Building a Diverse and Welcoming Community." You'll find it near the front, before the research and education sections.

Rarely will a monthly department heads meeting go by without some diversity matter surfacing or even being the focus of the meeting. We have had best practices sessions in which each department reported on its practices in recruitment and retention of diverse faculty and students; we have had our college and department administrators meet with our black alumni to learn about — and from — their experiences here. We have also had presentations on the campus environment for gays and lesbians in which I learned about issues of sexual orientation and gender identity sufficiently to get me involved in an effort that eventually led to the establishment of a GLBT center.

This brings me from the topic of diversity to that of inclusion.

What about Inclusion?

Much of the preceding material has focused on *recruitment* to build a diverse workplace and "training" to prepare women and underrepresented minorities for leadership. However, facilitating diversity in the leadership ranks also requires attention to the creation of an environment of mutual respect and individual success that enhances *retention* and invites the kind of open interactions implicit in the diverse learning environments and group dynamics settings I have mentioned. As shorthand, I am saying that we must also pay attention to "inclusion," and it must be intentional. A recent overview of inclusion in higher education is offered in an article titled "Are We There Yet?" (Rules of Engagement: Defining the Terms, Creating the Future, 2014) in a joint special supplement to the *Chronicle of Higher Education.*

There are some really small things that can be done that both make a real difference for those affected and raise awareness in others.

For example, when I came into the deanship, there was a long history of the monthly department heads meetings starting at 8 a.m. Despite some moaning, we pushed that back to 9 a.m. to allow participation by people with morning child-care responsibility. This meeting includes department heads, faculty and staff senators, and others in leadership positions. The ability for people with child-care responsibilities to take on those kinds of leadership roles could be compromised by an early start.

As another example, in the face of slow movement at the university level to confront child care for employees and students, we empanelled a college-level task force to collect data to document the need and to make recommendations on such work–life issues. This prompted the eventual establishment of the university childcare center.

Some other examples of work–life and other inclusion issues in which we have been involved are making the case for lactation rooms on campus; for example we were able to get one into the design of SAS Hall, a new building that would house our statistics and mathematics departments. That was 2009. Today NC State has 16 across campus. In addition, despite restrictive state rules about numbers of bathrooms, we were able to include a single-user, lockable, gender-neutral bathroom to serve the transgendered community. If you have not seen the documentary "Toilet Training," Sylvia Rivera Law Project, 2014, it is eye opening. This seemed to break a logjam, and many more such facilities have followed (Gender Neutral Bathrooms, 2014).

This brings me back to the GLBT Center I mentioned earlier. At the opening of the center, I was offered the opportunity to make some remarks.

I share some of them here in the spirit of addressing the editors' charge to suggest "strategies that organizational representatives can use...."

As a faculty member and administrator, I appealed especially to other faculty, staff, and administrators to learn about the issues facing the GLBT community. I urged them to begin by committing three hours to participate in an available training program and then to encourage others to do the same. I urged them to help others learn by bringing speakers to faculty and staff meetings, so that they can learn the facts surrounding suicide rates among gays and lesbians and about the size of the largely hidden GLBT population among students, faculty, and staff. They should learn about the trepidation that a transgendered student faces in approaching a campus restroom door.

They should learn about the closeted gay faculty member who yearns to come out so that he can offer the support of a successful role model to his gay and questioning students, but who fears for the impact it would have on his relationships with his colleagues ... and perhaps even his professional future. To the burden of living two lives and the nagging daily fear of discovery, add the guilt of the missed opportunity to support his students.

They should learn about the differences in benefits for spouses versus same sex partners. And they should learn that there are faculty and staff among their colleagues who may themselves be straight, but who are grappling with how to help their children who are questioning their sexuality or gender identity. (Do you know about PFLAG? (Parents, Families, Friends, and Allies United with GLBT People to Move Equality Forward, 2014))

What about Tomorrow?

Before women can move into leadership positions in statistics and other sciences, they have to join these careers. So we have tried to get beyond bemoaning demographic unavailability and act to help fill the science and mathematics pipeline at all levels.

Our K–12 outreach center, The Science House, offers a suite of programs for students and teachers, some of which are explicitly aimed at diversifying the pipeline. For example, they have hosted many Spring break programs in which 400–500 7th or 8th grade girls and their science teachers spend a day doing hands on science with female scientists from across the Research Triangle area. There is also a weeklong program for about 50 African-American and Latino 3rd through 6th graders in the community who do hands-on science led by NC State students and other scientists of color as role models. That program started in 1998.

Our academic associate dean and diversity leader–exemplar, mathematician Jo-Ann Cohen, together with a counterpart in Engineering, established in 2003 the Women in Science and Engineering Living and Learning Village for NC State undergraduates and has now exported the model to high school chapters. And we have a range of activities for mentoring multicultural undergraduate and graduate students to ensure their retention and graduation.

And the departments and university have similarly motivated programs to mentor or attract diverse graduate students, postdoctoral fellows and faculty such as the Building Future Faculty program described earlier.

Has Any of This Worked?

We have achieved a few successes in both diversity and inclusion. Here are some examples. In statistics, 44% of our undergraduates and 52% of graduate students are female. The latter is high for a statistics (vs. biostatistics) department. NC State has awarded 15 PhDs in mathematics in the last ten years to black students, a significant percentage of the national output. (Not including statistics and operations research, in 2012 there were 13 mathematics PhDs awarded to black students in the US (Doctorate Recipients, by Citizenship, Race, Ethnicity, and Subfield of Study: 2012, National Science Foundation, 2014).) And during the 2012–13 and 2013–14 fiscal years, of 19 new tenured or tenure-track faculty who joined our college, 9 are female.

In support of inclusion, we have added a child-care center, a GLBT Center, more family-friendly parental leave and tenure-clock policies, and an active spouse and partner hiring program.

To update some of the stories told earlier in this chapter, I would note that our "grow your own" faculty member who went on "sabbatical" her first year before starting her career on our faculty is now a high-impact full professor. And remember, the physics department that sought to change the conversation by hiring a female senior researcher? That was 1998. As this is written, no research active physics department in the US has more tenured or tenure-track women on the faculty. (And that senior researcher is now associate dean for research for the entire College of Sciences.)

But if there is a depressing side to this, it is that even these successes are achieved with very small numbers of diverse faculty or students. For example, that leadership position in female physicists is accomplished with only 8 women in a faculty of 33.

Looking Ahead

So I come to the end of this chapter, begun on a Wednesday some 33 years ago. As it approaches its 75th year, Dr. Cox's department has about the same number of tenure-track faculty, but 10 are women. We also followed Snedecor's advice to "consider a woman" and found "no one better qualified." The department has its first female head since Gertrude.

Which reminds me, enhancing diversity is a voyage not a destination:

> Ensure that your successors are increasingly more effective than you were. I was followed as department head by Tom Gerig, Sastry Pantula and now Montse Fuentes, each of whom ratcheted up the successes of the head they followed.

And heed the widely known NC State mantra, "Don't give up. Don't ever give up" (Valvano, 1993).

Bibliography

"Age and Retirement," NSF Science and Engineering Indicators January (2012), accessed July 11, 2014, http://www.nsf.gov/statistics/seind12/c3/c3s4.htm#s6.

Antonio et al., (2004). "Effects of Racial Diversity on Complex Thinking in College Students," *Psychological Science* 15: 507–510.

Braskamp, Larry A. and Wergin, Jon F. "Inside-Out Leadership," *Liberal Education* Winter (2008): 30–35.

Byers, William. (2014), "Ambiguity and Paradox in Mathematics," *The Chronicle Review* 53 (2007): B12, accessed July 11, http://chronicle.com/article/ AmbiguityParadox-in/17982.

Charlan, J.N. and Joel, W. (1983), Creative Problem Solving as a Result of Majority vs Minority Influence. *European Journal of Social Psychology*, 13: 45–55. doi: 10.1002/ejsp.2420130103.

E.g., see "Faculty Search Toolbox: Resources for Recruiting Diverse Faculty," North Carolina State University, accessed July 11, 2014, http://oied.ncsu.edu/advance/resources/faculty-search-toolbox-resources-for-recruiting-diverse-faculty/. Also see Appendix 2 at "Handbook for Faculty Searches and Hiring, Academic Year 2009-10," University of Michigan, accessed July 11, 2014, http://www.advance.rackham.umich.edu/HandbookFacultySearchesHiring_091609EP.pdf.

Examples cited in Antonio et al., (2004). "Effects of Racial Diversity on Complex Thinking in College Students," *Psychological Science* 15: 507–510.

"Gender Neutral Bathrooms," Gay, Lesbian, Bisexual, and Transgender Center at North Carolina State University, accessed July 11, 2014, http://glbt.ncsu.edu/resources-services/gender-neutral-bathrooms/.

Greenberg, Bernard G. et al., (1978). "Statistical Training and Research: The University of North Carolina System." *International Statistical Review* 46: 171–207.

"Inclusive Excellence — ADVANCE Grant," The University of Arizona, accessed July 11, 2014, http://diversity.arizona.edu/advance-grant. Also "WISELI: Promoting Participation and Advancement of Women in Science and Engineering," University of Wisconsin-Madison, accessed July 11, 2014, http://wiseli.engr.wisc.edu/.

"INTERRUPTING BIAS in the Faculty Search Process," University of Washington, accessed July 15, 2014, http://www.engr.washington.edu/lead/biasfilm/.

Land of Plenty: Diversity as America's Competitive Edge in Science, Engineering and Technology. Congressional Commission on the Advancement of Women and Minorities in Science, Engineering and Technology Development (CAWMSET), p. 1 [September 2000].

National Science Foundation. "ADVANCE: Increasing the Participation and Advancement of Women in Academic Science and Engineering Careers," accessed July 11, 2014, http://www.nsf.gov/funding/pgm_summ.jsp?pims_id=5383.

National Science Foundation. "Doctorate Recipients, by Sex and Subfield of Study: 2012," accessed July 11, 2014, http://www.nsf.gov/statistics/sed/2012/pdf/tab16.pdf.

National Science Foundation. "Doctorate Recipients, by Citizenship, Race, Ethnicity, and Subfield of Study: 2012," accessed July 19, 2014, http://www.nsf.gov/statistics/sed/2012/pdf/tab22.pdf.

North Carolina State University. "Developing Diverse Departments (D3) at NC State," accessed July 11, 2014, http://oied.ncsu.edu/advance/.

PFLAG. "Parents, Families, Friends, and Allies United with LGBT People to Move Equality Forward," accessed July 11, 2014, www.pflag.org.

Pascarella, E., Martin, G., Hanson, J., Trolian, T., and Gillig, B. (2001). "For Example, Do Diversity Experiences Influence the Development of Critical Thinking?" *Journal of College Student Development* 42: 257–271.

Project Implicit, https://implicit.harvard.edu/implicit/. Accessed July 11, 2014.

Proust, Marcel. "The Verdurins Quarrel with M. De Charlus" in *Remembrance of Things Past*, trans. C. K. Scott Moncrieff, vol. 5, chapter 2 (London: Knopf, 1929). (as cited in http://www.insidehighered.com/views/2011/04/08/robbins#ixzz33VcpHhMh)

"Racial/Ethnic Differences in the S&E workforce," NSF Science and Engineering Indicators January (2012), accessed July 11, 2014, http://www.nsf.gov/statistics/seind12/c3/c3s4.htm#s2.

"Rules of Engagement: Defining the Terms, Creating the Future," *The Chronicle of Higher Education/Diverse: Issues in Higher Education* (2012), accessed July 11, 2014, http://www.saabnational.org/docs/convergence2012.SAAB_Feature.pdf. Also available at Angela Dodson, "Are We There Yet? A Dialogue on Inclusion, *The Chronicle of Higher Education/Diverse: Issues in Higher Education* (2012): 44–49, http://ofd.ncsu.edu/wp-content/uploads/2012/12/Are-We-There-Yet-A-Dialogue-on-Inclusion.pdf.

Sega, Ronald. Department of Defense (DOD) spokesperson, in a speech before the Congressional Black Caucus, referenced in Shirley Malcolm, et al. *Standing Our Ground: A Guidebook for STEM Educators in the Post-Michigan Era*, (Washington, DC: AAAS and National Action Council for Minorities in Engineering, 2004), 10.

"Sex Differences in the S&E Workforce," NSF Science and Engineering Indicators January (2012), accessed July 11, 2014, http://www.nsf.gov/statistics/seind12/c3/c3s4.htm#s1.

Solow, Robert (1987), *Growth Theory: An Exposition* (New York: Oxford University Press) and Alan Greenspan, "Technology and the economy," (Remarks to the Economic Club of New York, New York City, NY, Jan. 13, 2000), both cited in: Shirley Malcolm, et al., *Standing Our Ground: A Guidebook for STEM Educators in the Post-Michigan Era*, (Washington DC: AAAS and National Action Council for Minorities in Engineering, 2004), 10.

"State and Country Quick Facts," US Census Bureau, accessed July 11, 2014, last modified July 8, 2014, http://quickfacts.census.gov/qfd/states/00000.html.

Steinpreis, Rhea E., Anders, Katie A., and Ritzke, Dawn (1999). "The Impact of Gender on the Review of the Curricula Vitae of Job Applicants and Tenure Candidates: A National Empirical Study," *Sex Roles* 41: 509–528, accessed July 11, 2014 via http://advance.cornell.edu/documents/ImpactofGender.pdf.

Stout, Jane G. et al., (2011). "STEMing the Tide: Using Ingroup Experts to Inoculate Women's Self-Concept in Science, Technology, Engineering, and Mathematics (STEM)," *Journal of Personality and Social Psychology* 100: 255–270, accessed July 11, 2014, http://psycnet.apa.org/journals/psp/100/2/255/.

"Toilet Training," Sylvia Rivera Law Project, accessed July 11, 2014, http://srlp.org/resources/toilettraining/.

Trix, Frances and Psenka, Carolyn (2003). "Exploring the Color of Glass: Letters of Recommendation for Female and Male Medical Faculty," *Discourse & Society* 14: 191–220, accessed July 11, 2014 via http://advance.uci.edu/images/Color of Glass.pdf.

"U.S. Citizen and Permanent Resident Doctorate Recipients, by Race, Ethnicity, and Major Field of Study: 2012," National Science Foundation, accessed July 11, 2014, http://www.nsf.gov/statistics/sed/2012/pdf/tab24.pdf.

"Using Intellectual Standards to Assess Student Reasoning," Foundation for Critical Thinking, accessed July 11, 2014, *http://www.criticalthinking.org/pages/using-intellectual-standards-to-assess-student-reasoning/469*.

Valian, Virginia. *Why so Slow? The Advancement of Women* (Cambridge: MIT Press, 1998), 170.

Valvano, Jim (1993). ESPY Speech (New York City, NY, March 4, 1993), http://www.youtube.com/watch?v=HuoVM9nm42E.

Wergin, Jon F. (2004). "Leadership in Place." *The Department Chair* 14: 1–3.

27

Fostering the Advancement of Women in Academic Statistics

Judith D. Singer

Harvard University

It is 1 p.m. Tuesday April 4, 2008. (I know this detail because I took notes; an important habit to develop for anyone aspiring to a leadership role.) I am in my faculty office, revising slides for my 2 p.m. class, a second-level regression course at the Harvard Graduate School of Education (HGSE), where I have taught since 1984. The phone rings, a woman identifies herself as the executive assistant to Harvard's provost and asks, "Do you have a moment to speak with the provost?" Although I knew the provost reasonably well from an earlier stint as academic dean at HGSE from 1998 to 2004 (including one year as acting dean), I was surprised to get a personal call, since I was no longer in an administrative role.

If the call itself was a small surprise, the substance of it was a bigger one — would I be interested in being considered a candidate for the role of senior vice provost for Faculty Development and Diversity (FD&D), a post created at Harvard in 2005 in the wake of then President Larry Summers' now infamous comments on women in science. I was not being offered the position; I was being asked if I would be interested in being considered. (I have learned over the years that not everyone realizes that an invitation to apply is not the same as being offered the job. And I have also learned the importance of outreach - active recruitment is key to diversifying the faculty and leadership). To say I was intrigued was an understatement. Two weeks later, I agreed to be considered, and after an extensive search process, I was offered and accepted the job.

I am now in my seventh year as senior vice provost at Harvard, a role I do not think I would have imagined possible during graduate school or my early years as a faculty member (to be fair, positions like this did not exist back then). In this role, I oversee and guide Harvard's policies and practices in all areas of faculty affairs — including searches, promotions, retentions, and quality of life — with the three goals of insuring excellence, increasing accountability, and fostering measurable progress in important domains. I finished my PhD in statistics at Harvard in 1983, advised by Fred Mosteller, one of the 20th century's greatest statisticians, and an institutional citizen par excellence. In my first year of graduate school, Fred gave me advice that he lived by (and I have followed over the years): If someone asks you to do something, either say "yes" (Fred still holds the record for chairing more departments at Harvard than any other faculty member — four) or suggest who else could do the job. I did not know it then, but that advice was my first lesson in academic leadership.

In this chapter, I describe six concrete strategies that institutions and individuals — both women and men — can use to support the careers of female statisticians in academia. I begin by setting the context with some national statistics on women in statistics (some uplifting; some depressing). As a statistician in a leadership role, I will emphasize the important role that statistical thinking can play in identifying issues, galvanizing attention, and implementing concrete strategies for improvement. Although institutional contexts certainly affect what can or will work in different environments, I believe that these lessons have direct implications across academia as well as other knowledge based organizations.

Some Statistics on Women in Statistics

Statistics departments in faculties of arts and sciences have always been male-dominated, and while their gender composition has changed substantially in the last fifty years, parity remains very far away.

Figure 27.1 presents three sets of statistics on women in statistics — some suggest progress; others less so. The leftmost graph presents data from Monti (2014), who analyzed the gender breakdown of new members of the American Statistical Association (ASA). Before 1975, women accounted for only 16% of new ASA members. With each passing decade, women account for a larger percentage of new members, doubling from 20% in the period between 1975 and 1984 (I joined in 1976, the year I started graduate school) to 40% in the period between 2005 and 2014.

The middle graph presents data from the National Science Foundation's Survey of Earned Doctorates (2012) on the percentage of new PhD recipients in statistics who are female. In 1982, only 18% of the recipients were female; ten years later, in 1992, that percentage had barely shifted to just 21%. More recent data are both more promising (the percentages are much higher) and more depressing (they are stagnant). In 2012, the most recent year for which complete data are available, 37% of the PhD recipients in statistics were female, a figure unchanged from 2002.

The rightmost graph presents data from the American Mathematical Society (AMS, 2014), which tracks the percentage of women faculty in University departments that offer a PhD in statistics. AMS estimates that in 2012, 19% of the tenured faculty, and 37% of the tenure-track faculty, in these departments are women. Focusing on those newly hired into these

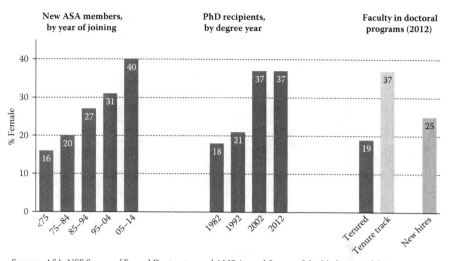

Sources: ASA, NSF Survey of Earned Doctorates, and AMS Annual Survey of the Mathematical Sciences

FIGURE 27.1
Some statistics on women in statistics. **See color insert.**

departments in 2012, 25% are female (reflecting a mix of tenure-track and tenured faculty).

As statisticians, we know that when the denominator is relatively small — a common occurrence given the size of many statistics departments — percentages are driven all too much by that small denominator instead of the size of the numerator, the statistic we really care about. We also recognize that percentages aggregated across many departments may mask interesting department-to-department variation. To address these two concerns, Figure 27.2 presents the composition — by gender and rank — of nine of the "top ten" statistics departments as designated by the 2010 National Research Council study of US PhD programs (the website for the tenth department, University of Wisconsin, had no faculty photographs, making it difficult to accurately assess the gender composition). I have sorted the departments by the number of tenured women faculty — Berkeley, Penn State, and University of Washington each have four, Michigan has three, the University of North Carolina at Chapel Hill, University of Chicago, Stanford, and Cornell each have only one. My alma mater and employer, Harvard, has none. Across all 9 departments, only 19 women have tenure and only 10 are on a tenure track. With numbers this small across 9 of the top US departments, the percentages — also shown in Figure 27.2 — barely matter.

These grim statistics suggest that despite increasing numbers of women earning PhDs in statistics, the representation of women on the faculties of top arts and sciences departments remains depressingly poor. Of course, statistics is hardly the only discipline in which women are underrepresented. The same

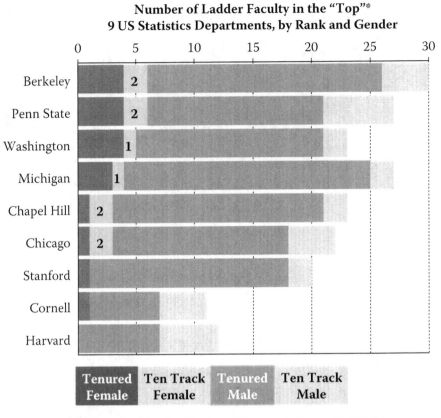

FIGURE 27.2
Number of ladder faculty in the top 9 US statistics departments, as ranked by
the National Research Council (2010), by rank and gender. **See color insert.**

can be said of other STEM (science, technology, engineering, and mathematics) disciplines: mathematics, physics, chemistry, astronomy, and engineering.

I am now in my seventh year as senior vice provost for Faculty Development and Diversity at Harvard. I am proud to say that, each year, the gender diversity of my university improves slightly. Taking a longer view, over the past 10 years, the number of tenured women has grown by 60%, from 164 to 263. At the same time, the number of tenured men has grown by only 13% (from 743 to 843). As a result of these differential growth rates, the percentage of women among the tenured faculty has increased 32% (from 18% to 24%). These increases are especially remarkable, given that the demographics of the tenured faculty at any university change relatively slowly, because most tenured faculty stay on the faculty for decades. An even more optimistic statistic comes from the entering cohort of new tenure track faculty at Harvard in 2014–15: It is 49% female.

With these data as backdrop, I now turn to six lessons I have learned in my current role as I work to foster the advancement of women in academia. To preview the highlights, here I present the lessons in brief, expanding upon them in successive sections:

1. Leadership and alliances matter (and can change much more quickly than the faculty)

2. Continually improve appointments policies and practices (do not accept "we have always done it this way...")

3. Share research relevant to the advancement of women (there's a huge literature, especially showing we're not always right!)

4. Concretely support work–life programs (rhetoric is great, but the institution has to act)

5. Assess and hold the institution accountable (anecdotes are nice, but data speak truth to power)

6. Help junior faculty take charge of their careers (institutional actions are necessary, yet not sufficient)

Leadership and Alliances Matter (and Can Change Much More Quickly Than the Faculty)

This volume is designed to highlight the importance of increasing the number and visibility of female statisticians in leadership positions. I would broaden this imperative to include the importance of increasing the number of women who are visible leaders across the spectrum — in statistics, yes, but in other domains of an organization as well. In my case, it certainly helps to have Drew Gilpin Faust, the first female president of Harvard, at the helm. So, too, it helps that Harvard has other visible female leaders — several of our deans are now female, including the dean of our School of Engineering and Applied Sciences. These female deans — and, of course, our female president — send a clear signal to the entire University community that women can and should be leaders, and that these women have real authority. And because leaders turn over more frequently than tenured faculty, it's easier to change the demographics of an organization's leadership more rapidly (not necessarily more easily!) than the tenured faculty as a whole.

In addition to the message that strong visible leadership sends, experience teaches me the importance of building strong alliances on the faculty — galvanizing stakeholders, both women and men, who believe strongly in the need to advance women. In a university setting, faculty champions and advocacy committees are vital. As statisticians we know: There is strength in numbers. Identify men in your organization who care about gender diversity and insure that they play a role in the discussion. Gender diversity is not just a "women's issue." An organization that seeks and promotes women will have an unfair advantage over one that overlooks and stymies talent.

If your organization has a committee or special interest group dedicated to supporting women, join it. If no committee or group exists, enlist colleagues and collectively ask your organization's leadership team to create one. At the fall 2014 meeting of the Society for Research on Educational Effectiveness (an organization on whose board I serve), a group of female methodologists approached the organization's executive director asking for time on the conference program when women in quantitative methodology could gather "to discuss career support for women conducting research on quantitative methods in education." They adopted the name WomenQuants. The response from the membership was so overwhelming that the relatively large room booked for the gathering was too small to accommodate all who wanted to attend.

Continually Improve Appointments Policies and Practices (Do Not Accept "We Have Always Done It This Way...")

As the adage goes, "If you keep on doing what you've always done, you'll keep on getting what you've always got." Most organizations are slow to change, and a common response to any proposed change is "but we've always done it this way." I suppose it's no surprise that Ivy League institutions like Harvard are resistant to change, but conversations with colleagues around the world tell me that we're not unique: Most institutions are slow to change.

As statisticians, we embrace W. Edwards Deming's philosophy of continuous improvement; I believe this mantra applies equally to academic settings: Times change, and we must change with them. I suggest that all institutions would be wise to examine their appointments policies and practices, and ask whether they should be revised to reflect the new realities of the academic marketplace, where the talent sought by your institution has many other excellent options.

I find it helpful to distinguish between two types of change: "wholesale" changes that come when we institute new policies and "retail" changes that come when we institute new practices. Policy changes can be more efficient, but in a university setting, I find that it's equally important to change practices — for example, how faculty search committee meetings are conducted and what happens when job candidates come to visit campus.

When thinking about improving the advancement of women, it becomes particularly important to examine the policies and practices that colleges and universities use to hire, evaluate, and promote faculty. At Harvard, we have recently developed new materials that we originally planned to share only with faculty serving on search committees. In the end, we decided to share it with the entire Harvard faculty. Our guide has many practical recommendations that may apply to your organization. Here are a few suggestions:

1. Position descriptions should be as broad as possible; because of pipeline issues, narrowly defined searches can inadvertently exclude highly qualified women who may not work in specific subfields.

2. Think carefully about which qualifications are "required" and which are "preferred." Kay and Shipman (2014) review the literature on confidence, and suggest that women are likely to apply only when they meet all the required qualifications whereas men are likely to apply even if they meet only half.

3. Avoid characterizing any search as a "replacement" for a departed or retired colleague (invariably a man). Searches are an opportunity to look forward, not backwards.

4. Examine the diversity of your peer institutions. If you're surpassing them, great. If you're not doing as well, start a discussion about why this might be so.

5. As you embark on a new search, audit the results of your last several searches. Did you pass over individuals who went on to great success? Discuss how this might have happened and think about how this could be avoided in this new search.

6. Actively recruit candidates to apply. There may never really have been a time when "post and pray" worked, but it certainly does not work now.

7. Beware of committee members who dominate meetings. This is not just the chair's responsibility; it's the responsibility of the entire committee.

Space prevents me from presenting the complete set of recommendations. I encourage interested readers to download our search guide: Best Practices for Conducting Faculty Searches, available for download at http://www.faculty.harvard.edu/appointment-policies-and-practices/resources-conducting-faculty-search. Many other universities have online guides well worth examining; I particularly recommend the University of Wisconsin Search Handbook, available for download from wiseli.engr.wisc.edu.

Share Research Relevant to the Advancement of Women (There is a Huge Literature, Especially Showing We Are Not Always Right!)

Recent years have seen an explosion in the quantity and quality of research relevant to the advancement of women. Social scientists from a variety of disciplines — particularly economics, sociology, and psychology — have conducted hundreds of laboratory and field studies on the effects of gender. Because the literature is far too vast to survey in this brief chapter, here are a few studies I find particularly compelling:

1. Goldin and Rouse (2000) found that when symphony orchestras moved from "open" to "blind" auditions — that is, having a musician's identity concealed by a screen — women were significantly more likely to be both hired and promoted.

2. Moss-Racusin, Dovidio, Brescoll, Graham, and Handelsman (2012) found that when faculty were sent resumes for a lab manager position — randomly varying the gender of the applicant — male and female participants were significantly more likely to view the male applicant (with identical credentials) as more competent and offer him a higher starting salary.

3. In a randomized trial, Blau, Currie, Croson, and Ginther (2010) found that female economists who were mentored — as part of an NSF-funded initiative under the auspices of the Committee on the Status of Women in the Economics Profession, a standing committee of the American Economic Association — had significantly higher publication rates and greater success rates in grant competitions.

4. In a randomized trial at the US Air Force Academy, Carrell, Page, and West (2010) found that professor gender had such a strong effect on the performance of female students — especially high achieving students — that they argue the gender gap in student achievement at the academy could be eliminated entirely simply by assigning women students to female faculty.

5. Trix and Psenka (2003) analyzed the language found in more than 300 letters of recommendation for faculty at a large US medical school and found significant differences — favoring men, in both length and tone — in the descriptions of job applicants.

6. Banaji and Greenwald (2013) synthesize the literature on implicit bias, providing convincing evidence that most of us have unconscious assumptions that can lead us to act in ways that are counter to our expressed beliefs. Further information about their work can be found at Project Implicit www.implicit.harvard.edu.

The accumulated evidence documents that women do experience discrimination in hiring processes, that mentoring can improve the academic outcomes for women, that at least some of the gender differentials that exist in academic performance can be ameliorated were there more women faculty, and that implicit and unconscious assumptions, regardless of gender or race, can affect our decisions. These conclusions are not anecdotes. They are based on serious, well-designed quantitative research.

Concretely Support Work–Life Programs (Rhetoric Is Great, but the Institution Has to Act)

Historically, female faculty members have been less likely to have children, but in recent years, the gender differential has declined. These secular trends are apparent in data my office has collected from Harvard faculty as part of a recent faculty climate survey: In 2013, among tenured professors, women were twice as likely as men to not have children (26% vs. 13%); among tenure-track faculty, the percentages were indistinguishable (39% and 42%).

Despite this smaller gender gap among tenure-track faculty in the likelihood of having children, a large statistically significant gender gap persists in reported levels of stress tenure-track faculty members experience due to responsibilities stereotypically associated with women: child care, children's schooling, and care of an ill child. Restricting analyses to faculty with children, women are significantly more likely than their male colleagues to report stress in each of these three domains, with child care being the single largest source of stress among our tenure-track women with children: fully 90% report "somewhat" or "extensive" stress in this domain.

These gender differentials in child care and other household responsibilities are not just a question of stress; they are also a question of time. Figure 27.3 presents the median amount of time male and female tenure-track faculty at Harvard report spending on work (the top portion of the graph) and household duties (the bottom portion of the graph). Since family structure logically predicts the amount of time spent on household duties, the graph divides the faculty into four groups: (1) single faculty with no children; (2) partnered faculty with no children; (3) faculty with children and a nonworking partner; and (4) faculty with children, who have a working partner or who are single parents (the single largest group, accounting for more than 50% of both the men and women). Note that for each group, there is no gender difference in the hours reported working. But when it comes to reporting the hours spent on household responsibilities, women spend twice as much time — a median of 40 hours a week — in comparison to only 20 hours a week for similarly situated men. This differential is not just statistically significant it is huge in practical terms, translating into almost an extra 3 hours each day.

These gender differentials in stress and time demonstrate the need for enlightened work–life policies. Women faculty cannot be as productive if they effectively have another full-time job taking care of their families and households. Enlightened work–life policies are not "just" for the women; they are for the organization. Such policies cost money, to be sure, but they bring increased satisfaction and productivity. Leaders seeking to enhance the opportunities for women in their workplaces need to consider strategies such as:

- Providing high quality, on-site child care, so that faculty can bring their children to campus and know that they will be well-cared for.

Work/Life Balance
Gender gaps in time spent on household duties for
Harvard faculty members with children

Analysis restricted to faculty who provided data on both hours worked and hours
spent on household duties. Source: Harvard Faculty Development & Diversity Office

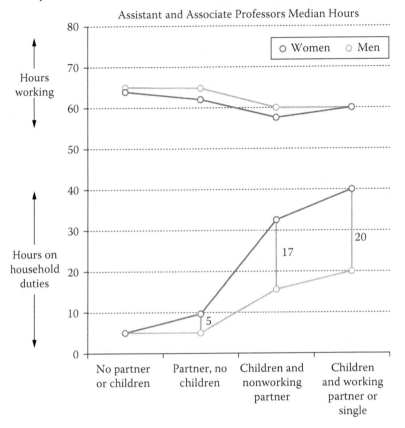

FIGURE 27.3
Gender differentials in work–life balance. **See color insert.**

- Subsidizing the costs of child care, so that younger faculty can afford
  to stay in the workforce. Many major research universities — including
  Harvard — now subsidize child-care costs, which in the Boston area can
  run as high as $20,000 a year.

- For faculty on tenure clocks, consider tenure clock extensions and parental
  teaching relief. At Harvard, we have made these policies automatic and
  gender neutral. They're automatic because of concerns that women will not
  ask, fearing that they will look like they are not dedicated to their careers.

And they are gender neutral to encourage our male faculty to participate in childrearing, which is good both for them and their partners/spouses.

The best women statisticians have many options. If you want your organization to attract the best candidates, family-friendly policies can give you a leg up over your competition.

Assess and Hold the Institution Accountable (Anecdotes Are Nice, but Data Speak Truth to Power)

This recommendation should be music to statisticians' ears. Effective leaders have to be good storytellers, but having data to support those narratives and anecdotes is essential — especially when the people you are trying to persuade are also statisticians.

Each year, my office issues an Annual Report on the Faculty (available for download at www.faculty.harvard.edu), which presents current and longitudinal data on the gender and racial/ethnic composition of the Harvard faculty, by school (and division within our Faculty of Arts and Sciences). Publicizing the data is essential to keeping diversity issues on the institutional agenda. In addition to sharing the data annually with the faculty (something Harvard started to do in 2006), I make regular presentations to our two governing boards: the Corporation and the Overseers. I also work with the student newspaper and alumni magazine to insure that these data are shared widely with the broader community.

Let me reiterate the value of our climate survey, which we first fielded in 2007 and which we repeated in 2013. It takes enormous institutional and political will to survey the faculty — I am reminded of the maxim "do not ask the question if you do not want to know the answer." Yet we find the climate survey to be especially effective for getting institutional leaders and faculty members to understand the more subtle issues of climate in the schools and departments. Several of the innovative work–life policies mentioned in the previous section were implemented in the wake of the 2007 climate survey, when faculty bitterly complained about not having parental teaching relief, tenure clock stops, and the high costs of child care.

Of course, a survey is only useful if the response rate is sufficiently high. Reviewing surveys conducted at other universities, I am surprised to find response rates so low as to render the data nonconvincing. I am pleased with our response rate of 72%, which took enormous effort to achieve. I also suspect that one reason people were willing to respond this second time was that it was clear that the institution used the results of the first survey to change policies.

Another example of the power of climate survey data is given in Figure 27.4, which presents, by gender, the responses to two questions that speak directly to differing perceptions of the atmosphere in faculty members' school/department: "The School/Department makes genuine efforts to recruit

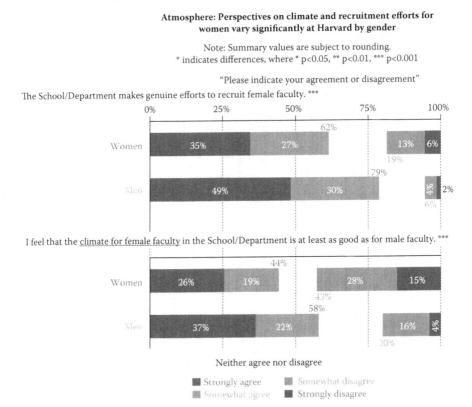

FIGURE 27.4
Climate survey data. **See color insert.**

female faculty" and "I feel that the climate for female faculty in the school/department is at least as good as for male faculty." In comparison to men, female faculty are three times as likely to disagree with the first statement and twice as likely to disagree with the second. Because women and men are not equally allocated across our schools and departments, these differentials become even more striking when we present them in even more specific smaller subgroups. We are using these data across the campus to spark conversations about the different experiences male and female faculty have, and what might be done to close the gap.

Help Junior Faculty Take Charge of Their Careers (Institutional Actions Are Necessary, Yet Not Sufficient)

No discussion of strategies for fostering the advancement of women in academic statistics would be complete without recommendations directed to junior faculty themselves. As much as I believe that institutional actions are necessary, it is clear that they must take responsibility for their own success. A faculty member is essentially a "small business owner," who must steer her own course in an academic setting. Below is some advice for statisticians — mostly equally applicable to women and men — both those who are interested in finding and growing in faculty role, perhaps with an eye downstream to a leadership role.

- **Develop your own personal career plan.** You are the steward of your career. Decide who you want to be in 5 years, 10 years, and 20 years. To concretize the exercise, think about a set of people senior to you who collectively — in whatever weighted average that makes sense to you — represent the kind of statistician you want to be down the road. Be careful not to assume that they got to where they are by doing what they are doing now (that's rarely the case). Ask yourself what you need to do to accrue that mix of activities — research projects, funding and grant opportunities, and additional training or certifications – that will help you realize your aspirations.

- **Plan your publications.** Statisticians have so many offers for collaboration that it's easy to become reactive, not proactive. Ask yourself what kinds of papers do you want to write? For which types of journals? Read those journals carefully and internalize those journals' expectations. Volunteer to be a peer reviewer for these journals. If you feel uncomfortable volunteering, ask a senior colleague to recommend you. Serving as a peer reviewer — and taking the responsibility seriously — helps you get known and establishes your reputation (see the next bullet).

- **Get known — in your department, across your institution, and in your field.** Do not underestimate the importance of getting known by colleagues. The academic world, in particular, depends heavily on reputation. When you go to a conference, send an email in advance to someone you would like to know your work and ask that person out for coffee. (If you do not meet someone new at every conference you attend, you are not fulfilling the purpose of conference attendance). Ask constructive questions at conferences and meetings. Volunteer to be a discussant. You can be doing fabulous work, but if people do not know who you are, your opportunities will be more limited.

- **Learn to say "no."** Time is our most precious resource and the one resource we are all allocated in equal amounts. To balance the many requests I receive, I have a policy of never (OK, rarely) committing immediately to any request that would require a substantial amount of time. Ask yourself what of your current responsibilities you would be willing to let go of if you were to take on this new commitment. Vesterland, Babcock, and Weingart (2014) have done a series of laboratory experiments documenting that women, on average, are much less willing to decline what they label "nonpromotable tasks" — tasks that do not lead to career advancement.

- **Balance collaborative research with independent contributions.** At various points in your career, others will make decisions about you predicated — at least in part — on your individual contributions. In this era of big science, the difficulty of separating out individual contributions from those of coauthors is being noticed across disciplines, but in my experience evaluating thousands (!) of faculty dossiers, statisticians are particularly vulnerable. I am a strong believer in collaboration, and all of my writings with my long-time collaborator, John Willett, include the footnote: "the order of the authors was determined by randomization." Yet experience teaches me that statisticians need to be especially sensitive to this issue and to figure out how to establish their specific contributions. Given the implicit biases mentioned earlier, women are particularly vulnerable to assessors concluding that a woman's contribution to a collaborative paper was less important than that of a male coauthor. A preponderance of middle authorships without a clear sense of your contribution can stymie your career.

- **Find mentors; preferably more than one.** If your organization has a formal mentoring program, participate. If not, seek out your own mentors (nota bene: Do not ask, "Will you be my mentor?" Instead share a draft manuscript with a senior colleague and ask for feedback. The mentoring will follow. And if it does not, send the draft paper to someone else.) Higgins and Kram (2001) explain the logic behind what they call "developmental networks," the set of individuals to whom you will turn for advice. The advice should resonate with anyone who has ever received differing feedback from referees or proposal reviewers. Different people will give you different advice. Not all advice is right for you. And as you advance in your career, your questions will change increasing the need to widen your developmental network.

- **Consider peer mentoring.** Do not underestimate the value of creating mentoring networks with peers. Many people feel more comfortable asking questions of peers than of senior colleagues, some of whom may be in an evaluative position at a later date. Peers can provide important reality checks (should I really do what so-and-so just asked me to do?), can provide support, and can read and comment on your work. I am certain I have gotten as much good advice from peers as I have gotten from senior mentors over the years.

Postscript: Advice for Women Statisticians Contemplating a Leadership Position

Women seeking a leadership position have many "how to" books, articles, and TED talks to guide them. Sheryl Sandberg's *Lean In* (2013) was Amazon's

No. 2 best seller in 2013. Amy Cuddy's 2012 TED talk on how "your body language shapes who you are" has received more than 20 million views. Kay and Shipman's (2014) article in *The Atlantic Monthly* and their subsequent book, *The Confidence Code: The Science and Art of Self-Assurance — What Women Should Know*, document gender differences in confidence and offer advice to women seeking to overcome them.

So the question becomes, what is unique about being a woman statistician contemplating a leadership position: To be sure, the lack of women in leadership positions in statistics means that we face a world in which we see far too few women role models. All of my mentors over the years have been men (for the simple reason that there were no women available).

As John Tukey famously noted, "The best thing about being a statistician is that you get to play in everyone's backyard." This attitude strikes me as an ideal attribute for a leadership position. Statisticians learn to collaborate early on in their training, and that ability to listen and take the other's perspective is essential for good leadership.

Bibliography

American Mathematical Society (2014). 2012 Annual Survey of the Mathematical Sciences Fall 2012 Departmental Profile Report. *Notices of the AMS*, 61(2), February.

Banaji, M.R. and Greenwald, A.G. (2013). *Blindspot: Hidden Biases of Good People*. Delacorte Press.

Blau, F.D. Currie, J.M., Croson, R.T.A. and Ginther, D.K. (2010). "Can Mentoring Help Female Assistant Professors? Interim Results from a Randomized Trial." *American Economic Review*, 100(2): 348–52.

Carrell, S.E., Page, M.E., and West, J.E. (2010). "Sex and Science: How Professor Gender Perpetuates the Gender Gap," The Quarterly Journal of Economics, MIT Press, MIT Press, vol. 125(3), pages 1101–1144, August.

Goldin, C. and Rouse, C (2000). "Orchestrating Impartiality: The Impact of 'Blind' Auditions on Female Musicians." *American Economic Review*, 90(4): 715–742.

Harvard University (2014). *Best Practices for Conducting Faculty Searches*. Available for download at http://www.faculty.harvard.edu/appointment-policies-and-practices/resources-conducting-faculty-search.

Higgins, M.C. and Kram, K.E. (2001). "Reconceptualizing mentoring at work: A developmental network perspective," *Academy of Management Review*, 26(2): 264–88.

Kay, K. and Shipman, C. (2014). "The Confidence Gap," *The Atlantic Monthly* April 14, 2014. Available for download at http://www.theatlantic.com/features/archive/2014/04/the-confidence-gap/359815/

Kay, K. and Shipman, C. (2014). *The Confidence Code: The Science and Art of Self-Assurance—What Women Should Know.* HarperBusiness.

Monti, K (2014). The ASA Fellow Award: 2014 Update. *Amstat News* Available for download at http://magazine.amstat.org/blog/2014/08/01/fellow-award-2014/.

Moss-Racusin, C.A., Dovidio, J.F., Brescoll, V.L., Graham, M., and Handelsman, J. (2014). Science faculty's subtle gender biases favor male students. *Proceedings of the National Academy of Sciences*, 109, 16474–16479.

National Research Council (2010) *A Data-Based Assessment of Research Doctorate Programs in the United States.* National Academy of Sciences.

National Science Foundation (2012) Survey of Earned Doctorates. Available for download at http://www.nsf.gov/statistics/surveys.cfm.

Sandberg, S. (2013). *Lean In: Women, Work, and the Will to Lead.* Knopf.

Trix, F. and Psenka, C. (2003). Exploring the color of glass: Letters of recommendation for female and male medical faculty. *Discourse & Society*, 14(2): 191–220.

University of Wisconsin (2012) Searching for Excellence & Diversity: A Guide for Search Committees. Available for download at wiseli.engr.wisc.edu.

Vesterland, L., Babcock, L., and Weingart, L. (2014). Breaking the glass ceiling with "no": Gender differences in declining request for non-promotable Tasks. Unpublished manuscript, University of Pittsburgh.

Contributors

Arlene S. Ash earned her PhD in 1977 from Dartmouth College and is professor and chief of the division of Biostatistics and Health Services Research in the Department of Quantitative Health Sciences at the University of Massachusetts Medical School in Worcester, Massachusetts. As a methods expert on risk adjustment in health services research, Dr. Ash has pioneered tools for using administrative data to monitor and manage health care delivery systems, including those now relied upon by the US Medicare program. In 1996, she cofounded DxCG, Inc., a company with more than 350 national and international clients to promote "fair and efficient health care" via predictive software. Many of her more than 150 research publications reflect her longstanding interests in women's health; gender, age, and racial disparities; and quality, equity, and efficiency in health care financing and delivery. She is also actively involved in improving US electoral integrity.

Lynne Billard earned her PhD in 1969 from University of New South Wales, Australia. She is currently a professor at University of Georgia. Her previous positions include professor and head, Department of Statistics and Computer Science, University of Georgia, 1980–1984; professor and head, Department of Statistics, University of Georgia, 1980–1989, associate to dean, University of Georgia, 1989–91; assistant/associate professor at University of Waterloo, Florida State University, and various visiting positions at Stanford, Berkeley, SUNY at Buffalo, and Birmingham (UK). She also served as president, Eastern North American Region, International Biometric Society (IBS); IBS International president; president, American Statistical Association, 1996; program secretary, Institute Mathematical Statistics; treasurer, Bernoulli Society; and council, International Statistical Institute. Dr. Billard has received numerous awards, including the Samuel S. Wilks Award, the Florence Nightingale David Award and the Elizabeth L. Scott Award from the Committee of Presidents of Statistical Societies, the Janet Norwood Award, and the American Statistical Association Founders Award. She has served on numerous National Research Council Committees and Panels and was secretary of Commerce's Census 2000 Advisory Board.

Cynthia Z.F. Clark earned her PhD in 1977 from Iowa State University. Prior to her retirement in 2014, she was administrator, National Agricultural Statistics Service (NASS) (2008–2014), the first female head of the agency, overseeing the agency's efforts to collect and disseminate data on every facet of

US agriculture. Dr. Clark is an accomplished statistician who is highly respected both in the United States and overseas for her expertise in survey research and development. Before joining NASS, she directed statistical research and survey methodology for the United Kingdom's Office for National Statistics and, prior to that, at the US Census Bureau. She has also served in numerous professional positions with the American Statistical Association (ASA), the International Statistical Institute (ISI), and the International Association of Survey Statisticians. Additionally, she is a fellow of the ASA, an elected member of ISI, and a fellow of the Royal Statistical Society. Her focus has been on research in survey and census methodology, confidentiality and privacy, statistics education, quality management, computer-assisted survey methodologies, and official statistics (including economic, environmental, demographic, agricultural). She earned a bachelor's degree in mathematics from Mills College in California, a master's degree in mathematics from the University of Denver, and master's and doctoral degrees in statistics from Iowa State University. She was recognized in 2014 by Iowa State University as a Distinguished Alumnae.

Charmaine B. Dean earned a PhD in 1984 from the University of Waterloo and is professor and dean of science at Western University. She was the 2007 president of the Statistical Society of Canada, 2002 president of the International Biometrics Society, Western North American Region, has served as president of the Biostatistics Section of the Statistical Society of Canada, and has given ten years of service to the Natural Sciences and Engineering Research Council of Canada, including two as chair of the Statistical Sciences Grant Selection Committee and one as chair of the Discovery Accelerator Supplement Committee for the Mathematical and Physical Sciences. She has served on the Michael Smith Foundation for Health Research Research Advisory Council and on selection panels for that foundation. She serves on the NIH Biostatistics Grant Review Panel; on the board of directors of the Pacific Institute for the Mathematical Sciences; on the board of directors of the Banff International Research Station; and is a member of the College of Reviewers of the UK Engineering and Physical Sciences Research Council and of the MITACS College of Reviewers. She has served on several editorial boards and is currently associate editor of *Biometrics*, of *Environmetrics*, and of *Statistics in Biosciences*, and senior editor of *Spatial and Spatio-temporal Epidemiology*. In 2003, Dr. Dean was awarded the CRM-SSC prize; in 2007, she was named a fellow of the American Statistical Association; and in 2007 was awarded the University of Waterloo Alumni Achievement Medal; in 2010 she was named a fellow of the American Association for the Advancement of Science; and in 2012 received a Trinidad and Tobago High Commission Award.

Katherine Bennett Ensor is professor of statistics at Rice University, where she served as the chair of the Department of Statistics from 1999 through 2013. She is currently director of the Center for Computational

Finance and Economic Systems (CoFES). Dr. Ensor develops statistical techniques to answer important questions in science, engineering, and business, with specific focus on the environment, energy, and finance. She is an expert in multivariate time series, categorical data, spatial-temporal and general stochastic processes, and sampling. She is an elected fellow of the American Statistical Association, the American Association for the Advancement of Science, and has been recognized for her leadership, scholarship, service, and mentoring. She holds a BSE and MS in mathematics from Arkansas State University and a PhD in statistics from Texas A&M University.

Lori A. Evarts earned her BA in Economics and MPH in biostatistics from the University of North Carolina at Chapel Hill, USA. She maintains certifications as a project management professional (PMP), as awarded by the Project Management Institute, and as a certified public health (CPH) professional from the National Board of Public Health Examiners. Evarts' professional experience spans more than 30 years, beginning with the UNC Biostatistics Department, Blue Cross & Blue Shield of North Carolina, and nearly 20 years at Quintiles, Inc. prior to rejoining UNC in 2005. She is a clinical assistant professor and the director of graduate studies of the Public Health Leadership Program at the Gillings School of Global Public Health. She teaches semester-long graduate courses, offers lectures, and develops workshops for campus-wide participation and other organizations interested in project management, leadership, quality improvement, team dynamics and development, and strategies to maximize team effectiveness.

Rongwei (Rochelle) Fu earned her PhD in plant science in 2000 and her PhD in statistics in 2003, both from the University of Connecticut. She collaborates extensively with investigators of Oregon Health and Science University (OHSU) with a very productive publication record of more than 100 peer-reviewed journal articles and 30 peer-reviewed reports. She has been the lead biostatistician for the Center for Policy and Research in Emergency Medicine, Pacific Northwest Evidence-Based Practice Center (EPC), and Research Center for Gender-Based Medicine. She led the efforts to develop methodological guidelines for quantitative synthesis for comparative effectiveness review (CER) for the Agency of Healthcare Research and Quality (AHRQ), and has been working in the leadership role for biostatistics curriculum development and improving diversity in the Oregon Master of Public Health Program.

Mary W. Gray earned a PhD from the University of Kansas, a JD from Washington College of Law, and is professor of mathematics and statistics at American University, Washington, DC. She is the founding president of the Association for Women in Mathematics, past president of the Caucus for Women in Statistics, and past vice president of the American Mathematical Society. Her more than 100 publications are in the fields of economic equity, human rights, applications of statistics, and the history of mathematics.

Professor Gray has served as a consultant to such agencies as UNICEF, the US Department of State, the Equal Employment Opportunities Commission, and many universities, corporations, and private individuals. She is a fellow of the American Statistical Association, the American Mathematical Society, the American Association for the Advancement of Science, and the Association for Women in Science, and the recipient of a Presidential Award for Excellence in STEM Mentoring.

Marcia L. Gumpertz is assistant vice provost for Faculty Diversity and professor of Statistics at North Carolina State University. Her charge is to promote faculty diversity, facilitate campus actions to enhance climate, and support faculty success. From 2009–2012, Dr. Gumpertz served as principal investigator on the NSF-funded ADVANCE Developing Diverse Departments (D3) at NC State project. Based on the understanding that the commitment of campus leadership is the major driver of change, this project focused on developing female faculty leaders and on developing strong allies among current campus leadership. The D3 project initiated a workshop series for mid-career faculty, called the Leadership for a Diverse Workshop series, and a Climate Workshop Series for department heads, and developed a core group of faculty who serve as change leaders promoting faculty diversity at NC State. In statistics, Dr. Gumpertz's interests lie in the areas of applied statistics, spatial data analysis, and design of experiments. Dr. Gumpertz is coauthor of the textbook *Planning, Construction and Statistical Analysis of Comparative Experiments* and is a fellow of the American Statistical Association.

Lee-Ann Collins Hayek, a fellow of ASA and RSS, is the chief mathematical statistician and a senior research scientist at the Smithsonian Institution. She has more than 150 publications in a wide range of peer-reviewed journals and has written 5 books and monographs, among which is her latest: *Surveying Natural Populations: Quantitative Tools for Biodiversity Assessment*, called a "classic" in the fields of paleoecology, ecology, and conservation biology. Contributing to both theoretical and application-based literatures, she has published major advances and problem solutions in fields as diverse as anthropology, art conservation, auditing, biodiversity, geology, ecology, forensic science, marine science, paleontology, scientific reproducibility, and zoo and wildlife medicine, among others. Dr. Hayek has a long list of professional organization participation as a member and leader in more than 20 organizations. A select list of her current involvement includes board member of the DC Chapter of AWIS, and spokesperson for the IYSTAT2013 for AWIS National, ASA Science and Public Policy Advisory Committee, Data Sharing Committee, Organizing Committee for New Annual Conference on Statistical Practice (CSP), and Software and Graphics Presentation Committee for the CSP.

Nancy Heckman is professor in the Statistics Department at the University of British Columbia and has served as chair of that unit since 2008. Her research areas are smoothing methods and functional data analysis, with application to a variety of areas including environmetrics, energy consumption and, most notably, evolutionary biology, via a long-standing interdisciplinary and international collaboration with evolutionary biologists and animal geneticists. She is a fellow of the IMS and ASA and a member of the ISI. She earned her PhD in mathematics from the University of Michigan, Ann Arbor.

Jude Heimel is a professional certified coach and accomplished organization development consultant. For the past 10 years, she has been coaching and consulting with university faculty and administrators. She has built her ontological coaching practice upon a 30-year foundation as an organization development practitioner, consultant, and evaluator. In addition to higher education organizations, she has worked with for-profit and not-for-profit organizations of global, national, and local scope. Heimel is currently studying neuroscience and its application to personal change. "I am fascinated by the power of the supposedly simple acts of compassionate detachment, deep listening, and nonjudgmental reflection. I apply these principles to better understand myself and with my clients in my coaching and consulting practice. The results have been significant insights and lasting breakthroughs."

Jacqueline M. Hughes-Oliver is professor of statistics at North Carolina State University. She has held positions at George Mason University and the University of Wisconsin Madison, and served as a faculty fellow at the Statistical & Applied Mathematical Sciences Institute (SAMSI). She also served as director of graduate programs for the Department of Statistics at North Carolina State University. Dr. Hughes-Oliver's research areas include prediction and classification, variable and model selection with dimension reduction, design of experiments, and spatial modeling. Her application areas include cheminformatics and drug discovery, ontology-guided genomics and metabolomics, and effect of point sources. Dr. Hughes-Oliver earned a bachelor of arts degree in mathematics from the University of Cincinnati and a PhD in statistics from North Carolina State University. She is a fellow of the American Statistical Association.

Sallie Keller received her PhD in 1983 from Iowa State University and is professor of statistics and director of the Social and Decision Analytics Laboratory within the Virginia Bioinformatics Institute at Virginia Tech. Formerly, she was professor of statistics at University of Waterloo and academic vice president and provost, director of the IDA Science and Technology Policy Institute in Washington, DC, professor of statistics and the William and Stephanie Sick Dean of Engineering at Rice University, head of the Statistical Sciences group at Los Alamos National Laboratory, professor and director of graduate studies in the Department of Statistics at Kansas State University, and statistics program director at the National Science Foundation. Dr. Keller has served as a member of the National Academy of Sciences Board on Mathematical Sciences and Their Applications, has chaired the Committee on Applied and Theoretical Statistics, and is currently a member of the Committee on National Statistics. She is a national associate of the National Academy of Sciences, fellow of the American Association for the Advancement of Science, elected member of the International Statistics Institute, and member of the JASON advisory group. She is also a fellow and past president of the American Statistical Association.

Jon R. Kettenring joined Drew University in 2004 as a fellow in the Charles A. Dana Research Institute for Scientists Emeriti, known as RISE, and presently serves as its director. From 1969–2003, he worked in industrial research at Bell Laboratories, Bellcore, and Telcordia Technologies. His research focus has been primarily on new methods for analyzing multivariate statistical data. From 1984 through 2003, he managed a variety of research groups in statistics, economics, computer science, and information analysis at Bellcore and Telcordia. He has held visiting appointments at the University of Washington, the University of Minnesota, Stanford University, and the University of Michigan. All of his degrees are in statistics: BS and MS from Stanford University and PhD from the University of North Carolina. He served as president of the American Statistical Association in 1997 and as chair of the board of trustees of the National Institute of Statistical Sciences from 2000 to 2004.

Yun Li earned her PhD in biostatistics in 2008 from the University of Michigan and is a research associate professor in the Department of Biostatistics at the University of Michigan. She is a member of the University of Michigan Comprehensive Cancer Center, the Arbor Research Collaborative for Health, and the Kidney Epidemiology and Cost Center. Her current methodology focuses on causal inference, missing data issues, surrogate and auxiliary data, mediation and unmeasured confounding. She collaborates primarily in health sciences, particularly in the areas of cancer, kidney disease, and cardiovascular disease. She currently serves as the University of Michigan principal investigator on the Dialysis Outcomes and Practice Patterns Study and the codirector of the Design and Methodology Core on a P01, with the focus on the challenges of individualizing treatments for breast cancer. She is also involved in many other grants as a coinvestigator. She has authored or coauthored numerous publications.

Laura J. Meyerson is vice president, Biogen Idec. Meyerson currently leads the Biostatistics, Clinical Data Sciences, and Medical Writing Organizations at Biogen Idec. During her 12-year tenure, she has led organizations spanning multiple therapeutic areas (neurology, immunology, and oncology) and geographies (US and UK.) The number of employees in her organization has grown dramatically under her direction. She started with less than 50 employees, and her current staff consists of approximately 200 employees. Prior to joining Biogen Idec, Dr. Meyerson held several management positions in clinical biostatistics, including senior director of biostatistics at Quintiles and director, biostatistics at Janssen Pharmaceuticals. Dr. Meyerson started her career in the pharmaceutical industry as a clinical project statistician at Marion Merrell Dow (a legacy company of Sanofi). She has more than 30 years of experience in the analysis of clinical trial data for the pharmaceutical industry and has participated as a leader in various professional organizations, including American Statistical Association (ASA) and the Eastern North American Region of the Biometrics Society (ENAR.) Dr. Meyerson was an assistant professor of statistics in the Department of Operations Research at the Naval Postgraduate School in Monterey prior to joining the pharmaceutical industry. She earned her PhD in biostatistics in 1983, an MA in biostatistics in 1980, and an AB in biological sciences in 1977.

Motomi (Tomi) Mori is a professor and head of the Division of Biostatistics, Department of Public Health & Preventive Medicine, Walter & Clora Brownfield Endowed Professor of Cancer Biostatistics, and the director of the Biostatistics Shared Resource of the Knight Cancer Institute at Oregon Health & Science University (OHSU). Dr. Mori earned a BA in psychology from the University of Montana, and MS in statistics and PhD in biostatistics from the University of Iowa. Prior to OHSU, she held a faculty position at the Fred Hutchinson Cancer Research Center, University of Washington, and the University of Utah. She served on the Cancer Centers of the National Cancer Institute Initial Review Group (2006–2010) and the Executive Committee of the Section on Teaching Statistics in Health Sciences (TSHS) of the American Statistical Association (2010-2012). She is a successful biostatistician collaborator with more than 25 years of experience in cancer research.

Sally C. Morton is professor and chair of biostatistics in the Graduate School of Public Health, and director of the Comparative Effectiveness Research Center in the Health Policy Institute, at the University of Pittsburgh. Previously, she was vice president for statistics and epidemiology at RTI International; and head of the RAND Corporation Statistics Group. Her research focuses on evidence synthesis, particularly meta-analysis. She is a member of the Patient-Centered Outcomes Research Institute (PCORI) Methodology Committee, and the Agency for Healthcare Research and Quality (AHRQ) Evidence-Based Practice Center Methods Steering Committee. She was the 2009 president of the American Statistical Association (ASA) and the 2013 chair of Section U (Statistics) of the American Association for the Advancement of Science (AAAS). She is a fellow of the ASA and AAAS. She earned a PhD in statistics from Stanford University.

Bhramar Mukherjee is currently a professor of biostatistics at the School of Public Health, University of Michigan. She earned her PhD in statistics from Purdue University in 2001 and spent the next 4 years as an assistant professor of statistics at the University of Florida. Her research interests are gene–environment interactions, Bayesian methods in genetic and environmental epidemiology, and statistical analysis under outcome dependent sampling. Dr. Mukherjee has authored more than 100 articles in reputed statistics, biostatistics, and epidemiology journals. She is leading multiple NIH and NSF grants on gene–environment interaction analysis. She is the director of the biostatistics core in two large center grants related to environmental health. She is a member of the University of Michigan Comprehensive Cancer Center and associate director of the Cancer Biostatistics Training Grant. She is presently serving on the editorial board of several journals in statistics, biostatistics, and epidemiology. Dr. Mukherjee has won multiple grants and awards for her teaching accomplishments, including the 2012 School of Public Health Excellence in Teaching Award at the University of Michigan. She has received an outstanding alumna award from Purdue University and outstanding young researcher award in statistical applications category from the International Indian Statistical Association. She is an elected member of the International Statistical Institute and is a fellow of the American Statistical Association.

Sastry G. Pantula is a professor of statistics and the dean of the College of Science at Oregon State University. His goals are to strive for excellence, enhance diversity, foster harmony, and to build future leaders in science. From 2010–2013, he served as the director of the Division of Mathematical Sciences at the US National Science Foundation. In 2010, he served as the president of the American Statistical Association (ASA). After earning his PhD in 1982 from Iowa State University, he spent about 30 years at North Carolina State University (NCSU). He is a fellow of ASA and AAAS and a member of the NCSU Academy of Outstanding Teachers at NCSU, Sigma Xi, Phi Kappa Phi, and Mu Sigma Rho. He received his B.Stat in 1978 and M.Stat in 1979 from the Indian Statistical Institute, Kolkata, India.

Sowmya Rao is an associate professor in the Quantitative Health Sciences Department at the University of Massachusetts Medical School, Worcester, and the Center for Healthcare Organization and Implementation Research at the Veteran's Administration in Bedford, Massachusetts. Her research interests include survey methods, missing data, disparities, and statistical methods applied to epidemiology, health care and health policy research. Dr. Rao also donates her expertise to address issues in international health as the secretary of Statistics Without Borders, an outreach group of the American Statistical Association, and the International Indian Statistical Association, a satellite organization of the American Statistical Association. After earning a PhD in biostatistics from Boston University, Boston, Massachusetts, and completing a postdoctoral fellowship at the National Cancer Institute, Rockville, Maryland, Dr. Rao held positions at Abt Associates, Harvard Medical School, and Massachusetts General Hospital.

Nancy Reid is a university professor of statistical sciences at the University of Toronto. She has served in a number of leadership roles during her career, including chair of the department, president of the Institute of Mathematical Statistics, president of the Statistical Society of Canada, vice president of the International Statistical institute, editor of the *Canadian Journal of Statistics*, and associate editor of *The Annual Reviews of Statistics and Its Applications*, *Bernoulli*, *Biometrika*, *JRSS B*, and *Statistical Science*. She has also served on a number of committees of the American Statistical Association, the International Statistical Institute, the Bernoulli Society, the Institute of Mathematical Statistics and the Statistical Society of Canada. Dr. Reid earned her PhD from Stanford University in 1979, working under the supervision of Rupert G. Miller, Jr. Her MSc degree is from the University of British Columbia, and she earned her BMath from the University of Waterloo in 1974. Her research interests are the theory of statistical inference and its interplay with applications. She won the COPSS Presidents' Award in 1992 and the FN David award in 2008.

Marilyn M. Seastrom currently serves as the chief statistician at the National Center for Education Statistics in the US Department of Education. As such, she provides technical assistance and advice to programs throughout NCES, and is responsible for the NCES technical review of all NCES products. Dr. Seastrom directs the IES data confidentiality program and serves as a consultant on matters of privacy and data confidentiality throughout the department. Dr. Seastrom served on the OMB interagency committee that drafted the regulations for the implementation of the e-Gov Confidential Information Protection and Statistical Efficiency Act. Dr. Seastrom is a recognized statistical authority on a range of topics relevant to federal statistics, as is evident by her selection for membership on the OMB Federal Statistical Committee on Methodology, as a fellow of both the American Statistical Association and the American Association of Educational Research, her service on a number of government-wide standard setting committees, and her role as chair of the OMB Committee on Privacy in Statistical Data Collections. Dr. Seastrom has served in a number of elected and appointed positions in the American Statistical Association, including currently as the education officer for the Survey Research Methods Section, a council of sections representative for the Social Statistics Section, a member of the Presidential Committee on Statistical Leadership, and chair of the Deming Award Selection Committee. Areas of specialty include survey methodology, survey design, data analysis, measurement errors, data confidentiality, statistical standards and guidelines, and technical review. Dr. Seastrom majored in biology and sociology as an undergraduate, holds master's degrees in both fields, and completed her doctoral degree in demography and applied social statistics.

Judith D. Singer is senior vice provost for Faculty Development and Diversity and James Bryant Conant Professor of Education at Harvard University.

An internationally renowned statistician, Dr. Singer's scholarship focuses on improving the quantitative methods used in social, educational, and behavioral research. Her contributions on multilevel modeling, survival analysis, and individual growth modeling have made these and related methods accessible to empirical researchers. She has published numerous papers and chapters as well as three coauthored books: *By Design: Planning Better Research in Higher Education, Who Will Teach: Policies that Matter* and *Applied Longitudinal Data Analysis: Modeling Change and Event Occurrence.* Dr. Singer is a fellow of both the American Statistical Association and the American Educational Research Association and is an elected member of the National Academy of Education. In 2012, her nomination by President Obama to serve as a member of the board of directors of the National Board of Education Sciences was confirmed by the US Senate. In 2014, she received the Janet L. Norwood Award for Outstanding Achievement by a Woman in the Statistical Sciences. After earning her PhD in statistics from Harvard in 1983, she was appointed assistant professor at the Harvard Graduate School of Education (HGSE) in 1984. She was promoted to associate professor in 1988 and professor in 1993. From 1999 to 2004, Singer served as academic dean of HGSE and acting dean from 2001 to 2002.

Sim B. Sitkin is professor of management and director of the Center on Leadership and Ethics at the Fuqua School of Business, and director of the Behavioral Science and Policy Center at Duke University. He has also been academic director at Duke Corporate Education, on the faculty of the University of Texas at Austin and the Free University of Amsterdam. He was elected a fellow of the Academy of Management in 2010 based on his research on leadership, trust, and control systems, and their influence on how organizations and their members become more or less capable of risk-taking, learning, change and innovation. He is editor of the *Academy of Management Annals*, founding editor of *Behavioral Science and Policy*, consulting editor of *Science You Can Use*, advisory board member of the *Journal of Trust Research*, and previously served as senior editor of *Organization Science* and associate editor of the *Journal of Organizational Behavior*.

William A. Sollecito earned his doctor of public health degree in 1982 from the University of North Carolina and is a clinical professor in the Public Health Leadership Program at the UNC Gillings School of Global Public Health. He holds a bachelor of business administration degree from Baruch College, a master of science degree from the University of Pittsburgh, and a doctor of public health degree in biostatistics from the University of North Carolina at Chapel Hill. After working for ten years in health services research, Dr. Sollecito worked in the contract clinical research industry at Quintiles Transnational Corporation from 1982–1996, where, as president of Quintiles Americas, he was responsible for all clinical operations in North and South America. He was appointed to the faculty of the UNC School of Public Health in 1997 and was director of the Public Health Leadership Program from 2000 through 2009. He currently teaches graduate courses, lectures, and publishes on topics that span team and organizational leadership, project management, and continuous quality improvement.

Daniel L. Solomon is professor of statistics and dean of the College of Sciences at North Carolina State University. He began his career in 1968 at Cornell University, heading the Biometrics Unit there until 1981 when he moved to NC State as head of the Department of Statistics. He became dean of the College of Physical and Mathematical Sciences in 2000 and inaugural dean of the College of Sciences in 2013. Over his career, Dr. Solomon has participated in or led an array of activities in support of diversity and inclusion, ranging from programs to diversify the STEM pipeline beginning at grade 3 through efforts to diversify the faculty and enhance work–life balance. Dr. Solomon is a fellow of the American Statistical Association, a 2010 winner of its Founders Award, and an elected member of the International Statistical Institute. He served as editor of *Biometrics*. Dr. Solomon chairs the governing board of the Statistical and Applied Mathematical Sciences Institute and was instrumental in the founding of the National Institute of Statistical Sciences.

Duane L. Steffey earned his PhD in 1988 from Carnegie Mellon University and is principal scientist and director of the Statistical and Data Sciences practice at Exponent, Inc., an engineering and scientific consulting firm. In that capacity, he leads a group of statisticians, mathematicians, computer scientists, and reliability engineers specializing in the application of quantitative methods in projects involving product development, manufacturing, regulatory, and safety issues. His experience as a consulting statistician spans more than 25 years and a breadth of applications in engineering, health, environmental science, and civil justice. Prior to joining Exponent, Dr. Steffey was a tenured professor at San Diego State University, where he cofounded and codirected the university's statistical consulting center, and also served as coordinator of the Division of Statistics. Dr. Steffey is an elected fellow of the American Statistical Association (ASA). Within ASA, he has held elected offices as a chapter president, district vice chair, and section program chair, in addition to committee and working group appointments addressing professional ethics, visibility, and impact in policymaking.

Gary R. Sullivan earned his PhD in 1989 from Iowa State University under the mentorship of Wayne W. Fuller, and is the senior director of Non-Clinical Statistics at Eli Lilly and Company, a major pharmaceutical manufacturer headquartered in Indianapolis, Indiana. Dr. Sullivan joined Lilly in 1989 and has 13 years of experience as a project statistician, where he collaborated with pharmaceutical formulators, chemists, biologists, and engineers on formulation design, process optimization and modeling, assay development and characterization, and production monitoring. He has spent the last 12 years in various managerial positions with responsibilities for statisticians collaborating in manufacturing, product development, discovery research, and biomarker research. His personal passions include leadership development, quality management, experimental design, process optimization, and statistical process control.

Ronald L. Wasserstein has served in leadership roles for more than two decades, first at a university, then as head of the world's largest professional association for statisticians. At Washburn University, he served as associate vice president for eight years, then as vice president for Academic Affairs for seven years, the second highest leadership position at the university. For the past seven years, he has served as executive director of the American Statistical Association (ASA). Other leadership roles include service as a member of the ASA Board of Directors, chair of the MATHCOUNTS Foundation Board of Directors, president of Kappu Mu Epsilon National Mathematics Honor Society, and president of the North American Association of Summer Sessions. He has served as board chair for several civic organizations as well, including Housing and Credit Counseling, Inc., Ballet Midwest, and Cornerstone Family School. Wasserstein earned a PhD in statistics from Kansas State University.

Roy E. Welsch is Eastman Kodak Leaders for Global Operations professor of management and professor of statistics and engineering systems in the Sloan School of Management and the Engineering Systems Division at the Massachusetts Institute of Technology and director of the MIT Center for Computational Research in Economics and Management Science. He serves as area head for the Management Science Area of the Sloan School. He earned his AB (1965) in mathematics from Princeton and his MS (1966) and PhD (1969) from Stanford, also in mathematics. Dr. Welsch joined the MIT faculty in 1969 as an assistant professor, became an associate professor four years later and was promoted to professor in 1979. He held the Leaders for Manufacturing Chair from 1988 until 1993. From 1973 until 1979, he was also a senior research associate at the National Bureau of Economic Research, where he participated in the development of the Troll econometric, financial, and statistical modeling system. He was codirector or director of the MIT Statistics Center from 1981 until 1989. Dr. Welsch is widely recognized for his book (with Edwin Kuh and David Belsley) on regression diagnostics and for his work on robust estimation, multiple comparison procedures, nonlinear modeling, and statistical computing. He is currently involved with research on robust process control and experimental design, credit scoring models and risk assessment, diagnostics for checking model and design assumptions, variable selection in bioinformatics, and volatility modeling in financial markets. Dr. Welsch has worked with major exchanges to design and test algorithms for detecting unusual trading patterns in prices and volumes based on intraday and interday data. He uses real time and historical data to formulate benchmarks and standards for "normal behavior and patterns" relative to a given market. Dr. Welsch is a fellow of the Institute of Mathematical Statistics, the American Statistical Association, and the American Association for the Advancement of Science.

Kelly H. Zou earned her PhD from the University of Rochester in 1997 and is senior director and statistics lead at Pfizer Inc. She was a director of biostatistics and an associate professor of radiology at Harvard Medical School (HMS), as well as an associate director of Rates at Barclays Capital. She is a fellow of the American Statistical Association and an accredited professional statistician. She has served several leadership roles: chair, Committee of Presidents of Statistical Societies Awards; vice chair, Committee on Applied Statisticians; secretary, Health Care Policy Statistics Section; vice chair, Committee on Statistical Partnerships among Academe, Industry & Government; awards committee cochair, Women in Statistics Conference; cochair, International Conference on Health Policy Statistics; faculty taskforce member, Joint Committee on the Status of Women, HSM and Harvard School of Dental Medicine; and member, Women's Leadership Network, Pfizer Inc. She has authored monographs titled, *Statistical Evaluation of Diagnostic Performance: Topics in ROC Analysis* and *Patient-Reported Outcomes: Measurement, Implementation and Interpretation,* both published by Chapman & Hall/CRC Press.

Index

A

Abortion issues, 234–235
Abramson, Jill, 230
Abrashoff, D. Michael, 267
Academic advisers, 389–390
Academic leadership, *See also*
　　Women leaders in statistics
　advancement, *See* Career
　　advancement
　building positive work culture,
　　347–348
　challenges, 339, 365–366
　critical mass of women faculty,
　　354
　cultural contexts, 111–115
　diversity leadership, *See*
　　Diversity leadership;
　　Gender diversity
　　enhancement
　Elizabeth L. Scott, 31–49
　encouraging junior faculty,
　　424–426
　engagement of faculty, 91–93
　engaging predecessors, 93–94
　entrepreneurship and
　　innovation, 367
　examples of personal journeys,
　　360–362
　examples of statistical leaders, 5
　focus and time management,
　　342–343
　grant funding parity, 107,
　　116–117
　Harvard vice provost for Faculty
　　Development and Diversity,
　　413–414
　improving gender diversity,
　413–427, *See also* Gender
　　diversity enhancement
　institutional responsibilities, 351
　leadership programs, 380–381
　need for women in leadership
　　roles, 353–356
　negotiation, 346
　networking, 346–347
　opportunities, 339–340
　practical advice for female
　　faculty, 365–369
　professional champions,
　　376–380, *See also*
　　Sponsorship
　professional development
　　training, 344–345
　research versus, 355–356
　status of women, 86
　student supervision and
　　training, 345
　team building, 341–342
　use of power, 368
　vision and, 340–341
　women as leaders, 94–98
　women university presidents,
　　101–102
　working with the organization,
　　342
Academy Health, 337
Accreditation, 107–108
Active listening, 288
Acumen, 205–213, 271
Administrative data utilization,
　179–182, 186–187
Administrators as mentors, 391–392
ADVANCE grant, 359, 392, 401–403
Advocacy leadership, 35–36, 47–49,

See also Gender diversity
    enhancement
Affirmative action, 36, 46–47, *See
    also* Gender diversity
    enhancement
Affordable Care Act, 234
Agora, 13
Agriculture Resource Management
    Survey (ARMS), 70–71, 79,
    81
Allen, Rich, 384
Alliance for Graduate Education and
    the Professoriate (AGEP),
    375–376
American Association of University
    Professors (AAUP), 96
American Council on Education
    (ACE) Fellows Project, 360
American Mathematical Society
    (AMS), 414
American Statistical Association
    (ASA), 247
    accreditation levels, 107
    Conference on Statistical
        Practice, 152
    development opportunities, 6–7,
        39, 59–61
    ethical guidelines, 156–157
    Founders Award winners, 6, 63
    leadership definitions, 275
    membership gender statistics,
        414
    Scientific and Public Affairs
        (SPA) Committee, 335
    Statisticians Without Borders,
        243, 307–308
    strategic planning, 210–211
    student membership, 249
    women presidents, 8, 61,
        255–258, 354, 393
Apollo 13, 259
Argumentation skills, 358
Aristotle, 11, 13
Asian American career advancement,
    111–115

Asian cultural contexts, 195–198
Asian students in the US, 193–194
Assigned leadership, 276
Association for Women in Science
    (AWIS), 247–248
Authentic self, 331
Authorship positions, 136, 140
Awards, 6, 32–33, 47, 63, 66

B
Bamboo ceiling, 111–112
Behavior domains of leadership, *See*
    Leadership behavior
    domains
B.E.L.I.E.F. Model, 202
Bell Telephone Laboratories, 57–58
Bernanke, Ben, 12
Big Telescopes Story, 42–44
Biostatistician personality type,
    200–201
Board on International Comparative
    Studies in Education
    (BICSE), 170
Bock, Mary Ellen, 7
Boehmer, Susan, 165
Bohman, Mary, 165
Boston University, 334–335
Brain structure, 111
Budgeting program change, 77–78
Bureau of Labor Statistics (BLS),
    165, 166
Bureau of Transportation Statistics
    (BTS), 165
Business and organizational acumen,
    205–213, 271

C
Cain, Susan, 268
Career advancement, 7, 344–345,
    365–369, *See also*
    Leadership development
    "accidental leader's" experiences
        and observations, 325–338
    accreditation, 107–108
    cultural contexts, 111–115,

193–202, *See also* Cultural
contexts
finding your niche, 335–338
gender-related inequities, *See*
Gender inequities
improving
business/organizational
acumen, 212–213
junior faculty, 424–426
organizational contexts, 115–118
sponsorship and, *See*
Sponsorship
successful strategies and skills,
356–359
top three jobs, 120
work-life balance, *See*
Family-work balance
Caucus for Women in Statistics, 335
Census of Agriculture, 70–71, 74, 79,
80–81, 82
Census Quality Management (CQM)
program, 72, 81
Center for Drug Evaluation and
Research (CDER), 5
Champions, professional, *See*
Professional champions
Champions for change, 76
Charisma, 24, 89, 120
Charles A. Dana Research Institute
for Scientists Emeriti
(RISE), 58–59
Child care, 238, 349, 356, 365–366,
405, 421
Chuang-Stein,Christy, 5
Cipsea, 185
Clark, Cynthia, 165
Class action lawsuits, 240
Clinical research teams (CRTs),
278–280
Coaching, 75, 110, 218–221, *See also*
Mentors and mentoring
Collaborative project leadership,
131–132, *See also* Teams
and teamwork
authorship positions, 136, 140

best practices, 136–138
challenges for women leaders,
138–141
getting started, 132
rising to leadership roles,
134–136
statistician roles, 133–134
staying involved, 132–134
women's underrepresentation,
134–136
Committee on National Statistics
(CNSTAT), 161–162, 165,
167, 168, 183–184
Committee on Women in Science,
Engineering, and Medicine,
64
Communicating change, 79
Communication plan, 289
Communication skills
argumentation and public
speech, 358
attaining acumen, 208
being heard, 334, 367–368
challenges for women, 328
critical competencies, 271
development programs, 7
elevator speech development,
329–330
expert witnessing, 329, 333–334
language barriers, 114–115
listening, 226, 288, 331
mastery of speech, 120–121
responding to sexist behavior,
331
virtual communication, 119–120
working in teams, 137–138,
287–289
Computer science, 158
Confidence, 26, 106–107, 117, 139,
357–358
Confidentiality protections, 171–174,
178, 185
Conflict resolution, 294, 358–359
Consultants, *See* Statistical
consulting

Contextual leadership behavior, 21, 24–25
Continuing education, 250
Continuous improvement, 359, 418
Convention on the Elimination of All Forms of Discrimination Against Women (CEDAW), 229, 241–242
Council for Advancement and Support of Education (CASE), 41
Council of Professional Associations on Federal Statistics (COPAFS), 171–172
Counseling, 218–220
Cox, Gertrude, 8, 395
Credibility, 22
Crisis leadership stories, 221–225
Critical incident technique, 157–158
Critical mass of women faculty, 354
Critical thinking, 399–400
Cross-functional skills, 118–119, 276
Cuddy, Amy, 427
Cultural contexts, 111–115, 193–195, *See also* Diversity leadership
    B.E.L.I.E.F. Model, 202
    building collaboration skills, 198–199
    gender considerations, 194
    leadership competencies, 203
    performance evaluation issues, 195–197
    prioritizing work, 197–198
    teams and leadership development, 282–285
    work-life balance, 199–200
Cultural intelligence, 201–202
Current Population Survey (CPS), 168, 182

D
Data confidentiality, 171–174, 178, 185
Data mining skills, 10

Davidian, Mary, 8, 354, 355
Decision making, critical competencies, 271, 292–293
Defense of Marriage Act (DOMA), 231
Delaney, Ann, 262
Delegation, 213, 297–298, 325–326, 358
Deming, W. Edwards, 152, 418
Discrimination, *See* Gender inequities
Disparate treatment or impact, 237
Diversity and leadership development, 282–285, *See also* Cultural contexts
Diversity leadership, 396–397, *See also* Gender diversity enhancement
    administrative roles, 397
    diversity imperatives, 398–401
    inclusion policies, 405–406
    K-12 outreach, 406
    meaning of demographic diversity, 397–398
    NC State successes, 407
    search committees, 402–404
    strategies and practices, 401–405
Drew University, 58–59
DxCG, Inc., 337

E
Economic Research Service (ERS), 165
Efficacy and supportive leadership, 27
Egan, Carmel, 271
Elevator speech, 329–330
Eli Lilly, 261–264, 271
Elizabeth L. Scott Award, 32–33, 63
E-mail, 93
Emergent leadership, 276
Emotional intelligence, 282
Employment law, 236–241
Empowerment, 297
Enthusiasm, 26

Entrepreneurship, 367
Equal Pay Act of 1963, 230
Equal Protection Clause, 14th
 Amendment, 232
Equal Rights Amendment (ERA),
 232
Ethics and integrity
 ASA guidelines, 156–157
 consulting considerations,
  155–158
 critical competencies, 271
 critical incident technique,
  157–158
 large-scale collaboration project
  leadership, 138
 principle-centered leadership,
  155
 research teams and, 299–300
 responsible leadership behavior,
  27–28
 transformational leaders and, 88
 unethical leadership, 89–90
 violations of professional norms,
  225–226
Ethos, logos, and pathos, 11–13
Executive presence, 14–15
Experiential learning theory (ELT),
 274, 277–278
Expert power, 276–277
Expert witnessing, 329, 333–334, 337

F
Faculty meetings, 91–93
Fair Labor Standards Act, 238
Family and Medical Leave Act, 238
Family-work balance, 140–141, 238,
 356
 "accidental leader's" experiences
  and observations, 324–325
 cultural contexts, 199–200
 maintaining personal ties,
  350–351
 supporting in academic contexts,
  348–351, 365–366, 421–423
Federal Committee on Statistical

Methodology (FCSM), 162,
 181, 185–186
Federal statistical system and
 agencies, *See also specific*
 agencies
 administrative data use,
  179–182, 186–187
 author's experience, 182–187
 change in official statistics
  programs, 69–84, *See also*
  Government official
  statistics programs, leading
  change in
 confidentiality protections,
  171–174, 178, 185
 examples of statistical leaders,
  5, 163–182
 gender considerations, 83–84
 organization, 162–163
 "Principles and Practices,"
  161–162
 US chief statistician, 162,
  163–164, 171
Flexibility, 62, 357
Florence Nightingale David Award,
 63
Focus, 342–343, 357
Fourteenth Amendment, Equal
 Protection Clause, 232
Franklin, Christine, 6
Fuentes, Montserrat, 354, 355–356
Fuller, Wayne, 389

G
Gardner, Martha, 6
Gates, Robert, 266
Geduldig v. Aiello (1974), 235
Geller, Nancy, 8
Gender and leadership, 111
Gender composition of statistics
 discipline, 414–417
Gender differences in leadership
 styles, 131, 139–140
Gender diversity enhancement,

413–427, *See also* Diversity
leadership
alliances, 418
appointments policies and
practices, 418–419
E. L. Scott and affirmative
action, 36, 46–47
Harvard improvements, 413–414,
416–417
high school statistics exposure,
61
institutional accountability,
423–424
junior faculty career paths,
424–426
search committees, 402–404,
418–419
sharing relevant research,
419–420
statistics on women in statistics,
414–417
supporting work-life balance,
348–350, 356, 365–366,
421–423
visible leadership, 417–418
Gender inequities
awards, 47
the glass ceiling, 103–104
Goldberg effect, 96–97
grant funding parity, 107,
116–117
health-related, 234–236
international cultural contexts,
194
parity in education, 104–105
research opportunities, 95–97
responding to sexist behavior,
331
review of women's rights,
230–232
salary differences, *See* Pay
equity
Scott's Big Telescopes Story,
42–44

STEM disciplines
underrepresentation, 56
UN Convention (CEDAW), 229
women statisticians' role,
242–243
General Electrics (GE), 6
Genetics and leadership qualities,
110–111
Gerig, Tom, 391–392
Gertrude M. Scott Scholarship, 63
Gilford, Dorothy, 167, 169–171,
184–185
Gladwell, Malcolm, 267
Glass ceiling, 103–104, *See also* Pay
equity
Goldberg effect, 96–97
Government Accountability Office
(GAO), 181–182
Government official statistics
programs, leading change
in, 69–84, *See also* Federal
statistical system and
agencies
analysis of change process, 80–82
examples, 70–73, 179, 186
significant factors, 73–80
women's roles, 83–84
Grant funding parity, 107, 116–117
Grant review committees, 344
Gravitas, 14–15
Groshen, Erica, 165, 166
Gumpertz, Marcia, 354, 360–361, 388

H
Harvard University, 41, 413–414,
416–417
Hat wearing, 331–332
Health-related issues, 234–236
medical school careers, 332–333
policy work, 336–338
volunteer statistical service and,
316
Heredity and leadership qualities,
110–111
Hewlett, Sylvia Ann, 377

High School Longitudinal Study, 169, 185
High school statistics curriculum, 60–61, 65
Hu, Patricia, 165
Hughes-Oliver, Jacqueline, 354, 361–362
Human capital investment, 330
Hurricanes, 223

I
Image and gravitas, 14–15
Implicit bias, 13–14, 112–114, 403–404, 420
Influence
    core inner circle, 355
    defining leadership, 20–21
    leadership is based on, 34–38, 276–277
    LEAP framework, 11–13
    personal observations, 16–17
    persuasion skills, 9–14
    types of power, 91, 276–277
    using your power, 368
Informal organization, 209
Innocence Project, 243
Innovation, 367
Inspirational leadership behavior, 21, 25–26
    storytelling, 42–45
Institute for Educational Management (Harvard), 41
Institute for the Operations Research and Management Sciences (INFORMS), 107–108
Institute of Mathematical Statistics (IMS), 247, 248, 255–258
Integrity, *See* Ethics and integrity
Intel Corporation, 6
Interagency Council on Statistical Policy (ICSP), 162, 163–164, 180, 185
Internal Revenue Service (IRS), 165
International Association of Education (IEA), 170

International Biometric Society (IBS), 247
International Statistics Institute (ISI), 247
International students, 193–194
Internet resources, 270

J
Joint Statistical Meetings (JSM), 47, 104, 374
Judicial sector, statisticians and, 316–317

K
Kaiser Permanente, 336
Kay, K., 427
King, Martin Luther, Jr., 12
Kutner, Michael, 379

L
Laissez-faire leaders, 89
Language barriers, 114–115
Latent organizational chart, 209
LaVange, Lisa, 5
Leader-follower relationship, 23
    transformational leaders and, 88, 91
Leadership, 85–86
    assigned versus emergent, 276
    gender differences in styles, 131, 139–140
    management versus, 34–35
    principle-centered, 155
    research balance, 355–356
    situational, 298–299
    trait theory, 153
    transactional, 87, 110
    transformational, 86–94, 110
    unethical, 89–90
    without official title, 355
Leadership, defining, 206, 261, 275–277, 313–314
    influence behaviors, 20–21
    roles and positions, 20

Leadership behavior domains, 19,
    21–22, 154
  contextual, 21, 24–25
  gender considerations, 28–29
  inspirational, 21, 25–26
  personal, 21, 22
  relational, 21, 23–24
  responsible, 22, 27–28
  supportive, 22, 26–27
Leadership competencies, 38–42,
    262–264, 270–271, *See also*
    Leadership skills
  decision-making, 271, 292–293
  international cultural contexts,
    193–204
  recruitment and hiring, 330
Leadership development, 259, *See
    also* Career advancement;
    Mentors and mentoring
  academic contexts, 339–352, *See
    also* Academic leadership
  "accidental leader's" experiences
    and observations, 323–338
  becoming a leader, 273–274
  competencies determine
    effectiveness, 38–42
  defining leadership, 261, 275–277
  developing areas of expertise,
    333
  diversity and, 282–285
  expert witnessing, 329
  finding your niche, 335–338
  focusing on competencies,
    262–264, 270–271
  formal education programs, 278,
    360, 380–381
  institutional responsibilities, 351
  keeping motivated, 269–270
  large-scale collaboration
    projects, 134–138
  lifelong adaptive process, 274
  media resources, 270
  Millennium Leadership
    Institute, 42
  observing/interacting with
    leaders, 265–266
  practice and experience,
    264–265, 274, 277–278,
    366–367
  preparation, 107–108
  professional societies and, 6–7,
    39, 59–61, 247–254, 329, 335
  recommended literature,
    266–268, 426–427
  research team experience, 273,
    277–285
  strategy for developing future
    leaders, 65–66
  successful strategies and skills,
    356–359
  volunteer service, 307–318,
    328–329
Leadership in statistics, *See
    Statistics leadership;
    Women leaders in statistics
Leadership qualities
  cultural contexts, 111–115
  gender-specific roles, 111
  heredity, 110–111
  personality, 108–110
  universal successful qualities,
    106–108
Leadership skills, 4, 285–299, *See
    also* Communication skills;
    Leadership competencies
  "accidental leader's" experiences
    and observations, 327–328
  competencies determine
    effectiveness, 38–42
  cross-functional, 118–119, 276
  development opportunities, 6–7,
    See also Leadership
    development
  effective scientific leaders,
    285–286
  executive presence, 14–15
  meetings, 290
  organizational and business
    acumen, 205–213

patience, 16
people skills, 295–299
persuasion, 9–14, *See also*
    Influence
problem-solving, 291–292
requirements for women in
    government, 83
trait-based perspectives, 153
universal successful qualities,
    106–108, 201
working in teams, 118–119,
    285–299
Leadership stories, 217
    life-changing events, 224–225
    outstanding leadership qualities,
        271
    unforeseen catastrophes,
        221–224
    violations of professional norms,
        225–226
LEAP framework for persuasion,
    11–13
Leave policies, 348–350
Legal issues, 229–230, 232
    class action suits, 240
    disparate treatment versus
        disparate impact, 237
    educational discrimination,
        233–234
    employment law, 236–241
    expert witnessing, 329, 333–334,
        337
    health-related inequities,
        234–236
    litigation
        advantages/disadvantages,
            239–240
    review of women's rights,
        230–232
    sexual harassment, 239–240
    UN CEDAW, 229, 241–242
    whistleblower protections, 239
    women statisticians' role,
        242–243

Legal/judicial sector, statisticians
    and, 316–317
LGBT Center, 405–406
Lifelong adaptive process, 274
Lilly Ledbetter Fair Pay Act of 2009,
    238
Listening, 226, 288, 331
Logos, 11, 12
Los Alamos fire, 221–223
Louise Hay Award, 63

M
Madigan, David, 5
Management, leadership versus,
    34–35
Martin, Margaret, 8, 167–168,
    182–183
Martinez, Rochelle (Shelly), 167, 178
Maslow's hierarchy of needs, 296
Maternity leaves, 348
Mathematics Association of America,
    248
McConnell, John, 268
Medical doctors, statistician
    interactions with, 332–333
Medical school careers, 332–333
Medicare purchasing program, 336
Meetings, 290
Meng, Xiao-Li, 5
Mentors and mentoring, 4, 65,
    218–221, 359, 369, 373–376,
    426, *See also* Coaching
    academic advisers, 389–390
    academic outcomes and, 420
    administrators, 391–392
    colleagues, 390–391
    consultants, 151–152
    defining, 384–385
    large-scale collaboration
        projects, 135–136
    leveraging, 269
    parents, 385–386
    Pathways to the Future
        workshops, 94–95, 374
    peers, 386–388, 426

a personal story, 385–394
professional societies and, 251
RISE and, 58–59
sponsorship, 117–118, 359,
    376–380
VIGRE, 375–376, 387
volunteer service, 313
Military leadership, 39
Millennium Leadership Institute, 42
Mission-oriented meetings, 290
Morton, Sally, 7
Mosteller, Fred, 413–414
Motivation, 295–296
Motsinger-Reif, Alison, 354, 356
Myers Briggs Type Indicator,
    108–110

N
National Academy of Science
    Committee on National
    Statistics (CNSTAT),
    161–162
National Agricultural Statistics
    Service (NASS), 70–82, 165
National Center for Education
    Statistics (NCES), 161,
    169–170, 171, 179, 183–184,
    186–187
National Center for Health Statistics
    (NCHS), 179
National Institute of Statistical
    Sciences (NISS), 61–62
National Institutes of Health (NIH),
    234
National Labor Relations Board
    (NLRB), 334
National Public Health Leadership
    Institute, 278
National Science Foundation (NSF),
    359, 375, 387, 392, 401
Needs hierarchy, 296
Negotiation, 139–140, 271, 346
Networking, 120, 346–347
    informal interactions, 368–369
    professional champions, 373,

376–380, *See also*
    Sponsorship
Neyman, Jerzy, 35, 146
Nightingale, Florence, 64
North Carolina State University,
    354–356, 359–361, 392,
    395–407

O
Obama, Barack, 11
Office of Management and Budget
    (OMB), 5, 162, 167–168,
    173–174, 178–182, 185
On-Call Scientists program, 243
Optimism, 25
Organizational and business acumen,
    205–213, 271
Organizational chart, 209
Organizational contexts in career
    advancement, 115–118
Orren, Gary, 11, 13

P
Palmer, Gladys, 168
Pantula, Sastry, 5
Parental leaves, 348–350
Parents as mentors, 385–386
Path-goal theory, 154
Pathos, 12
Pathways to the Future workshops,
    94–95, 374
Patience, 16
Paycheck Fairness Act, 229–230
Pay equity, 56
    Elizabeth L. Scott and, 34,
        45–47
    the glass ceiling, 103–104
    Lilly Ledbetter Fair Pay Act of
        2009, 238
    research opportunities, 95–96
    UN Convention (CEDAW), 229,
        241–242
    US legislation, 229–230, *See also*
        Legal issues

women statisticians' role, 243–244

Pearl Meister Greengard Prize (PMG), 63

Peck, Roxy, 6

Peer mentoring, 386–388, 426

Pensions, 240–241

People skills, 295–299, 359

Performance evaluation, 195–197, 212, 326–327

Personality types, 108–110, 200–201

Personal leadership behavior, 21, 22

Persuasion skills, 9–14, *See also* Influence

Peters, Tom, 10–11

Pfizer, 5

Positive work culture, 347–348

Postdoctoral appointments, 65

Potok, Nancy, 167, 174–178, 186

Power, 91, 276–277, 368, *See also* Influence

Presence, 14–15

Principle-centered leadership, 155

Prioritization of work, 197–198

Problem solving, 291–292, 400–401

Process-oriented meetings, 290

Professional champions, 373, 376–380, *See also* Sponsorship

Professional development, *See* Career advancement; Leadership development

Professional societies, 247–248, *See also* American Statistical Society

codes of ethics, 156–157

development opportunities, 6–7, 39, 59–61, 247–254, 329, 335

mentoring opportunities, 251

mid-career membership, 251–252

senior-career membership, 252–253

student/early-career membership, 249–251

Project management, *See also*

Collaborative project leadership

Census Bureau and, 174–176

communication plan, 289

program change and, 78

teams and, 300–301

Project Management Institute, 301

Public Health Leadership Program, 278

Public speaking skills, 358

Q

Quality improvement (QI), 300

R

Recruitment and hiring skills, 330

Referential power, 276–277

Relational leadership behavior, 21, 23–24, 45–47

Reproductive rights, 234–235

Respect, 23

Responsible leadership behavior, 22, 27–28

Retirement benefits, 240–241

Rice University, 374–375

RISE, 58–59

Rodriguez, Robert N., 3–4, 116

Roe v. Wade (1973), 234

Royal Statistical Society (RSS), 247, 249

S

Salary equity, *See* Pay equity

Same-sex marriage, 231

Sandberg, Sheryl, 204, 229, 266–267, 377, 426–427

San Diego State University, 36–37

Scheaffer, Richard, 6

Schwarzkopf, Norman, 267

Science and mathematics education, 170

Scientific and Public Affairs (SPA) Committee, 335

Scott, Elizabeth L., 31–49, 95

advocacy leadership, 35–36,
47–49
Big Telescopes Story, 42–44
military background, 38–39
Search committees, 402–404, 418–419
Security and supportive leadership,
27
Self-assessment, 108
Self-awareness, 213
Self-confidence, *See* Confidence
Sex discrimination, *See* Gender
inequities; Legal issues
Sexist behavior, responding to, 331
Sexual harassment, 239–240
Shah, Aarti, 262–263
Shipman, C., 427
Sierra Leone, 307–312, 318
Singer, Judith, 5
Situational leadership, 298–299
Skills, *See* Leadership skills
Snedecor, George, 146, 395
Social media, 120
Social networks, *See* Networking
Social policy, statisticians and,
315–316
Social Security Administration
(SSA), 187
Solomon, Dan, 5, 391–392
Southern Regional Council on
Statistics (SRCOS), 379
Sponsorship, 117–118, 351, 359, 373,
376–380
Statistical and Applied
Mathematical Sciences
Institute (SAMSI), 61–62
Statistical and Science Policy
Branch, 162
Statistical Community of Practice
and Engagement (SCOPE),
162–163
Statistical confidentiality protections,
171–174, 178, 185
Statistical consulting, 145
in academia, 146–147

consultant/client relationship,
149
education in, 151
effective consultant qualities,
149–150
ethical considerations, 155–158
in industry and government,
147–148
interdisciplinary issues, 158
leadership considerations,
148–150, 153–155
mentoring, 151–152
professional activities, 152
sole proprietorships, 148
volunteer service, 313–314
Statistical Society of Canada (SSC),
247
Statisticians Without Borders
(SWB), 243, 307–308, 315
Statistics education, K-12, 6, 60–61,
65
Statistics leadership, 3–5, 31, 207,
See also Women leaders in
statistics
in academia, *See* Academic
leadership
in government agencies, *See*
Federal statistical system
and agencies
health care sector and, 316
in industry, 5–6
K-12 education, 6
need for, 10
opportunities, 4
personal perspective of robust
leadership, 55
skills development, *See*
Leadership development;
Leadership skills
social sector and, 315–316
Statistics of Income Division, IRS
(SOI), 165
STEM disciplines (science,
technology, engineering,
and math), 401–403

diversity imperatives, 398–399
international students in US,
      193–194
NSF ADVANCE grants, 359,
      392, 401–403
representation of women, 56,
      402, 416
tenure issues, 365
Storytelling, 42–45
Strategic planning, 210–211
Student membership in professional
      associations, 249–251
Student supervision and training,
      345
Summers, Lawrence H., 104–105
Supportive leadership behavior, 22,
      26–27
Survey methodology, 310
Survey methodology research, 72, 74,
      81
SWOT analysis, 211
Szygy, 13

T

Task relevant maturity, 298
Teaching statistics, K-12, 6
Teaching statistics, volunteer, 312
Team grants, 345
Teams and teamwork, 154–155, *See
      also* Collaborative project
      leadership
   academic leadership and,
      341–342
   best practices, 136–138
   communication skills, 137–138
   conflict resolution, 294
   critical leadership skills, 285–299
   cross-functional skills, 118–119
   diversity considerations, 282–283
   leadership development
      opportunities, 273, 277–285
   meetings, 290
   people skills, 295–299
   problem-solving and
      decision-making, 291–293

   project management skills,
      300–301
   scientific integrity and quality,
      299–300
   statistician roles, 133–134, 279
   virtual teams, 280
   vision for change, 74–75
   volunteer service, 313
Thought leadership, 20
Time management, 342–343, 425
Transactional leadership, 87, 110
Transformational leadership, 87–94,
      110
Trust, 24, 226–227, 281–282

U

UK Office for National Statistics
      (ONS), 70, 72–73, 76, 78,
      81–82
United Nations Convention on the
      Elimination of All Forms of
      Discrimination Against
      Women (CEDAW), 229,
      241–242
University leadership, *See* Academic
      leadership
US Census Bureau, 70–72, 75, 179
   Census of Agriculture, 70–71,
      74, 79, 80–81, 82
   Census Quality Management
      (CQM) program, 72, 81
   women leaders in, 174–178, 186
US chief statistician, 162, 163–164,
      171
Utlaut, Theresa, 6

V

Verisk Health, 337
Vertical Integration of Graduate
      Research and Education
      (VIGRE), 375–376, 387
Vietnam, 325
Violence Against Women Act
      (VAWA), 236
Virtual communication, 119–120

Virtual teams, 280
Vision, 73–76, 210–212, 286–287,
    340–341
Volunteer service, 243, 307–318,
    328–329
  collaboration, 313
  consulting, 313–314
  mentoring opportunities, 313
  potential impacts, 315–317
  in Sierra Leone, 307–312, 318
  teaching, 312

W
Walker, Helen M., 8
Wallman, Katherine, 5, 163–164,
    167, 171–174, 185
Walmart, 240
Weems, Kim, 354
Whistleblower protections, 239
Wilson, Alyson, 354, 356
Women in statistics, statistics on,
    414–417
Women leaders in statistics, 8, 161,
    163–182, 353, *See also*
    specific persons
  in academia, 86, 395, *See also*
    Academic leadership
  "accidental leader's" experiences
    and observations, 323–338
  ASA presidents, 8, 61, 255–258,
    354, 393
  balancing collaborative and
    independent research, 426
  examples, 5–8, 163–182
  federal statistical agencies,
    83–84, 161–188, *See also*
    Federal statistical system
    and agencies
  large-scale collaboration
    projects, *See* Collaborative
    project leadership
  legal system and, 242–243
  need for women in leadership
    roles, 353–356

  recognitions and awards, 6,
    32–33, 63, 66
  role models and mentors, 64,
    141, *See also* Mentors and
    mentoring
Women's rights, 230–232, *See also*
    Legal issues
Women university presidents,
    101–102
Work culture, 347–348
Work-life balance, *See* Family-work
    balance
Works Progress Administration
    (WPA), 168

X
Xun, Lu, 121